Tropical Fruit Tree Diversity

Farmers have developed a range of agricultural practices to sustainably use and maintain a wide diversity of crop species in many parts of the world. This book documents good practices innovated by farmers and collects key reviews on good practices from global experts, not only from the case study countries but also from Brazil, China and other parts of Asia and Latin America.

A good practice for diversity is defined as a system, organization or process that, over time and space, maintains, enhances and creates crop genetic diversity, and ensures its availability to and from farmers and other users. Drawing on experiences from a UNEP-GEF project on 'Conservation and Sustainable Use of Wild and Cultivated Tropical Fruit Tree Diversity for Promoting Livelihoods, Food Security and Ecosystem Services', with case studies from India, Indonesia, Malaysia and Thailand, the authors show how methods for identifying good practices are still evolving and challenges in scaling-up remain. They identify key principles effective as a strategy for mainstreaming good practice into development efforts. Few books draw principles and lessons learned from good practices. This book fills this gap by combining good practices from the research project on tropical fruit trees with chapters from external experts to broaden its scope and relevance.

Bhuwon Sthapit is an *in situ* conservation specialist at Bioversity International based in Pokhara, Nepal.

Hugo A.H. Lamers is an associate scientist at Bioversity International based in New Delhi, India.

V. Ramanatha Rao has worked in plant genetic resources at ICRISAT (Hyderabad, India) and at Bioversity International (Italy, Singapore and Malaysia).

Arwen Bailey is science editor at Bioversity International based in Rome, Italy.

Issues in Agricultural Biodiversity

Series editors: Michael Halewood and Danny Hunter

This series of books is published by Earthscan in association with Bioversity International. The aim of the series is to review the current state of knowledge in topical issues associated with agricultural biodiversity, to identify gaps in our knowledge base, to synthesize lessons learned and to propose future research and development actions. The overall objective is to increase the sustainable use of biodiversity in improving people's well-being and food and nutrition security. The series' scope is all aspects of agricultural biodiversity, ranging from conservation biology of genetic resources through social sciences to policy and legal aspects. It also covers the fields of research, education, communication and coordination, information management and knowledge sharing.

Published titles:

Crop Wild Relatives
A Manual of *in situ* Conservation
Edited by Danny Hunter and Vernon Heywood

The Economics of Managing Crop Diversity On-Farm
Case Studies from the Genetic Resources Policy Initiative
Edited by Edilegnaw Wale, Adam Drucker and Kerstin Zander

Plant Genetic Resources and Food Security
Stakeholder Perspectives on the International Treaty on Plant Genetic Resources for Food and Agriculture
Edited by Christine Frison, Francisco López and José T. Esquinas

Crop Genetic Resources as a Global Commons
Challenges in International Law and Governance
Edited by Michael Halewood, Isabel López Noriega and Selim Louafi

Community Biodiversity Management
Promoting Resilience and the Conservation of Plant Genetic Resources
Edited by Walter S. de Boef, Abishkar Subedi, Nivaldo Peroni and Marja Thijssen

Diversifying Food and Diets
Using Agricultural Biodiversity to Improve Nutrition and Health
Edited by Jessica Fanzo, Danny Hunter, Teresa Borelli and Federico Mattei

Community Seed Banks
Origins, Evolution and Prospects
Edited by Ronnie Vernooy, Pitambar Shrestha and Bhuwon Sthapit

Farmers' Crop Varieties and Farmers' Rights
Challenges in Taxonomy and Law
Michael Halewood

Tropical Fruit Tree Diversity
Good practices for *in situ* and on-farm conservation
Edited by Bhuwon Sthapit, Hugo A.H. Lamers, V. Ramanatha Rao and Arwen Bailey

Tropical Fruit Tree Diversity

Good practices for *in situ* and on-farm conservation

Edited by
Bhuwon Sthapit, Hugo A.H. Lamers, V. Ramanatha Rao and Arwen Bailey

Routledge
Taylor & Francis Group
LONDON AND NEW YORK

earthscan
from Routledge

First published 2016 by Routledge

2 Park Square, Milton Park, Abingdon, Oxfordshire OX14 4RN
52 Vanderbilt Avenue, New York, NY 10017

Routledge is an imprint of the Taylor & Francis Group, an informa business

First issued in paperback 2018

British Library Cataloguing-in-Publication Data
A catalogue record for this book is available from the British Library

Library of Congress Cataloging in Publication Data
Names: Sthapit, Bhuwon.
Title: Tropical fruit tree biodiversity : good practices for in situ and on-farm conservation / edited by Bhuwon Sthapit, Hugo A.H. Lamers, V. Ramanatha Rao and Arwen Bailey.
Description: New York : Routledge, 2016. | Includes bibliographical references and index.
Identifiers: LCCN 2015039779| ISBN 9781138781276 (hbk) | ISBN 9781315758459 (ebk)
Subjects: LCSH: Fruit trees—Conservation—Tropics.
Classification: LCC SB359 .T679 2016 | DDC 634—dc23
LC record available at http://lccn.loc.gov/2015039779

ISBN: 978-1-138-78127-6 (hbk)
ISBN: 978-0-367-17278-7 (pbk)

Typeset in Bembo
by Florence Production Ltd, Stoodleigh, Devon

Contents

Illustrations

Plates

Figures

Tables

Boxes

Contributors

Achmad Rafieq is an Agricultural Extension Worker at South Kalimantan Assessment Institute for Agricultural Technology, Indonesian Agency for Agricultural Research and Development (IAARD), and Ministry of Agriculture. He is an anthropologist whose main focus is tropical fruit tree diversity conservation, also with experience in upland rice diversity research. He has significant experience in partnership activities in rural agribusiness development.

Adhitya Marendra Kiloes is a Socio-Economist and Policy Analysis Researcher with the Indonesian Center for Horticultural Research and Development (ICHORD), Indonesian Agency for Agricultural Research and Development (IAARD) and Ministry of Agriculture, and conducts research on agricultural social economics and policy including agricultural development, competition and market preferences.

Adriana Alercia studied Informatics at the Universidad Blas Pascal, Argentina. She has led the development of descriptors for more than 100 crops and has co-authored international documentation standards, including *Descriptors for Farmers' Knowledge of Plants* and the *Core Descriptors for In situ Conservation of CWR*. She worked as a Germplasm Information Specialist at Bioversity International and is currently working at The International Treaty of FAO.

Alfredo Kingo Oyama Homma, PhD, is a distinguished Economist specializing in rural agricultural and forest economics of Brazil. He conducts leading research on regionally and internationally valued non-timber forest products and has worked for four decades at the Brazilian Enterprise for Agricultural Research (EMBRAPA), Belém, Brazil.

Anne Sennhenn is a Scientist at the University of Goettingen, Department of Crop Production Systems in the Tropics, Germany, where she obtained her PhD in 2015. Her main research interest is on plant physiology as well as agrobiodiversity and crop modelling with focus on smallholder farming systems in semi-arid areas.

Antonio Jose Elias Amorim de Menezes, PhD, is an Agronomist with a Doctorate in agricultural production systems of family farms studying the economics and management of highly valued Amazonian fruit species. He is an Analyst at the Brazilian Enterprise for Agricultural Research (EMBRAPA), Belém, Brazil.

Arwen Bailey is an ex-language teacher with an MSc in Systems Thinking in Practice (Open University, UK). She works at Bioversity International as a Science Editor, helping scientists to express complex ideas through the written word. She is interested in how change happens through knowledge-sharing, and has experience in facilitation, project and people management.

Auttapon Rukkaphan, MSc graduate in Botany from Kasetsart University, Bangkok, is an Agricultural Researcher at Srisaket Horticultural Research Center, Thailand. He has experience in the identification of herbs and local fruit trees. He is interested in research on physiology and morphogenesis of tropical fruit crops for the enhancement of horticultural production.

Awtar Singh, PhD (Plant Breeding, CCS Haryana Agricultural University, Hissar, India), works on the breeding of citrus and litchi. His expertise includes germplasm characterization and improvement of these crops. He is currently working as a Principal Scientist at the Indian Agricultural Research Institute, Indian Council of Agricultural Research, New Delhi.

Barsati Lal, PhD (Extension Education) from Indian Veterinary Research Institute, Izatnagar, has more than 20 years' involvement in technology transfer and demonstration, training, participatory research, assessment and refinement with more than 30 publications. He is currently working as a Senior Scientist at the Central Institute for Subtropical Horticulture, Lucknow, India.

Bhuwon Sthapit has a PhD in Plant Biology (plant breeding) from the University of Wales, UK. He has 16 years of experience in implementing projects in Brazil, China, India, Indonesia, Malaysia, Nepal, South Africa, Thailand, Vanuatu and Vietnam. His expertise is in participatory plant breeding, seed systems, community seedbanks, home gardens and tropical fruit conservation. His work focuses on community biodiversity management for enhancing livelihoods of resource poor farmers through conservation and use of biodiversity. He has been with Bioversity International since 1997 as *In Situ* Conservation Specialist and is currently based in the Nepal Office. He has authored/co-authored more than 320 publications.

B. Marambe is a Professor of Crop Science at the Faculty of Agriculture, University of Peradeniya, Sri Lanka. He is the Director of the Agriculture Education Unit of the Faculty and was the Dean of the Faculty from 2003 to 2009. He is an experienced researcher on weed science and vulnerability assessment of home gardens.

B.M.C. Reddy, PhD, is Vice-Chancellor of YSR Horticultural University, Andhra Pradesh. While with ICAR, he worked on conservation of tropical and underutilized and lesser-known fruits and crop production technology of banana, pineapple, grape, citrus and mango while holding various positions, including Director of the Central Institute for Subtropical Horticulture, Lucknow. He was the National Project Coordinator – India of the UNEP/GEF TFT project.

B.V.R. Punyawardena is Senior Climatologist at the Natural Resources Management Centre, Department of Agriculture, Sri Lanka. He focuses his research in the area of climatology, climate change and agro-ecology. He was instrumental in developing the agro-ecological region map of Sri Lanka.

Charles R. Clement, PhD, has worked on the development of native fruit crops since 1976. As a Biologist at the National Institute of Amazonian Research (INPA), he studies the historical ecology of the Amazonian biome, especially the origin, domestication and dispersal of native crops, and the location and extent of anthropogenic soils and forests.

Chatchanok Noppornphan, PhD (Agronomy and Soil Science, Louisiana State University, USA), is a retired Agricultural Scientist in the Department of Agriculture, Thailand. His past experiences dealt with research in management of soil fertility, fertilizers and plant nutrition to sustainably improve productivity and quality of horticultural crops.

Chunlin Long is a Botanist working for Minzu University of China. He specializes in ethnobotany research, and in particular landraces associated with traditional knowledge and ethnic culture.

Chunmin Lu is a Rice Breeder working for Yunnan Agricultural University (YAU), China. He specializes in local rice landrace collecting, characterizing and evaluation.

Dalva Maria da Mota, PhD Sociology, is a Researcher of the Brazilian Enterprise for Agricultural Research (EMBRAPA) and Lecturer at the Universidade Federal do Pará. She has worked on peasant agriculture and rural communities on sociability, gender, social division of labour and livelihoods in the northeast and north of Brazil.

Danny Hunter, PhD, is Senior Scientist at Bioversity International and currently holds adjunct staff posts at the National University of Ireland, Galway (NUI Galway) and Charles Sturt University (CSU), Australia, where he contributes to postgraduate courses in plant genetic resources, climate change, sustainable agriculture and food security.

Devra Jarvis, PhD, is a Botanist and Principal Scientist at Bioversity International. She specializes in the assessment, conservation and use of crop genetic diversity in cultivated and wild ecosystems.

D.K.N.G. Pushpakumara is Professor of Crop Science, Faculty of Agriculture, University of Peradeniya, Sri Lanka. He is also the country liaison scientist with the World Agroforestry Centre (ICRAF). He is an experienced researcher focusing on tree domestication, up-scaling of agroforestry, conservation and use of agrobiodiversity and pteridophytes, with experience in home garden characterization and improvement programmes.

Enlai Zhang is a Rice Breeder working for Yunnan Academy of Agricultural Sciences (YAAS), China. He is specialized in local rice landrace collection, characterization and evaluation.

Froukje Kruijssen, PhD, an Agricultural Development Economist, was with the Asia-Pacific regional office of Bioversity International for three years working on issues related to market-based approaches for improved on-farm management of tropical fruit diversity for livelihoods. After getting her doctorate she moved to the World Fish Center, where she leads the research on value chains.

G.L.L.P. Silva is Professor of Animal Science in Faculty of Agriculture, University of Peradeniya, Sri Lanka. Her research areas include characterization of farm animal genetic resources, population diversity analysis and conservation of indigenous farm animal genetic resources.

H.M.S. Heenkenda is Additional Director General of Agriculture (Research), Department of Agriculture, Sri Lanka. He is an experienced agricultural scientist on pomology and home garden analysis and improvement. He has also served as a consultant horticulturist and master trainer of fruit production project, Metta Foundation, Myanmar.

Huaxian Peng is a Plant Pathologist working for Sichuan Academy of Agricultural Sciences (SAAS), China. She has been working on using maize diversity for pest and disease control for several years.

Hugo A.H. Lamers has worked since 2009 for Bioversity International as Associate Scientist and contributed to research projects on on-farm and *in situ* conservation of fruit trees in India, Indonesia, Malaysia and Thailand. His expertise is on value chain development, the marketing of fruit diversity, community biodiversity management and impact assessments with a special interest in participatory methods and research tools. He holds an MSc in Development Economics from Wageningen University in the Netherlands.

Idha Widi Arsanti is a Senior Scientist in the field of agribusiness, agricultural economics and policy. She has been working on the tropical fruit trees research and development project in Indonesia for the last three years. She is the Head of Programme Evaluation at the Indonesian Centre for Horticulture Research and Development (ICHORD), Indonesian Agency for Agricultural Research and Development (IAARD), Ministry of Agriculture, Indonesia.

Indra Pal Singh, PhD, a Principal Scientist at the National Research Centre for Citrus, Nagpur, India, has 29 years of research experience on citrus genetic resources. Trained at Yale University and the University of Florida (USA) and the University of Puerto Rico, he has released two acid lime varieties and led 21 citrus germplasm collecting missions in India.

Jamaluddin Lani, BSc (Horticulture, Utah State University) is currently the Agriculture Officer, Papar District, Sabah, Malaysia with eight years of experience in agriculture extension. Previously, he was a Lecturer with the Agriculture Institute, Department of Agriculture Sabah. His work with communities involves advisory services, subsidy programmes and development of rice cultivation.

Jens Gebauer is a Professor of Sustainable Production Systems with Special Focus on Horticulture and the Head of the Tropical Greenhouse and Study and Showpiece Gardens at Rhine-Waal University of Applied Sciences, Germany. His research interests include home gardens and wild fruit trees such as the African baobab.

Jie Wu is a Researcher working for Sichuan Academy of Agricultural Sciences (SAAS), China. She focuses on plant protection, especially using biodiversity to improve maize resistance to pests and disease.

Jie Yuan is a Plant Pathologist working for Guizhou Academy of Agricultural Sciences (GAAS), China. She has focused on integrated pest and disease management for several years.

José Edmar Urano de Carvalho is an Agronomist with a Masters in vegetable production, working for close to three decades to discover innovative techniques to improve the production and quality of native Amazonian fruit trees. He is a Researcher at the Brazilian Enterprise for Agricultural Research (EMBRAPA), Belém, Brazil.

Josué Francisco da Silva Júnior, MSc (Tropical Fruit Crops, Universidade Federal da Bahia, Brazil), has more than 20 years of involvement in genetic resource conservation of tropical fruits, underutilized species and *in situ* conservation by traditional communities, mainly in Northeast Brazil. Currently, he is a Researcher at the Brazilian Enterprise for Agricultural Research (EMBRAPA).

J. Rizvi is the Regional Director of the South Asia Programme of the World Agroforestry Centre (ICRAF). He is an Agricultural Scientist and worked for more than 13 years in ICARDA as the country programme manager.

J. Weerahewa is Professor of Agriculture Economics and Business Management in the Faculty of Agriculture, University of Peradeniya, Sri Lanka. Her research areas include quantitative policy analysis and their applications in agriculture.

Katja Kehlenbeck, PhD in International Agriculture (University of Goettingen, Germany), is a Scientist at the World Agroforestry Centre, Kenya. Her current focus is on improving nutrition through promoting food tree diversity, diversification of farming systems and improvement of home garden systems in Africa and Asia.

Keyu Bai, PhD, is an Ecologist working for Bioversity International, based in the Beijing Office. She specializes in agricultural resource management and using biodiversity to improve the resilience and sustainability of agricultural ecosystems.

Kuntoro Boga Andri, PhD, works as Associate Researcher in the Agricultural Policy Division, Assessment Institute for Agricultural Technology, East Java Agency for Agricultural Research and Development, Indonesian Ministry of Agriculture. He conducts research on agricultural economics and policy including supply chain management, agribusiness development, strategic marketing and regional development.

Lau Cheng Yuon, B Hort Sc (Hons) (Lincoln College, University of Canterbury), is currently an Assistant Director of Research at Agriculture Research Centre, Semongok. He has 34 years of work experience with the Department of Agriculture Sarawak.

Luyuan Dai, PhD, is a Botanist, who used to work for Yunnan Academy of Agricultural Sciences (YAAS). He is now the Deputy Director General of Yunnan Association for Science and Technology. His specialized expertise is on traditional knowledge documentation in plant germplasm resource conservation.

Margaret C. Yoovatana, PhD (Crop Science, Central Luzon State University, Philippines), currently holds the position of Senior Policy and Plan Specialist, International Agricultural Affairs Group, Planning and Technical Division, Department of Agriculture, and Head of the Multilateral and International Organization Collaborations Section.

Martin Wiehle, PhD in Sustainable International Agriculture, is the Coordinator of the Centre of International Rural Development at the University of Kassel, Germany. His main research interests include home gardens and agrobiodiversity of wild fruit trees of Africa.

Mohd Nizam Abdullah, Senior Assistant Agriculture Officer in the Department of Agriculture, Perak, Malaysia. He is responsible for extension and dissemination of agriculture technology in Bukit Gantang area. He is also responsible for tropical fruit and vegetable farms.

Montree Issarakraisila, PhD (Plant Physiology, the University of Western Australia, Australia), is involved in teaching and research of tropical fruit crops. Research areas of interest are effects of environment on yields and quality of fruit crops, diversity of tropical fruits and sustainable horticultural production in Thailand.

M. Kiran Kumar, MSc Agriculture (Crop Physiology), works as a Regional Coordinator at the DHAN Foundation, Madurai. His expertise is in building social capital in actions related to water tank conservation, watershed development and agriculture development and he focuses on farm ponds, microfinance for small and marginal farmers and agriculture commodity marketing.

M.P. Vasimalai, MSc (Agri.), MBA, is a leading expert in the development profession in India, with vast experience in building development institutions and models for pro-poor grassroots action. His expertise includes microfinance, agriculture and water resource development. He has trained around 1,000 development professionals in different disciplines in the 30 years of his development career.

M. Palanisamy, MSc (Agri.), has 22 years of experience in the development sector. His work focuses on promotion of farmers' organizations at multiple levels, renovation of minor irrigation tanks, training rainfed farmers in improved agricultural practices and grooming development professionals. He is on the Board of Trustees in various development institutions in India.

M.R. Dinesh, PhD (Fruit Breeding), has worked for the past 26 years on mango, guava and papaya breeding at the Indian Institute of Horticultural Research, Bengaluru, including field genebank maintenance of mango. In addition, he is heading the Division of Fruit Crops, managing the research projects in the Division.

Muhammad Sabran, PhD in Statistical Genetics, with research experience in plant breeding, participatory crop improvement, genetic resource management, integrated farming systems and genomic data analyses. He works at the Indonesian Centre for Agricultural Biotechnology and Genetic Resources Research and Development, Indonesian Agency for Agricultural Research and Development (IAARD), MoA. He is also involved in developing national and international policy on biodiversity and genetic resource management, particularly on access and benefit sharing.

Muhammad Shafie Md. Sah, BSc (Genetics and Molecular Biology, University of Malaysia), is a Researcher at the Malaysian Agricultural Research and Development Institute (MARDI). He has a long record of involvement in conservation biology especially in traditional vegetable species of Malaysia. He is involved in traditional knowledge documentation, field and seed genebank management as well as on-farm conservation of tropical fruit trees in Malaysia.

M. Winarno, PhD (Horticultural Science, University of Minnesota, USA), has served as first Director of the Indonesian Tropical Fruit Research Institute (ITFRI) and served in various positions from Solok, West Sumatra and Jakarta with the Indonesian Agency for Agricultural Research and Development (IAARD), Ministry of Agriculture. He was the Chair of a joint Indonesia–Australia task force on horticultural crops/industry. He also worked as National Project Manager of the GEF/UNEP TFT project in Indonesia.

Narasimha Hegde, MSc, General Manager of Life Trust, Sirsi, India, has been involved in promoting integrated and interdisciplinary approaches for conservation and management of non-timber forest products in the tropical forests by combining landscape/ecosystem knowledge and participatory methods. He conducts action research on ecological restoration of the tropical fresh water swamps.

Nono Sutrisno, PhD, has a long record of involvement in integrated farming systems, soil and water management and natural resource management. He also served as the National Project Coordinator of the TFT project. Currently, he is working at the Indonesian Agroclimate and Hydrology Research Institute, Indonesian Agency for Agricultural Research and Development (IAARD), Ministry of Agriculture.

Norhayati Md Haron, BSc (Bioindustry, University of Putra Malaysia), is a Senior Agriculture Officer at the Department of Agriculture, Titi Gantung, Perak, Malaysia. She has been involved in horticultural crops including tropical fruits for 11 years.

Paola De Santis is a Researcher working for Bioversity International. She mainly focuses on participatory diagnostic approaches to assess crop genetic diversity in agroecosystems.

Paul Quek, DVM (Veterinary Medicine, University Putra Malaysia) and Dip KE (Knowledge Engineering, National University of Singapore), has more than 20 years of experience in information management for genetic resource conservation, traditional knowledge documentation and use of information technology in public awareness. He is an Honorary Research Fellow for Bioversity International, Serdang, Malaysia.

Patricia Shanley, PhD, has worked with forest communities in the Brazilian Amazon for more than two decades, focusing on the effects of land use change on locally valued fruit and medicinal species. She is on the Steering Committee of People and Plants International and is the Programme Director of Woods & Wayside International. She also worked for CIFOR, Bogor, Indonesia.

Pearlycia Brooke, MSc (Biotechnology, University of Malaysia Sarawak), works as a Researcher at the Agriculture Research Centre Semongok, Sarawak. She has been involved in fruit agronomy work in Sarawak for the last six years. Her work focuses on developing technology packages for various fruit crops and providing agronomic advice and information to growers.

Percy Sajise, PhD (Plant Ecology, Cornell University, Ithaca), has written and edited numerous publications including books and is a member of editorial boards of several research journals. He headed the Bioversity International Office for Asia and the Pacific and the Southeast Asia Regional Center for Graduate Study and Research in Agriculture (SEARCA).

Pichit Sripinta, MSc (Horticulture, Kasetsart University), is a Horticulturist with long experience of research into breeding and cultural practices for temperate fruit and subtropical fruits. At present he works at Chiangmai Royal Agricultural Research Centre, Horticulture Research Institute, Department of Agriculture, Thailand.

Prama Yufdy, PhD, is the Director of the Indonesian Centre for Horticulture Research and Development (ICHORD), Indonesian Agency for Agricultural Research and Development (IAARD), MoA. He is also a Senior Scientist on soil fertility and biology. He has wide experience in agricultural management and policy.

Putu Bagus Daroini, Agricultural Extension, Indonesian Agency for Agricultural Research and Development, Ministry of Agriculture, has experience in counselling on agricultural institutional, agribusiness development and community strategic marketing. He currently works in agricultural extension at the Assessment Institute for Agricultural Technology, East Java.

Ratchanee Siriyan holds a Doctorate degree in Agriculture from the University of Khon Kaen, Thailand. Her expertise is in the area of plant virology and molecular biology. She is an Agricultural Research Officer at Sisaket Horticultural Research Center, Sisaket province, Thailand, working on papaya and tomato breeding.

Rateng Girid is an Agriculture Assistant at Serian District Agriculture Office. He has served with the Department of Agriculture Sarawak since 1983.

R.H.G. Ranil is Senior Lecturer at the Faculty of Agriculture, University of Peradeniya, Sri Lanka. He is a Naturalist working on biodiversity and agrobiodiversity conservation with special reference to assessments and development of conservation priorities.

Rosa Lía Barbieri, PhD (Genetics and Molecular Biology, Universidade Federal do Rio Grande do Sul, Brazil), has a long record of involvement in plant genetic resources, agricultural biodiversity, neglected and underutilized species and community biodiversity management in Southern Brazil. Currently, she is a Researcher at Brazilian Enterprise for Agricultural Research (Empresa Brasileira de Pesquisa Agropecuária, EMBRAPA).

Salma Idris, PhD (Botany, University of Putra Malaysia), has a long record of involvement in taxonomy, botany and agricultural biodiversity. Her research experience encompasses the genera *Durio*, *Mangifera* and *Nephelium* and wild tropical fruits. She is the National Project Coordinator of the TFT project. Currently, she is a Research Fellow at the Malaysian Agricultural Research and Development Institute (MARDI), Serdang, Malaysia.

Samroeng Changprasert, MSc (Bioindustrial Technology, King Mongkut's Institute of Technology Ladkrabang, Bangkok), is an Agricultural Researcher, Chanthaburi Horticultural Research Center, Horticulture Research Institute, Department of Agriculture, and is involved in research in processing products and postharvest technology of horticultural crops.

Sanjay Kumar Singh, PhD, is working as a Senior Scientist (Fruit Science) at the National Research Centre of Litchi, Muzaffarpur, Bihar. He has a record of working on the mango and pummelo diversity of Bihar and Jharkhand and its conservation, including motivating factors for sustainable maintenance to harness livelihood security. He is also working on physiology of shoots responsible for flowering of mango and litchi, pruning and paclobutrazol effects on off season flowering.

Shailendra Rajan, PhD, is Principal Scientist and Head of the Division of Crop Improvement and Biotechnology at the Central Institute for Subtropical Horticulture (CISH), Indian Council of Agricultural Research (ICAR), Lucknow, India. His research focuses on genetic resource management of mango and guava. He has a long record of exploration, characterization and conservation of genetic resources of *Mangifera* and *Psidium*.

Sombat Tongtao, PhD (Postharvest Technology, King Mongkut's University of Technology Thonburi, Thailand), is Director of Chanthaburi Horticultural Research Center, Horticulture Research Institute, Department of Agriculture, and he has experience in research of tropical fruit crops. Focus research areas are postharvest physiology, supply chain and quality management of fruit crops.

Songpol Somsri, PhD (Plant Breeding and Biotechnology, The University of Queensland, Australia), is the Senior Advisor in Plant Production and also Chief of Senior Expert Office, Department of Agriculture, Thailand and National Project Coordinator of the TFT project. His expertise is in plant breeding, plant germplasm conservation, molecular markers and plant production of tropical fruits and horticultural crops.

Suchitra Changtragoon, Doctor of Forestry, is the Forest Technical Expert on forest conservation research at Department of National Parks, Wildlife and Plant Conservation, Thailand. She has more than 25 years of research experience on genetic diversity and genetic resource conservation of tropical trees and partly on wildlife forensics. She has more than 75 publications.

Sudha Mysore, PhD (Agricultural Economics, University of Agricultural Sciences, Bengaluru), also a Fulbright-Nehru Senior Fellowship holder, has been associated with issues relating to agricultural economics, biodiversity conservation, technology transfer through commercialization and intellectual property management. She has several national and international publications.

Supattanakit Posawang, MSc (Posharvest Technology of Agriculture, Chiangmai University, Thailand) is an Agricultural Researcher at Chiangmai Royal Agricultural Research Center, Horticulture Research Institute, Department of Agriculture. He is involved in research on postharvest technology of tropical and subtropical fruit crops and herbs.

Susi Lesmayati is a Postharvest Scientist at South Kalimantan Assessment Institute of Agricultural Technology, Indonesian Agency for Agricultural Research and Development (IAARD), Indonesia. She is focusing on postharvest issues, both handling and processing. She has experience in guiding women farmers in the processing of agricultural products, especially fruits and vegetables.

Tawatchai Nimkingrat, MS (Horticulture, Kasetsart University, Bangkok), Senior Agricultural Scientist, Director of Srisaket Horticultural Research Center, Horticulture Research Institute, Department of Agriculture, has experience in research on cultivation of tropical fruit crops. He is very interested in germplasm collection of tropical fruits, vegetables and herbs.

T.M. Gajanana, PhD (Agricultural Economics, Indian Agricultural Research Institute, New Delhi), has been involved in production, marketing and impact assessment of horticultural crops. Currently he is working on agricultural biodiversity conservation and linking it to markets. He is Principal Scientist (Agricultural Economics) of the Indian Council of Agricultural Research (ICAR) at the Indian Institute of Horticultural Research, Bengaluru, India.

Vasudeva R, PhD (Plant Genetics, University of Agricultural Sciences, Bangalore), is a Professor of Forest Biology in the University of Agricultural Sciences, Dharwad, India. Working extensively in the Western Ghats of India for the last 20 years, he has rich experience in the conservation and use of forest genetic resources, tree domestication and community biodiversity management.

V. Dakshinamoorthy, MA, MPhil in Economics (University of Madras and Madurai Kamaraj University, respectively) has 34 years of experience in agro-economic research. He has carried out work on production, marketing, export and postharvest loss estimation pertaining to horticultural crops.

Vishal Nath, PhD, is working as Director, ICAR-NRC on Litchi, Muzaffarpur, India and has worked on various horticultural crops for the last 21 years, developed 11 varieties, identified 30 promising germplasm lines of various horticultural crops, and developed several agro-techniques such as modified stone grafting in mango and canopy architecture management in mango.

V. Ramanatha Rao, PhD in Plant Breeding and Genetics, has worked in plant genetic resources since 1975, at ICRISAT (Hyderabad, India) and at Bioversity International (Italy, Singapore and Malaysia). He has expertise in plant genetic diversity, conservation (both *ex situ* and *in situ*) and in using different crop and forest genetic resources for sustainable yields. He has contributed to several projects and activities dealing with agricultural biodiversity conservation issues in Brazil, Germany, China, India, Indonesia, Sri Lanka, Malaysia, Mongolia, Laos, Thailand, Papua New Guinea, Fiji and Vietnam. He has authored/co-authored more than 300 publications.

William W.W. Wong, MSc (Technology of Crop Protection, University of Reading, England), is Head of Research at the Agriculture Research Centre, Tuaran, Sabah, Malaysia, and has over 25 years of involvement in work on genetic resource conservation of tropical fruits and underutilized species. His main research area is fruit tree agronomy and germplasm conservation.

Yayun Yang is a Plant Pathologist working for the Yunnan Academy of Agricultural Sciences (YAAS), China. She has been working on using rice diversity for pest and disease control for several years.

Yunyue Wang is Vice Director General, Chinese Ministry of Education Key Laboratory of Agrobiodiversity and Plant Disease Management at the Yunnan Agricultural University (YAU), China. She has been working as a Plant Pathologist on using crop genetic diversity for pest and disease management for several years.

Zahimi Hassan, Senior Agriculture Officer, has a long record of involvement in horticultural crops including tropical fruits. He has long years of service with the Department of Agriculture, Malaysia in extension work in various locations in Peninsular Malaysia. He has been the Director of Integrated Agriculture Development Area (IADA) Kerian, Perak since 2012.

Zahirotul Hikmah Hassan is a Postharvest Scientist involved in several research activities. She spent eight years working in South Kalimantan Assessment Institute for Agricultural Technology (IAAT) before moving to the Indonesian Center for Agricultural Postharvest Research and Development (ICAPOSTRD), Indonesian Agency for Agricultural Research and Development (IAARD), MoA, Indonesia. Currently she is a PhD scholar at Wageningen University and Research Centre in the Netherlands.

Abbreviations and acronyms

AFLP	Amplified Fragment Length Polymorphism
ASI	Agribusiness Systems International
ATMA	Agriculture Technology Management Agency, India
BPTP	Balai Pengkajian dan Penerapan Teknologi Pertanian (Indonesian: Assessment Institute for Agricultural Technology)
CAP	Community Action Plan
CBD	Convention on Biological Diversity
CBFM	Community-Based Forest Management
CBM	Community Biodiversity Management
CBNRM	Community-Based Natural Resource Management
CBO	Community-Based Organization
CBR	Community Biodiversity Register
CBT	Community-Based Tourism
CCAFS	Research Program on Climate Change, Agriculture and Food Security
CCER	Centre-Commissioned External Review
CGIAR	Consultative Group of International Agricultural Research
CDI	Centre for Development Innovation
CFTRI	Central Food Technological Research Institute
CIAL	Comité de Investigación Agrícola Local (Spanish: Local Agricultural Research Committee)
CIFOR	Center for International Forestry Research, Indonesia
CIP	International Potato Center, Peru
CISH	Central Institute for Subtropical Horticulture, India
CLIP	Collaboration/Conflict, Legitimacy, Interests and Power
CSR	Corporate Social Responsibility
CWR	Crop Wild Relative
DFID	Department for International Development
DFS	Diversity Field School
DFSC	Danida Forest Seed Centre
DID	Department of Irrigation and Drainage
DLP	Diversity for Livelihood Programme

DNP	Department of National Parks, Wildlife and Plant Conservation, Thailand
DOA	Department of Agriculture
DOF	Department of Forestry
EMBRAPA	Empresa Brasileira de Pesquisa Agropecuaria (Brazilian Agricultural Research Corporation)
ESE	Ecosystem Services Economics
FA	Farmers' Association
FAMA	Federal Agriculture Marketing Authority, Malaysia
FAO	Food and Agriculture Organization of the United Nations, Italy
FCA	Four Cell Analysis
FFS	Farmer Field School
FG	Farmers' Group
FGD	Focus Group Discussion
FGR	Forest Genetic Resources
FPIC	Free, Prior Informed Consent
FSC	Forest Stewardship Council
FSMP	Forestry Sector Master Plan
GDA	Genetic Diversity Analysis
GEF	Global Environment Facility
GI	Geographic Indication
GIAHS	Globally Important Agricultural Heritage Systems
GMZ	Gene Management Zone
GPD	Good Practice for managing Diversity
HCA	Hydroxycitric Acid
HG	Home Garden
HH	Household
IARD	International Agriculture and Rural Development, Indonesia
IBPGR	International Board for Plant Genetic Resources (now Bioversity International), Italy
ICAR	Indian Council of Agricultural Research
ICCAs	Indigenous and Community Conserved Areas
ICHORD	Indonesian Center of Horticulture Research and Development
IDRC	International Development Research Centre, Canada
IFAD	International Fund for Agricultural Development, Italy
IFOAM	International Federation of Organic Agriculture Movements
IFPRI	International Food Policy Research Institute, USA
IFTS	Indigenous Fruit Tree Species
IGNOU	Indira Gandhi National Open University
IIED	International Institute for Environment and Development, London
IMO	Indigenous Microorganism
INPA	National Institute for Amazonian Research, Brazil

IPGRI	International Plant Genetic Resources Institute (now Bioversity International), Italy
IPM	Integrated Pest Management
IRD	Informal Research and Development
ITTO	International Tropical Timber Organization
IUCN	International Union for Conservation of Nature
JBES	Journal of Biodiversity and Environmental Sciences
JBIC	Japan Bank for International Cooperation
KHG	Kandyan Home Garden
KVK	Krishi Vigyan Kendra (India: Agricultural Science Centre)
LI-BIRD	Local Initiatives for Biodiversity, Research and Development, Nepal
LIC	Life Insurance Corporation of India
LMDH	Lembaga Masyarakat Desa Hutan (Indonesian: Forest Village Community institution)
MACAB	Marketing Approach to Conserve Agricultural Biodiversity
MADAS	Ministry of Agriculture Development and Agrarian Services, Sri Lanka
MARDI	Malaysian Agriculture Research and Development Institute
MENR	Ministry of the Environment and Natural Resources
MFE	Ministry of Forestry and Environment
MI	Margalef Index
MOGA	Maharashtra Orange Growers' Association, India
MOSTI	Ministry of Science Technology and Innovation
NARC	Nepal Agricultural Research Council
NBARD	National Bank for Agricultural and Rural Development, India
NBPGR	National Bureau of Plant Genetic Resources, India
NCSD	National Council for Sustainable Development
NGO	Non-Governmental Organization
NHB	National Horticulture Board
NRCC	National Research Centre for Citrus, India
NTFP	Non-Timber Forest Product
NUS	Neglected and Underutilized Species
OTOP	One Tambon One Product
PACS	Payment for Agrobiodiversity Conservation Services
PGRC	Plant Genetic Research Centre
PGS	Participatory Guarantee Systems
PMCA	Participatory Market Chain Approach
PMDH	Pembinaan Masyarakat Desa Hutan (Indonesian: Community Development Village Forest)
PNAS	Proceedings of the National Academy of Sciences
PPV&FRA	Protection of Plant Varieties and Farmers' Rights Authority, India
PRA	Participatory Rural Appraisal
PVS	Participatory Variety Selection

RAPD	Random Amplified Polymorphic DNA
RAU	Rajendra Agricultural University, India
RE	Retinol Equivalents
REDD	Reducing Emissions from Deforestation and Forest Degradation
REI	Resource Exchange International, Indonesia
REST	Responsible Ecological Social Tours Project
RFD	Royal Forestry Department, Thailand
SAAS	Sichuan Academy of Agricultural Sciences, China
SANPGR	South Asia Network on Plant Genetic Resources
SCMD	Society for Conservation of Mango Diversity, India
SCS	Seed Conservation Service
SEDC	Sarawak Economic Development Corporation, Malaysia
SHG	Self-help Group
SI	Simpson Index
SkTGA	Sarawak Tourist Guide Association, Malaysia
SSA	Sub-Saharan Africa
SSR	Simple Sequence Repeat
TFT	Tropical Fruit Tree
TFTGR	Tropical Fruit Tree Genetic Resources
TRIPS	Trade-Related Aspects of Intellectual Property Rights
UN	United Nations
UNDP	United Nations Development Programme
UNEP	United Nations Environment Programme
UPWARD	Users' Perspectives With Agricultural Research and Development
VCD	Value Chain Development
VFC	Village Forest Committee
WTO	World Trade Organization
YAAS	Yunnan Academy of Agricultural Sciences, China

Species

Note: C. = Citrus, G = Garcinia, M = Mangifera, N = Nephelium

Aroi aroi	*G. forbesii*
Asam gelugor	*G. atroviridis*
Assai	*Euterpe oleracea*
Avocado	*Persea americana*
Bacuri	*Platonia insignis*
Bael trees	*Aegle marmelos*
Banana	*Musa* spp.
Baobab	*Adansonia digitata*
Bitter beans (petai)	*Parkia speciosa*
Brazil nut	*Bertholletia excelsa*
Buriti	*Mauritia flexuosa*
Butiá or jelly palm	*Butia odorata*
Cacao	*Theobroma cacao*
Cashew	*Anacardium occidentale*
Cha muang	*G. cowa*
Champak	*Michelia champaca*
Chocolate berry	*Vitex payos*
Christ's thorn jujube	*Ziziphus spina-christi*
Cleopatra mandarin	*C. reshni*
Cupuassu	*Theobroma grandiflora*
Date palm	*Phoenix dactylifera*
Desert date	*Balanites aegyptiaca*
Domesticated madan	*G. schomburgkiana*
Durian	*Durio zibethinus*
Galgal	*C. pseudolimon*
Gingerbread palm	*Hyphaene thebaica*
Guava	*Psidium guajava*
Iron wood	*Hopea odorata*
Jujube	*Ziziphus mauritiana*
Kafir lime	*C. hystrix*
Kasturi	*M. casturi*

Kokum	*G. indica*
Kuini	*M. odorata*
Langsat	*Lansium domesticum*
Lemon grass	*Cymbopogon* spp.
Lemon	*C. limon*
Longan	*Dimocarpus longan*
Malacca teak	*Intsia palembanica*
Mandarin	*C. reticulata*
Mangaba	*Hancornia speciosa*
Mango	*M. indica*
Mangosteen	*G. mangostana*
Marula	*Sclerocarya birrea*
Nagpur mandarin	*C. reticulata*
Orange	*C. sinensis*
Papaya	*Carica papaya*
Passion fruit	*Passiflora edulis*
Pawpaw	*Carica papaya*
Pequi or Souari nut	*Caryocar brasiliense*
Pineapple	*Ananas comosus*
Pomelo	*C. grandis*
Pupunha	*Bactris gasipaes*
Rambutan	*N. lappaceum*
Rangpur lime	*C. limonia*
Red/pink-fleshed or acid lime	*C. aurantifolia*
Rose apple	*Syzygium aquem*
Rough lemon	*C. jambhiri*
Salak	*Salacca zalacca*
Som jeed or Som Mapit	*C. madurensis*
Sweet lemon	*C. medica*
Sweet orange	*C. sinensis*
Tamarind	*Tamarindus indica*
Tapang	*Koompassia excelsa*
Tucuma	*Astrocaryum aculeatum*
Uxi	*Endopleura uchi*
Waterberry	*Syzygium guineense*
White crossberry	*Grewia tenax*
Wild custard apple	*Annona senegalensis*
Wild madan	*G. fusca*

Acknowledgements

This book is a result of the project 'Conservation and Sustainable Use of Cultivated and Wild Tropical Fruit Diversity: Promoting Sustainable Livelihoods, Food Security and Ecosystem Services', implemented in India, Indonesia, Malaysia and Thailand from 2009 to 2014. The project was financed by the Global Environmental Facility (GEF) to meet the goals of the Convention on Agricultural Biodiversity (CBD) and was implemented by the United Nations Environment Programme (UNEP). The project was regionally executed by Bioversity International in collaboration with the Indian Council of Agricultural Research (ICAR), New Delhi, India; the Indonesian Centre for Horticulture Research and Development (ICHORD), Jakarta, Indonesia; the Malaysian Agricultural Research and Development Institute (MARDI), Kuala Lumpur, Malaysia; and the Department of Agriculture (DOA), Bangkok, Thailand.

The work presented here would not have been possible without the support of the following institutions in the participating countries.

India: Central Institute for Subtropical Horticulture (CISH), College of Forestry, Indian Institute of Horticultural Research, National Bureau of Plant Genetic Resources (NBPGR), National Horticulture Board, National Research Centre for Citrus (NRCC), National Research Centre for Litchi (NRCL), Protection of Plant Varieties and Farmers' Rights Authority (PPV&FRA), DHAN Foundation, Life Trust, EcoWatch, Maharashtra Orange Growers' Association (MOGA), Society for Conservation of Mango Diversity (SCMD), YSR Horticultural University.

Indonesia: Indonesian Agency for Agricultural Research and Development (IAARD), Indonesian Tropical Fruit Research Institute (ITFRI), Assessment Institute for Agricultural Technology (BPTP), East Java and South Kalimantan, Indonesian Centre for Agricultural Biotechnology and Genetic Resources Research and Development (ICABIOGARD), Resource Exchange International (REI) NGO.

Malaysia: Department of Agriculture Malaysia, Sabah Department of Agriculture, Sarawak Department of Agriculture, Semongok Agriculture Research Centre, Federal Agriculture Marketing Authority (FAMA), Peninsular Malaysia Department of Forestry, Sarawak Forestry Department, Sabah Forestry

Department, Forest Research Institute of Malaysia (FRIM), International Medical University (IMU) Malaysia, Universiti Putra Malaysia (UPM), Forestry Departments Peninsular Malaysia, Sabah and Sarawak.

Thailand: Department of Agriculture, Chiangmai Royal Agricultural Research Centre, Sisaket Horticultural Research Center, Trang Horticultural Research Center, Horticultural Research Institute, Department of National Parks, Wildlife and Plant Conservation, Walailak University and Kasetsart University.

We would like to acknowledge all key actors for their unshaken commitment and support during implementation of the project activities. At the country level, the following people played key roles in coordination and support:

India: H.P. Singh, N.K. Krishna Kumar, B.M.C. Reddy, V.A. Parthasarathy, Awtar Singh, Sanjay Kumar Singh, Shailendra Rajan, M.R. Dinesh, I.P. Singh, Vasudeva R, Narasimha Hedge, T.M. Gajanana, Dinesh Kumar, K.V. Ravishankar, Sudha Mysore, S. Vinoth, Prakash Patil, Amrik Singh Sidhu, V.J. Shivankar, M.S. Ladaniya, H. Ravishankar, Vishal Nath, C. Vasugi, M.P. Vasimalai, M. Palanisamy, M. Kiran Kumar, Barsati Lal, V. Dakshinamoorthy and M.B. Nayak.

Indonesia: Yusdar Hilman, Muhammad Prama Yufdy, Sudarmadi Purnomo, Anto Hardiyanto, Nono Sutrisno Saad, Idha Widi Arsanti, Winarno, Achmad Rahieq, Kunturo Boga Andri, Putu Bagus Daroini, Adhitya Kiloes, Dian Kurniasih, Linda Guno, Rima Setiani, Endro Gunawan and Z. Hikmah Hassan.

Malaysia: Sharif Haron, Salma Idris, Muhammad Shafie Md Sah, Abd. Shukor Rahman, Mohd Yusoff Abdullah, Mohd. Norowi Hamid, Zahimi Hassan, William W.W. Wong, Pearlycia Brooke, Shariah Umar, Yusof Talib, K Maya Izar Khaidizar, D. Garten, Mohd Syauqi Nazmi, Abdul Latif Ahmad Zabidi, Lau Cheng Yuon, Shukor Abdullah, Norhayati Md. Harun, Stephen Joseph Lim, Jamaluddin Lani and Termize Alui.

Thailand: Somchai Channarongkul, Jirakorn Kosaisawi, Dumrong Jirasutas, Anan Suwannarat, Vina Pongpattananon, Suwit Chaikiattiyos, Chamrong Daoruang, Songpol Somsri, Margaret Yoovatana, Chatchanok Noppornphan, Pichit Sripinta, Suchitra Changtragoon, Tawatchai Nimkingrat, Montree Issarakraisila, Nataya Dumampai, Samroeng Changprasert, Sombat Tongtoa, Thaveesak Sangudom, Kasemsak Palakorn, Satja Prasongsup, Suphalak Ariyaphuchai, Rachanee Siriyan, Auttapon Rukkaphan and Supattanakit Poswang.

Felipe S. dela Cruz, Dorothy Chandrabalan, Percy Sajise, Judith Thompson, Lee Choon Pei and Ambika Thapa helped us on the first part of the journey when we developed the rationale for the book, designed the case studies, drafted the analytical framework for the analysis of functions and services and collected the first reference materials. Their assistance is greatly appreciated.

The idea of publishing on good practices for diversity in the context of the series ‘Issues in Agricultural Biodiversity’, a collaboration between Bioversity International and Earthscan (Routledge), was triggered by encouragement from our colleagues Danny Hunter and Michael Halewood (series editors), for which we are very grateful.

Our special thanks go to the authors of six external case studies and 18 case studies by the project partners. We must also thank a number of reviewers for various case studies at different times. They include: Percy Sajise, Judith Thompson, David Williams, Per Rudebjer, Paul Quek, Danny Hunter, Mauricio Bellon, Marlène Elias, Riina Jalonen, VA Parthasarathy, Salma Idris, Jacob Zuckerburg, Claire Willey, Pitambar Shrestha, Ronnie Vernooy, Michael Halewood, Adriana Alercia, Ehsan Dulloo, Sajal Sthapit, Froukje Kruijssen, B.M.C. Reddy, Vasudeva R, Matthias Jaeger, Stefano Padulosi, Devendra Gauchan, Margaret Yoovatana, Sara Hom, Hannah Jackson, Camilla Zanzanaini, Toby Hodgkin, Devra Jarvis, Pablo Eyzaguirre, Gennifer Meldrum, Walter de Boef, Stephan Biggs, Abhiskar Subedi and M.S. Sabran.

We would also like to thank Marieta Sakalian from UNEP, Ramni Jamnadass from ICRAF, S. Ann Tutwiler, Stephen Weise, Laura Snook, Maurico Bellon, Devra Jarvis and Ehsan Dulloo for encouraging the idea of the publication.

We are very much grateful to the Department of Agriculture, Government of Thailand for supporting a bulk purchase of copies of the book from the publisher. Songpol Somsri and Margaret Yoovatana are gratefully acknowledged for this effort.

Francesca Giampieri in the Bioversity library was tireless in searching for many useful references and contacting people in her networks who knew someone or something useful for the book.

Tim Hardwick and Ashley Wright steered the publication process smoothly. We would like to express our special gratitude to the team who at various times coordinated and shouldered the whole process until the manuscript was submitted to the publishers: Maninder Kaur, Safal Khatiwada and Richa Gurung.

Most of all, we would like to specially mention the rural communities, custodian farmers, fruit tree farmers and orchardists for sharing their knowledge and time for global benefit freely and without any reservations.

Any errors, omissions and misleading statements are, of course, solely our responsibility.

The editors

Foreword

In my 30 years of experience in agricultural policy and development, I have seen time and time again how successful approaches need to be 'owned' by the people they are designed to help. Ownership comes with recognition that people are active agents of their own development, holders of knowledge about practices and skills honed over time that make sense in their context. Smallholder farmers around the world have intimate knowledge about the plants they tend. Working together with scientists, both can learn the principles behind successful agronomic, social and economic practices that can improve their livelihoods sustainably.

Half a year after the Sustainable Development Goals were adopted in New York, this book comes as welcome vision of what 'sustainable' can mean and how to get there. Most people are familiar with the concept of 'sustainability' as making progress simultaneously on the fronts of the environment, society and the economy. It is a delicate balance: too much focus on conservation and people may not gain the full economic benefits they might; too much emphasis on economic development and the environment might irreversibly suffer.

Emerging from Bioversity International's focus on effective genetic resource conservation and use, the editors' vision was to provide guiding principles and concepts strongly embedded in practical case studies for our work on how farmers use agricultural biodiversity to improve their livelihoods, sustainably, in different ways. While the book focuses on tropical fruit trees, the messages are applicable to all rural communities where genetic resources are available as natural resources to exploit and enhance for better livelihoods.

Historically fruit tree breeding has taken place as small, individual and isolated efforts. It is often a losing endeavour as financial return to investment is slow and uncertain. As a result, there has been low public, and even lower private, sector investment in fruit tree research and development, particularly in some of the low-income countries where Bioversity International works. CGIAR research on fruit trees is also limited. Yet, in the current context of a renewed focus on integrated agro-food systems, fruit tree research in smallholder home gardens and orchards provides critical contributions to address challenges such as malnutrition and hidden hunger, poverty, climate change and environmental degradation.

What is needed today is to understand and implement different and innovative approaches – far different from the conventional green revolution type of crop improvement. This book shows the way to make this a reality. It demonstrates how to integrate the innovation of smallholder farmers into large-scale research for development by selecting farmers' best fruit varieties, employing locally developed good practices and restoring degraded lands with diverse fruit species that at the same time sustainably improve the diets of the poor.

For Bioversity International, this book represents a milestone, presenting key concepts of good biodiversity management that the authors have matured over seven years of research. By embedding practice within theory, the authors illustrate and analyse concepts such as good practices in diversity management, custodian farmers and community biodiversity management—backed up with appropriate practical case studies. Based on their experiences across India, Indonesia, Malaysia and Thailand, the authors have extensively documented farmer-developed good practices for maintaining, marketing and safeguarding fruit trees, outlining a framework for on-farm conservation, drawn from the real ways that communities and farmers implement de facto conservation strategies through their everyday practices. Participation of communities leads to increased understanding of local traditions and knowledge that can in turn lead to increases in productivity and income.

The findings suggest that empowering farmers and their institutions, creating space for social learning and innovation and a dynamic system of small-scale innovation might be a sustainable way to mainstream good practices. Experience tells us that approaches, processes, methods and principles are more practical for scaling up than context-specific good practices. This kind of innovation holds promise to be productive, simple, low cost and sustainable in practice.

Tropical Fruit Tree Diversity: Good practices for in situ and on-farm conservation is the result of a successful collaboration between national research partners, including the Indian Council of Agricultural Research (ICAR), Indonesian Centre for Horticulture Research and Development (ICHORD), Malaysian Agricultural Research and Development Institute (MARDI), and Thailand's Department of Agriculture (DoA), as well as the invaluable assistance of over 80 custodian farmers who guided the editors in understanding the roots of good practices and shared their vision of scaling up.

This book is based on work under the Tropical Fruit Tree project 'Conservation and Sustainable Use of Cultivated and Wild Tropical Fruit Tree Diversity: Promoting Sustainable Livelihoods, Food Security and Ecosystem Services', launched in 2009 with funding from the Global Environment Facility (GEF), implemented through the United National Environment Programme (UNEP) and led by Bioversity International. We thank both GEF and UNEP for their valuable support.

Ann Tutwiler,
Director General, Bioversity International

Preface

This book is about knowledge and practices for sustainable management of agricultural biodiversity by farmers and scientists. What we see on the ground is that farmers often blend traditional and modern farming practices that best suit their livelihood needs and support their individual instincts for innovation and survival. This is particularly true among small-scale farmers and for underutilized, under-researched perennial tropical fruit crops. Farmers in many parts of the world have developed a range of good practices for managing diversity (GPD). These are good agricultural practices to sustainably use and maintain a wide diversity of crop varieties. Such practices tend to be simple, practical, cost efficient and sustainable under the given context. They maintain, enhance and create crop genetic diversity, and ensure its availability to and from farmers and other sources. These practices connect society to nature, biodiversity, ecosystem health and human well-being. The benefits of such good practices can be multiplied and scaled up to wider geographical, institutional and socio-cultural settings.

Policymakers, donors and institutional leaders are always interested in promoting good practices – often with external interventions – to improve agricultural intensification without environmental degradation. This book aims to understand practical challenges in scaling up good practices of *in situ* and on-farm conservation of tropical fruit tree diversity and to identify key principles for mainstreaming good practice into development efforts.

Drawing from experiences from a UNEP-GEF project on 'Conservation and Sustainable Use of Wild and Cultivated Tropical Fruit Tree Diversity for Promoting Livelihoods, Food Security and Ecosystem Services' in India, Indonesia, Malaysia and Thailand, this book documents good practices innovated or adopted by smallholder farmers from four countries. The book also includes experiences of experts from other countries, who share good practices of conservation and sustainable management of tropical fruits from their perspective and work. Few books are available that scientifically document the experiences of identifying and implementing good practices of on-farm management of perennial fruit trees. This book attempts to fill this gap by exploring the concept of good practices as a means to understand and improve *in situ* and on-farm management efforts by farmers, communities and states around the world, particularly in the context of poorly researched tropical fruit tree crops.

This is arguably the first book to offer a comprehensive overview and analysis of good practices that support the maintenance of tropical fruit tree diversity *in situ* and also contribute to sustainable livelihoods, food security and ecosystem services. The concept of good practices for diversity is still new and the methodology and approaches are still evolving. The book presents the current status of a conceptual framework, method and experiences of good practice resulting from 24 case studies. Hopefully they will offer new perspectives and approaches in this surprisingly poorly documented field. We hope that this work will stimulate the research community to explore further this new area of work on conservation of diversity through practices that also contribute to custodians' livelihoods. The book may be of interest to anyone working in the field of innovation, knowledge transfer, and mainstreaming good practices of agricultural biodiversity for joint conservation and livelihood benefits. The book also furthers the global understanding of good practice for diversity (GPD) management as a methodology which contributes to *in situ* conservation and implementation of farmers' rights regarding plant genetic resources.

Bhuwon Sthapit
Hugo Lamers
Ramanatha Rao
Arwen Bailey

Part 1

Setting context

1 On-farm and *in situ* conservation of tropical fruit tree diversity

Context and conceptual framework

Bhuwon Sthapit, Hugo A.H. Lamers, V. Ramanatha Rao, Arwen Bailey, Percy Sajise and Paul Quek

The origins of the book

Wild and cultivated tropical fruit tree diversity in Asia is threatened by rapid genetic erosion due to the destruction of natural ecosystems, commercialization, land use changes, global climate change and a variety of other socio-economic and cultural pressures. Despite these pressures, some farmers continue to successfully manage a range of tropical fruit tree diversity in different production systems, reaping benefits in terms of nutritional and food security, income-generating opportunities, ecosystem services or cultural identity. They identify, select, propagate, manage, use and promote fruit tree diversity through local social networks, market linkages, community rules and local customary practices. These unique management practices can serve as a valuable knowledge base for the sustainable management of tropical fruit tree genetic resources in other geographic areas with similar social, economic and ecological contexts today and in the future. This book aims to document these good practices for maintaining diversity so that farmers and other practitioners can fully and sustainably benefit from the unique diversity conserved in and near their lands. The identification of good practices can help researchers, government institutions and other farmer-support organizations to plan and conduct better informed and targeted interventions within their on-farm and *in situ* conservation programmes and projects.

From 2009 to 2014, Bioversity International coordinated a research-for-development project supported by the Global Environment Facility (GEF) with implementation support from the United Nations Environment Programme (UNEP), 'Conservation and sustainable use of cultivated and wild tropical fruit tree diversity: sustainable livelihoods, food security and ecosystem services', abbreviated as the 'TFTGR Project'. This project, implemented in India, Indonesia, Malaysia and Thailand, focused on livelihood and environment benefits that people could derive from the conservation of species and varietal diversity of *Citrus*, *Garcinia*, *Mangifera* and *Nephelium*. These common tropical fruits include mandarin (*C. reticulata*), pomelo (*C. grandis*), mangosteen (*G. mangostana*), mango (*M. indica*) rambutan (*N. lappaceum*) and their edible wild relatives.

Why a focus on conservation of tropical fruit trees?

Tropical Asian countries are the centre of origin and diversity of many globally important tropical fruit tree species and their wild relatives. These species contribute to the well-being of human communities by providing a source of supplementary food, sustaining healthy diets and enhancing both household incomes and national revenues (Arora and Ramanatha Rao, 1998).

Fruit trees have recalcitrant seeds (i.e. they die if dried or frozen), which means that they cannot easily be maintained in genebanks. This makes it important to find solutions for their on-farm and *in situ* conservation. Natural reproduction of tropical fruit trees requires hot and humid conditions for germination and depends largely on their interaction with animal pollinators and fruit and seed dispersers. This unique reproduction system together with their perennial nature raises specific challenges (long investment period before initial fruiting, difficulty in propagation, limited breeding potential) and opportunities (long productive lifespan, low labour input crop, staggered harvest timings). It also determines their specific role in the agroecosystem as host and food source for pollinators, shade provider, conserver of soil organic matter, nutrient recycling enhancer, food for wild fauna and retainer of water.

Tropical fruit tree species have traditionally been selected to suit the specific hot and humid environments in which they have been cultivated or to satisfy the particular needs and preferences of local growers and consumers. Through natural selection processes, wild tropical fruit species found in this region have developed specific adaptive traits and qualities. Farmers' varieties grown in fields and gardens are the products of domestication of wild trees taken from natural forest areas, then subjected to the historical process of farmers' continued selection for preferred traits over several generations and for their sustainable management practices in traditional systems: agroecosystems (such as buffer zones surrounding villages, borders between farm fields or along paths), horticultural (semi-commercial or commercial orchards) or agri-silvicultural (home gardens or agroforestry). These management practices have been defined as on-farm conservation (Altieri and Merrick, 1987; Bellon, 1996; 2004; Maxted *et al.*, 2002). See Chapter 2 for a more detailed account of on-farm conservation. Planting materials of tropical fruit tree diversity tend to be sourced via social networks, where certain individuals with superior enthusiasm, skills and knowledge ('custodian farmers') play a key role (Sthapit *et al.*, 2013). See Chapter 4 for discussion of the concept of custodian farmer.

Despite their cultural and economic importance in Asia, there has been a lack of extensive research on the cultivation and management of these perennial tropical fruit tree species. Global and national investment in tropical fruit tree research is meagre because of the high costs, length of time needed and the limited economic importance and critical research mass of these species in Western countries. In such a context, a cost effective and efficient method for research-for-development interventions is to identify 'good practices' from scientific research and from farmers' innovation, which can be tested on site, strengthened and tried out in new sites (Figure 1.1).

The rest of this chapter will (1) provide a conceptual framework for what constitutes a 'good practice' in the context of on-farm and *in situ* conservation of tropical fruit tree diversity; (2) outline the steps taken to identify good practices for diversity management of tropical fruit trees; and (3) outline the main findings about categories, types and contexts of good practices as an analytical tool. Then in Chapter 2 more detail is given on key concepts used in this book; in Chapter 3 the characteristics of Community Biodiversity Management (CBM) are discussed, an approach whose principles underpin this research; and in Chapter 4 one of the major findings of this research, the existence and role of 'custodian farmers', is considered. Part 2 consists of six chapters from experts from countries outside the project countries to expand the scope of the book and to give the current status and examples of good practices found elsewhere. In Part 3 we present 19 case studies (that were selected from 33 original case studies), each documenting, under four different

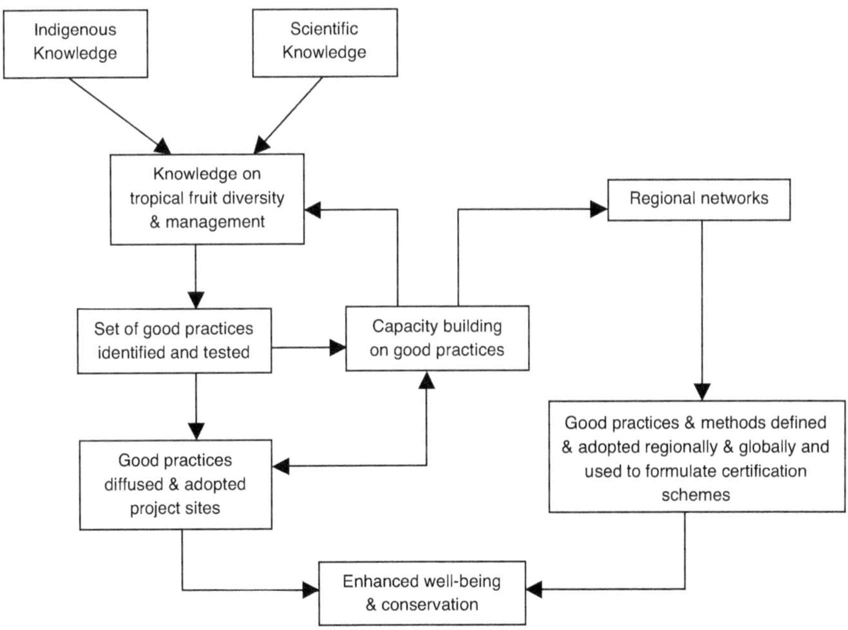

Figure 1.1 Impact pathway for enhanced community well-being and conservation of tropical fruit tree diversity.

categories, different good practices for diversity identified and tested during the project. Finally, in Part 4 we share reflections on identifying, documenting, piloting and mainstreaming good practices for diversity management and offer principles for identifying and using good practices for diversity as a combined livelihood and conservation tool, based on the lessons learned during the project.

Good practices

'Good practice' and 'best practice' are widely used terms in agriculture, manufacturing and the processing of products. Often they refer to a farm or production technique that can be adopted by farmers or companies to improve harvest yields or ensure the standardized quality of products. They are often measurable practices that include certification schemes to ensure compliance towards agreed standards. In the context of biodiversity conservation and poverty alleviation, good practices are considered by the Convention on Biological Diversity (CBD) to be an effective way to document and share tools, instruments and methods (CBD, undated; Gemmill, 2001). Sometimes the term 'best practice' is also used, but we consider 'good practice' more acceptable due to the complexity of on-farm conservation practices and the inability to measure or compare practices (GEF, 2001).

Practices are considered 'good' when they effectively work towards the achievement of certain objectives under a given set of conditions or contexts. A practice can be a technique, a method, a process, an institutional arrangement or any combination of these. Good practices should follow some criteria, as they should be practical, cost-efficient, sustainable, and have the potential for scaling up to wider geographical, institutional and socio-cultural spheres (Sthapit *et al.*, 2003).

A good practice for diversity management is defined as a system, organization or process that over time and space maintains, enhances and creates crop genetic diversity and ensures its availability to and from farmers and other actors for improved livelihoods on a sustainable basis. It is abbreviated as GPD (Sthapit *et al.*, 2003).

The argument for a focus on good practices is that it constitutes a low-cost approach for on-farm and *in situ* conservation. Instead of starting from scratch, it allows the practitioner (whether a farmer, an extension agent or a researcher) first to recognize and understand existing practices and then to build further on these. A wealth of documented case studies exist about the functions and values of agricultural biodiversity (Frankel and Soule, 1981; Brush, 2000; Maxted *et al.*, 2002; Bellon, 2004; Smale *et al.*, 2004; Heywood and Dulloo, 2005; Jarvis *et al.*, 2007; 2011); however, limited research has focused on what could be successful intervention strategies for on-farm and *in situ* conservation, how to strengthen or promote agricultural biodiversity efficiently and effectively on the ground, and what approaches can be systematically replicated that government agencies can support. This book strives to fill this gap by exploring the concept of good practice as a means to understand and improve on-farm and *in situ* conservation efforts of farmers, communities and researchers around the world. In particular, it looks at the good practices involved in managing and conserving tropical fruit tree diversity.

As many farmers' livelihoods depend on the success of their crops in any given year, good practices are key for perpetuating a long-term cycle of sustainable development and livelihood improvement. GPDs combine the achievement of both improved livelihoods and conservation in a given context. Most GPDs stem from traditional agricultural practices. In agricultural research, large-scale monocropping and modern agricultural practices have tended to be the main interventions considered for improving the livelihoods of farmers, thus contributing to loss of varietal and species diversity as well as traditional agricultural practices. Our contention is that these interventions are not the most appropriate in all contexts. Biological diversity and traditional systems can help many a poor smallholder farmer to improve their livelihood as they are better adapted to local socio-economic and environmental conditions and give farmers a range of options to manage climate or market risks. These risks are especially apparent in regions that can be characterized as remote, marginal and with limited market infrastructure, the areas where agricultural biodiversity is mostly still found and where high input-oriented monocropping systems often have had limited positive impacts on livelihoods. Farmers make their livelihood decisions based on multiple types of benefits in which home use and social and cultural factors play a major role besides private economic gains. What we see on the ground is that farmers often combine modern and traditional farming practices (and seed materials) that best suit their own interests and support their individual instinct for innovation and survival. In Chapters 3 and 4 we elaborate more on how conservation practices are embedded within sustainable livelihoods, and in Chapter 22 on the contribution of markets to conservation practices.

Thus, good practices can be farmers' own innovations or can be developed by formal research and development agencies and later adapted by farming communities in a variety of local contexts. In this book we focus first on farmer-innovated practices, as these tend to be overlooked by the research and government sectors. We also provide, however, examples of contexts where farmer-innovated practices did not exist and where the adoption of practices developed by the formal sector has allowed farmers to create livelihood gains and improve the management of plant genetic diversity on farm. Additionally we invited external experts to share their own experiences and perceptions of good practice to expand the scope of the book. See Table 1.1 for a summary.

Process of identifying good practices

The process used for identifying good practices for the conservation and sustainable use of tropical fruit tree diversity included the following steps:

1. A literature review to attain a better conceptual understanding of GPDs within the wider context of (a) on-farm and *in situ* conservation, (b) sustainable livelihoods, (c) Community Biodiversity Management (CBM) and (d) the specific context of cultivated and wild tropical fruit tree species.
2. A workshop with experts and partner institutions to understand and agree on the definition, types, evaluation criteria and methodology for the identification and promotion of good practices. Through discussions at this workshop, four broad topic areas were identified by which good practices maintain and enhance tropical fruit tree diversity:
   - propagation and planting material management
   - production and crop management
   - commercialization and home use
   - collective action and social networking
3. Inventory of potential good practices. Multiple sources of information and collection methods were used to inventory good practices: (1) review of scientific and popular literature, (2) review of case studies as encountered in the field, (3) formulation of additional case studies based on interviews and direct field observations with specific individuals and (4) experimental learning and innovation based on ongoing projects and case studies.
4. Development and refinement of evaluation criteria. A set of descriptors for good practices was developed from the good practice definition that served as a starting point for describing, characterizing and rapidly screening good practices to shortlist selected good practices for further research (Box 1.1).
5. Further development of conceptual framework. Given the multiple desired outcomes of livelihood and conservation benefits, it was agreed to use a sustainable livelihood framework when piloting and assessing good practices for diversity (Sajise, 2005; Sajise and Keizer, 2005). The sustainable livelihood framework recognizes five different forms of capital or assets of a household – human, social, natural, financial and physical – and explains how they are deployed by household members in livelihood activities to achieve certain livelihood goals and to deal with external shocks, uncertainties or policies (DFID, 1999). See Chapter 2 for more details on this framework. In the sustainable livelihood framework, agricultural biodiversity is embedded within the asset of natural capital – the seeds or tree crops that the farming households possess – but it is also part of the external environment in which the household lives, through the provision of ecosystem services such as lower pest and disease pressure, shade or the retention of water, which they enjoy from the direct local environment. Despite rich bio-wealth and traditional knowledge, poor and smallholder farmers often feel helpless when dealing with their current situations and, therefore, empowering such farmers to develop self-confidence through social capital building and livelihood enhancement is one essential part of good practice interventions.

Box 1.1 Descriptors for screening good practices for diversity (GPD) derived from the good practice definition

1. GPD in action over time and space
   Description: GPD that has been in use for more than one year and in more than one location
   0 = not sure, 1 = no, 2 = yes

2. GPD crop genetic diversity
   Description: Describe the GPD impact on crop genetic diversity
   –1 = reduces, 0 = no impact or not sure, 1 = maintains, 2 = enhances, 3 = creates

3. Genetic diversity availability to and from farmers
   Description: Indicate whether the crop genetic diversity used for implementing the practice is available to farmers and others
   0 = not available or not sure, 1 = available to farmers, 2 = derived from farmer knowledge, 3 = available to and derived from farmer knowledge

4. GPD improving livelihoods
   Description: Indicate if the GPD has any impact on improving the livelihoods of farmers
   0 = not sure, 1 = no, 2 = yes

5. GPD is economically sustainable
   0 = not sure, 1 = no, 2 = yes

6. Analysis of selected good practices to identify ways to strengthen the good practice on site and how to replicate it elsewhere. National partners were requested to follow a specific outline (3–5 pages long) and provide a few key tables to describe the good practices. Every good practice proposed was described and analyzed using the five assets of the sustainable livelihood framework to understand how the practice affects livelihoods and could be improved and implemented. In addition, all practices were analyzed by describing the driving forces and barriers favouring or hindering success.
7. Exploration of ways to strengthen existing good practices. The project used a CBM approach, which is a participatory methodology supporting community empowerment through diversified use of biodiversity resources. Many research-for-development tools have been developed in

the context of the CBM approach (e.g. Four Cell Analysis, Diversity Fairs, Diversity Kits, Participatory Diversity Selection, CBM fund), with the aim of strengthening and facilitating existing processes. This methodology and tools are described in more detail in Chapter 3.

8. Further analyses. Papers presented by the national partners and international experts were reviewed to draw lessons and principles from the key areas of research identified. Some of these are outlined in Part 4.

What we found

Contexts of good practice

Tropical fruit tree genetic resources were found to be traditionally managed in two overarching contexts:

- in communities interacting with natural forests or protected areas, mostly in remote hilly to mountainous landscapes with limited access to services and support systems including markets
- in on-farm or home garden systems or orchards in which communities are engaged in intensive horticultural practices in flat agricultural landscapes with good access to markets by substantial physical infrastructure such as power, roads and communications.

Context 1: Natural forests and protected areas

The majority of wild fruit tree species cannot be conserved *ex situ* in plantations or field genebanks because of biological, technical and resource limitations. Most so-called field genebanks for fruit tree species actually do not conserve genetic diversity but only elite materials (i.e. trees have undergone at least one cycle of selection at the time of sample collection) for fruit tree improvement researchers. Conserving a wide range of tropical tree species such as mango, durian, rambutan, breadfruit and jackfruits is practical in natural forests and protected areas as the size and canopy of trees is too large in home gardens and orchards. Therefore, the conservation of tropical fruit genetic resources relies heavily on *in situ* conservation efforts in public or semi-public lands (see Chapters 2 and 5 for more on *in situ* conservation). Guidelines for tree and forest genetic resource conservation and management *in situ* in managed natural forests and protected areas are already published jointly by FAO, DFSC and IPGRI (2001), but they do not recognize the role of human intervention in natural ecosystems. The roles farmers and communities play in managing wild tropical fruit tree species diversity are increasingly recognized but still limited and, as a result, traditional ecological knowledge and good practices for such management systems are eroding fast. Community roles in *in situ* conservation in buffer zones, community forests, sacred groves and religious forests require supportive national and local level policies, which are often lacking, weak or threatened because development-oriented land use policies usually ignore farmers' long-term needs.

Context 2: Home gardens, orchards and nurseries

Home gardens are plots located adjacent or in close proximity to a given homestead, in which a range of crops and trees are maintained for numerous culinary, medicinal and cultural purposes. Home gardens are a time-tested traditional practice throughout tropical countries, where combinations of trees, crops and animal and aquatic biodiversity are managed by family members for household food supply, income and well-being (Soemarwoto, 1987; Eyzaguirre and Linares, 2004; Kumar and Nair, 2006; Gautam *et al.*, 2009). Although the effective population size of target fruit trees in a single home garden can be limited for on-farm conservation, a landscape of home gardens tends to maintain a wide range of high-value, unique and rare fruit trees. These landscapes have been found to be a place for blending traditional knowledge with scientific knowledge through farmers' experimentation and innovations. Most of the custodian farmers identified (Sthapit *et al.*, 2013 and Chapter 4) have well-tended home gardens that harbour a rich diversity of tropical fruits. Networks of such home gardens might represent a bigger population size for maintaining reproductive biology of taxa. Home garden systems play an important role in the preservation of indigenous traditional knowledge, as they are often organized and defined by a set of traditional management mechanisms that do not always translate into large-scale or commercialized configurations such as orchards and nurseries.

Semi-commercial orchards and nurseries are two additional contexts in which innovative propagation and management techniques are displayed for using diversity. These settings differ from home gardens in several respects, though the main contrast is that they are often less diverse, focusing on commercial species and varieties that offer more direct economic incentives and benefits to the farmers who own them. Chapter 18 describes how varietal diversity is used by commercial farmers to minimize risk and improve yields, and how it is used to increase incomes through sales.

Categories of good practice

Propagation and planting material management

In many communities, a major constraining factor farmers face when attempting to expand the diversity and overall size of their seed portfolios is a lack of access to adequate planting materials. As such, developing and implementing good practices for propagation at the nursery level is key to encouraging the cultivation of a diverse seed portfolio.

Responsibly using naturally occurring wild fruit diversity and domesticating economically important fruit diversity from forests, combined with sharing knowledge and skills relating to propagation and management with local communities, plays an important role in tropical fruit tree conservation efforts. One example of this is seen in India, where communities in the proximity of natural or community forests harvest wild mango varieties such as Appemidi and Jeerige for pickle making, which contributes significant income to the livelihoods of local people. Farmers were able to collectively identify the large number of varieties of pickle mangoes that exist in the forest. They verified the best trees and standardized grafting techniques. Expert grafters now graft pickle mangoes in their friends' and families' home gardens or orchards (see Chapters 11 and 28).

A number of the case studies in this book relate to propagation and planting material management techniques for home garden systems. For instance, Chapters 11, 16 and 19 depict the manner in which home gardens in Indonesia, India and Malaysia integrate a variety of plant and animal species in an intensive farming arrangement as a way to conserve and enhance mango and citrus varietal diversity while generating an additional source of household income, all the while minimizing waste and providing a valuable set of ecosystem services. Farmers have developed context-specific propagation methods, for example, side grafting in dry areas of Thailand (Chapter 15) and marcotting of citrus in swampy peat land conditions in South Kalimantan (Chapter 14). Chapter 12 illustrates traditional practices of planting seedling mango as a thick boundary fence and assessing potential best varieties as a source of diversity.

Semi-commercial orchards and nurseries tend to be hotspots of innovative propagation techniques. For instance, Chapter 13 discusses maintenance of mother block seed production used by orchard owners in India that strengthens the local seed system, supplies healthy saplings, increases the yield and quality of fruit and improves overall orchard health and life span estimates. Similarly, Chapter 11 describes how a farmer has experimented with grafting methods to be able to introduce wild species into his semi-commercial orchard and Chapter 28 describes how a network of farmers with excellent propagation skills provide grafting services to their fellow villagers with home gardens and orchards. Sets of such practices help maintain diversity and increase crop productivity.

Production and crop management

Another broad topic this book emphasizes is good production and crop management practices for home gardens, orchards and other settings. Once a sustainable source of seed and planting material has been established, farmers must optimize their use of these resources when confronted with limited spatial, financial and technical inputs.

The greater land area and more direct economic connections of orchards and nurseries make them ideal sites for experimenting with and implementing distinctive management and production practices. Chapter 12 presents a study of historic heritage orchards in Malihabad, India, some of which contain as many as 135 different varieties of mango that are managed using unique, time-tested techniques passed down from generation to generation of farmers. Systems such as this often house a startling array of diversity due to a combination of social, cultural and economic interests, all of which tend to intersect in semi-commercial institutions like orchards and nurseries. One low-cost and efficient strategy is to identify the best trees available in the community, characterize and evaluate them and further multiply them for community benefits (Dinesh *et al.*, 2015). Chapter 17 shows how the needs of a traditional Hindu festival ensure the maintenance of a range of fruit species and in particular a genetically highly variable population of pomelo in home gardens in Bihar, India. Chapter 18 describes how varietal diversity is used in commercial orchards in Chittoor, India to lengthen harvest seasons, manage risk, avoid the glut season and improve productivity and pollination services. Chapter 20, from Thailand, demonstrates sustainable use of *Garcinia fusca* by holistic production and management practices, whereas Chapter 21 showcases the example of successful post-harvest management practices to support the livelihoods of aroi aroi (*G. forbesii*) farmers and genetic resources in home gardens and orchards in Subah, Malaysia.

Commercialization and home use

Several chapters in this book focus on how farmers establish market linkages and use commercialization of local biodiversity as a means to improve livelihoods. While historically the commercialization of agricultural systems had a negative impact on local agricultural biodiversity (see Chapter 22), a number of the good practices recorded in this book showcase different ways in which local communities have constructively engaged with markets in such a way as to support and promote the maintenance and enhancement of tropical fruit tree diversity by providing income, among other livelihood benefits in order to ensure a win–win situation of conservation and income generation. The nearly ubiquitous presence of markets in agriculture, even in remote, biodiverse regions, and the importance of providing economic incentives and benefits to farmers make it important to find ways to sustainably use and commercialize native biodiversity. Three primary strategies by which markets can contribute to the commercialization and conservation of tropical fruit tree diversity demonstrated in this book are: (1) by making the local agroecosystem and diversity an economically competitive good through community-based agrotourism (Chapters 22, 23 and 24); (2) by product development in conjunction with the creation of market links based on unique native fruit tree species and landraces (Chapters 24 and 25); and (c) by premiums or rewards paid by consumers or companies for conservation services conducted by local communities (Chapters 21 and 24 on the contribution of markets and Chapter 20 on *G. fusca* in Thailand).

Chapter 23 describes the establishment of a fruit diversity garden and trekking route for tourists in Sarawak, Malaysia. Chapter 25 summaarizes how a self-empowered women's group in Thailand successfully commercialized a unique local dish that uses a local species, *G. cowa*, for its particular flavour. Another example of how local agricultural biodiversity can contribute to the creation of added-value products is showcased in Chapter 28, describing the development of mango pickle made from carefully selected varieties of a unique aromatic type of mango found only in the Western Ghats of India. The establishment of several commercial activities at a community level, as presented in Chapter 24 by the community groups of Kiriwong village in Thailand, is a prime example of a practice that combines the strategies described above. This case study shows how several groups could organize themselves and sell a range of products from *Garcinia* species, along with using their protected local natural landscape to attract tourists. This type of value chain development often entails the provision of external support to local communities in the form of training in processing, financial loans or grants and help with economic and marketing expertise.

Collective action and social networks

The last theme this book covers relates to how good farmer practices form working modalities with a range of actors hailing from local, political and commercial spheres. Part of creating market linkages and providing livelihood benefits to rural and impoverished communities relies on the institutional and personal relationships they have with others capable of providing much needed capital assistance. As the case studies presented in this volume demonstrate, this social capital support can come in a number of different forms, from rallying political influence for policy ends to harnessing social and human capital in the form of collective action initiatives. Diversity fairs, CBM funds and income generation activities organized by farmers' self-help groups are good practices for social capital building (Chapters 3, 20, 24, 25 and 29). One manifestation of this theme can be observed in Chapter 23, a case study relating to the establishment of a tropical fruit tree agrotourism park in Malaysia. It illustrates the manner in which tapping a wide collection of sources for technical and material support can lead to a sustainable mechanism for preserving and expanding tropical fruit tree biodiversity in conjunction with a host of livelihood benefits. We observed that empowering farming communities and their local institutions for self-directed conservation and development goals by capitalizing tropical fruit tree diversity is challenging and a long-term investment. Developing skills and knowledge, raising awareness of the potential values of local biodiversity resources, exploring market linkages and setting up a CBO (community-based organization) do not require a long time, but institutionalizing practices and being able to work in a self-sustaining way require at least 8–10 years. Often external agencies, such as funding partners, have little patience for this persistent effort. Chapters 24 and 25 show examples where collective action has been formed in the villages of Kiriwong and Trok Nong, Thailand, with a view to facilitating this kind of institutional, transformative change, but it is early to assess whether the changes will be long lasting. We also noticed emerging collective action and leadership by the farmer organization in India called the Society for Conservation of Mango Diversity (SCMD) (Chapters 12 and 29) in a time frame of four to five years of social capital building.

Typology of good practices

Since a good practice can be a process, a method, a technique, an institutional arrangement or any combination of these, a typology was developed to better understand good practices in the context of a sustainable livelihood framework. Three key questions – what? (Techniques), how? (Processes) and in what ways? (Methods) – are posed to gain a conceptual grasp of good practices and how they can be better documented, so that piloting and scaling up of good practices can be practical, cost-effective and simple to implement.

Scaling up and out

There are challenges in adopting a GPD in a new context and place. During the piloting and implementation period, several workshops at various levels and cross-site visits by farmers, development workers and researchers were carried out in anticipation of mainstreaming some of the GPDs. Experience showed that GPDs cannot simply be 'copied and pasted' but instead need to be de-packaged by researchers or farmers and re-packaged to suit their own local context. For example, farmers and researchers from Sirsi in India saw a household-level cottage industry of soap, candle and shampoo preparation from *G. mangostana* and oilpalm among woman farmer entrepreneurs in Kuching, Sarawak but adapted the practice to produce candles made of kokum (*G. indica*). In Malaysia and Thailand, nursery owners display colour pictures of fruit along with young saplings in order to provide varietal information and facilitate customers in their purchase decisions. This idea was picked up by practitioners from the Sirsi site, who are now using information from the community fruit catalogues for marketing purposes. A side-grafting idea from Thailand has been tried for grafting multiple scions in the Malihabad site in India, while the multivarietal orchards idea from India was transferred as labelling diversity blocks in natural private land or community home gardens in East Java. The Thai packaging skills for value added products are well appreciated by all other countries, but it has been found difficult to transfer this skill set to other country contexts. Even within the same country, we cannot assume that good practices will be transferred without facilitation. Farmers and researchers from Sisaket, Thailand, learned tie dye fabric dyeing using natural dyes extracted from the by-products of tropical fruits from the Kiriwong community also in Thailand, through project activities (Chapters 20 and 24). This kind of social learning and local innovation can be created using a CBM approach. It is enhanced by organizing opportunities for farmer-to-farmer knowledge sharing and exchange visits.

Table 1.1 Summary chapters, practice, focus, context, type, origin and scope for scaling

Chapter	*Practice*	*Focus: (a) propagation and planting material management, (b) production and crop management, (c) commercialization and home use, (d) collective action and social networking*	*Context (i.e. wild or home garden, orchard)*	*Type (i.e. system, technique, process, method, institution)*	*Farmer or formally innovated*	*Scope for scaling up and out (+ little, ++ moderate, +++ high)*
3	Community Biodiversity Management as an approach for realizing on-farm management of agricultural biodiversity	Overarching, holistic	Both contexts	Combination of system, method, technique, institution, process	Blended	++
4	Custodians of tropical fruit tree diversity: Identifying and strengthening the roles and rights of custodian farmers	Overarching, holistic	On farm and home gardens	Method	Farmer and formal	+++
5	Good practices for conservation and sustainable use of crop wild relatives of tropical fruit tree diversity	(a) Propagation and planting material	Wild, protected, *in situ*	Process	*De facto*	+
6	Exotic and indigenous fruit tree diversity on farm and the conservation of tree genetic resources: Case studies from sub-Saharan Africa	(a) Propagation and planting material (b) Production and crop management	On farm, home gardens	Process	Farmer	+
7	Promoting community management of underutilized tropical and subtropical fruit genetic resources in Brazil	All four categories	On farm and wild	Method	Farmer	++
8	Good practice: Using intraspecific crop diversity to manage pests and pathogens in China	(b) Production and crop management; (d) Collective action and social networking	On farm	Method	Blended	+

9	Kandyan home gardens: A time-tested good practice from Sri Lanka for conserving tropical fruit tree diversity	All four categories integrated	On farm; agroforestry; buffer zone	System	Farmer	+++
10	Amazonian fruits: How farmers nurture nutritional diversity on farm and in the forest	(a) Propagation and planting material (b) Production and crop management (c) Commercialization and home use	Wild, agroforestry	System	Farmer	++
11	A suite of propagation and management techniques for *Garcinia* in the central Western Ghats region of Karnataka, India	(a) Propagation and planting material	Forest	Technique	Farmer	++
12	A set of interconnected practices which enhances and conserves mango diversity in Malihabad, India	(a) Propagation and planting material (b) Production and crop management	Home gardens, orchard	Processes and techniques	Farmer	++
13	Maintenance of mother blocks of *Citrus* rootstocks by farmers and nurseries for production of quality planting materials	(a) Propagation and planting material (b) Production and crop management (d) Collective action and social networking	On-farm mother block nurseries	System with technique	Farmer	+++
14	Marcotting as good practice in community nursery of citrus for maintaining diversity in south Kalimantan, Indonesia	(a) Propagation and planting material (b) Production and crop management	Home gardens, orchards	Technique	Farmer	++
15	Combination of side-grafting technique and informal germplasm exchange system in non-irrigated mango orchards in Thailand	(a) Propagation and planting material (b) Production and crop management	Home gardens, orchards, commercial orchards	Technique	Farmer	+++
16	Propagation technique and pruning techniques of *Garcinia atroviridis* (asam gelugor) in Bukit Gantang, Perak, Malaysia	(a) Propagation and planting material (b) Production and crop management	Home gardens, orchards	Technique	Farmer	+++

Table 1.1 continued

Chapter	*Practice*	*Focus: (a) propagation and planting material management, (b) production and crop management, (c) commercialization and home use, (d) collective action and social networking*	*Context (i.e. wild or home garden, orchard)*	*Type (i.e. system, technique, process, method, institution)*	*Farmer or formally innovated*	*Scope for scaling up and out (+ little, ++ moderate, +++ high)*
17	The role of a traditional festival, Chhath Puja, in the conservation and sustainable use of tropical fruits	(b) Production and crop management (d) Collective action and social networking	Home gardens	System	Farmer, culture	++
18	Multivarietal orchards – an age old conservation practice in mango	(b) Production and crop management	Home gardens and orchards	System	Farmer	+++
19	Integrated home gardens for maintaining diversity of mango and citrus, and for family well-being in East Java	All four categories	Home gardens	System	Farmer	+++
20	Management of *Garcinia fusca* for sustainable use	(a) Propagation and planting material (b) Production and crop management (d) Collective action and social networking	Wild and home gardens	System	Blended	++
21	Production and management of an underutilized fruit: aroi aroi (*Garcinia forbesii* King) in home gardens or orchards	(b) Production and crop management	Home gardens, orchards	Technique	Formal	++
23	Agrotourism in Kampung Kakeng, Serian: Development and challenges	All four categories	Natural forest and home gardens	Combination of system, method, technique, institution	Formal	++

24	Conserving tropical fruit tree diversity by using their products and promoting agrotourism: Lessons from an empowered community in Southern Thailand	All four categories	Home gardens, agroforestry, community forestry	Combination of system, method, technique, institution	Blended	+++
25	Value addition of a local food through collective actions and marketing by women's groups	All four categories	Home gardens, commercial orchards	Combination of system, method, technique, institution	Blended	+++
26	Value creation for *Garcinia gummi-gutta* and *Garcinia indica* through energy-efficient dryers and product differentiation in central Western Ghats region of Karnataka, India	(c) Commercialization and home use; (d) Collective action and social networking	Wild and home gardens	Technique, institution	Formal	++
27	Community forests: Utilization and informal regulation for tropical fruit tree conservation	(d) Collective action and social networking	Community forests	Combination of system, method, technique, institution	Blended	+
28	An informal network of grafting experts to help communities conserve and use wild pickle mango (*Mangifera indica*) diversity in the central Western Ghats region of Karnataka, India	(a) Propagation and planting material; (d) Collective action and social networking	Community forests, home gardens, on farm	Combination of system, method, technique, institution	Farmer	+
29	Social capital building for tropical fruit tree diversity management	(d) Collective action and social networking	On farm, agroforestry, home gardens	Process	Formal	+++

Table 1.1 summarizes all the case studies in terms of focus, context, types, origin and potential for scale up and scale out. Potential for scale out is based on subjective assessment of context, crops, categories, source of innovation, quality of participation amongst stakeholders and partners and authors' experiences of interventions.

Conclusions

The intention of this introduction is to clarify the conceptual understanding of the term 'good practice' for tropical fruit tree diversity (inter- and intraspecific diversity), maintenance and enhancement in the context of a sustainable livelihood framework. The rationale for identifying good practices is to build upon existing interests, activities and social structures while providing a range of options that farmers can experiment with and ultimately implement, which not only support dynamic on-farm and *in situ* conservation of tropical fruit tree species diversity but also confer social, economic and environmental benefits on the rural communities that actualize them. It is anticipated that this methodology and compendium of good practices can be used by research and academic institutions across South and Southeast Asia and the broader global community to identify, strengthen and promote locally available good practices. Additionally, the methodology could help national level on-farm and *in situ* conservation programmes to increase understanding and subsequently improve the effectiveness, efficiency and sustainability of their activities.

Experience suggests that 'good practices' are very context-specific and hence a better understanding of the particular circumstances and corresponding drivers that result in positive outcomes is required. Because of this context specificity, we have tried, rather than using a blueprint approach to replicate good practices, to focus on the identification of key principles that are embedded within a good practice of *in situ* and on-farm conservation of tropical fruit tree diversity. These principles are further explained and defined in Part 4, where we also discuss the challenges and lessons learned during the process of conceptual understanding, identifying, piloting and scaling up of good practices for diversity management. Using principles allows the easier application of the concept of good practices to other species and geographic locations. The concept of GPD is still new; there is a great need for more work in other locations and countries so that the methodology can be further refined through an experiential learning process. We hope that the case studies presented in this book will stimulate action by other interested multidisciplinary researchers.

It should be noted that the parameters and contours of what constitutes a good practice and the criteria by which it is evaluated are continuously evolving. Though a preliminary foundation such as that laid out above is necessary for orienting the following case studies within the confines of a broader field of study, experiential learning from these studies and those to follow is indispensable to refining the concept of good practices and how they relate to a diverse array of local contexts. The hope is that books and studies such as this will encourage the scientific community to grapple with these issues on a more universal scale, thus facilitating the global exchange of knowledge, data and lessons learned. Such a movement of ideas and research has the potential not only to make sustained progress in ongoing efforts towards conserving biodiversity, but also towards improving the quality of life for rural and impoverished communities worldwide.

References

Altieri, M.A. and Merrick, L.C. (1987) '*In situ* conservation of crop genetic resources through maintenance of traditional farming systems', *Economic Botany*, vol 41, pp. 86–96

Arora, R.K. and Ramanatha Rao, V. (eds) (1998) *Tropical Fruits in Asia. Diversity, Maintenance, Conservation and Use*, Proceedings of the IPGRI-ICAR-UTFANET Regional Training Course on the Conservation and Use of Germplasm of Tropical Fruits in Asia held at the Indian Institute of Horticultural Research, 18–31 May 1997, Bangalore, India; IPGRI, Rome, Italy

Bellon, M.R. (1996) 'The dynamics of crop intra-specific diversity: A conceptual framework at the farmer level', *Economic Botany*, vol 50, no 1, pp. 29–36

Bellon, M.R. (2004) 'Conceptualizing interventions to support on-farm genetic resource conservation', *World Development*, vol 32, pp. 159–172

Brush, S.B. (2000) *Genes in the Field: On-farm Conservation of Crop Diversity*, Lewis Publishers, Washington, DC

CBD (undated) Good practice guides, Convention on Biological Diversity; available at: www.cbd.int/development/training/guides/, accessed 2 April 2015

DFID (1999) DFID sustainable livelihoods guidance sheets, Department For International Development; available at: www.ennonline.net/resources/667, accessed 25 June 2015

Dinesh, M.R., Ravishankar, K.V., Sthapit, B.R., Parthasarathy, V.A., Sandya, B.S., Nischita, P., and Lavanya, B. (2015) 'Genetic diversity studies in certain indigenous mango (*Mangifera indica* L) varieties', *Indian Journal of Plant Genetic Resources*, vol 28, no 1, pp. 153–160. DOI 10.5958/0976-1926.2015.00019.4

Eyzaguirre, P. and Linares, O.F. (2004) *Home Gardens and Agrobiodiversity*, Smithsonian Books, Washington, DC, p. 296

FAO, DFSC, and IPGRI (2001) *Forest Genetic Resources Conservation and Management, Vol 2: In Managed Natural Forests and Protected Areas* (*in situ*), FAO, DFSC, and IPGRI, Rome, Italy, p. 87

Frankel O.H. and Soule, M.E. (1981) *Conservation and Evolution*, Cambridge University Press, Cambridge

Gautam, R., Sunwar, S., Subedi, A., Shrestha, P.K., and Sthapit, B.R. (2009) 'Home gardens and their roles in domestication of wild and uncultivated plant genetic resources in Nepal', *Acta Horticulturae 806: International Symposium on Underutilized Plants for Food Security, Nutrition, Income and Sustainable Development*, vol 2, pp. 677–681

GEF (2001) *Good Practices: Country Coordination and Global Environment Facility*, report, GEF/UNEP, Washington, DC

Gemmill, B. (2001) *Managing Agricultural Resources for Biodiversity Conservation: A Guide to Best Practices*, UNEP/GEF/UNDP, p. 87; available at: http://biotools.earthmind.net/files/elci-agbioguide.pdf, accessed on 24 April 2015

Heywood, V. and Dulloo, E. (2005) In Situ *Conservation of Wild Plant Species – A Critical Global Review of Good Practices*, IPGRI Technical Bulletin, no 11, FAO and IPGRI, Rome, Italy

Jarvis, D.I., Padoch, C., and Cooper, H.D. (2007) *Managing Biodiversity in Agricultural Ecosystems*, Columbia University Press, New York

Jarvis, D.I., Hodgkin, T., Sthapit, B.R., Fadda, C., and Lopez-Noriega, I. (2011) 'A heuristic framework for identifying multiple ways of supporting the conservation and use of traditional crop varieties within the agricultural production system', *Critical Reviews in Plant Sciences*, vol 30, no 1–2, pp. 1–49

Kumar, B.M. and Nair, P.K.R (eds) (2006) *Tropical Homegardens: A Time Tested Example of Sustainable Agriculture*, Springer, Dordrecht, The Netherlands, p. 387

Maxted, N., Guarino, L., Myer, L., and Chiwona, E.A. (2002) 'Towards a methodology for on-farm conservation of plant genetic resources', *Genetic Resources and Crop Evolution*, vol 49, pp. 31–46

Sajise, P. (2005) 'The sustainable livelihood framework in understanding good practices on on-farm conservation of tropical fruits', paper presented at the UNEP-GEF and IPGRI meeting on "Good practices for conservation and sustainable use of tropical fruits and its wild relatives", Bangkok, Thailand, 4–5 April 2005

Sajise, P. and Keizer, M. (2005) 'Sustainability of poverty reduction in coconut growing communities: Can it be attained?', in P. Batugal and J. Oliver (eds), *Poverty Reduction in Coconut Growing Communities Vol. III: Project Achievements and Impact*, IPGRI APO, Serdang, Malaysia

Smale, M., Bellon, M., Jarvis, D., and Sthapit, B. (2004) 'Economic concepts for designing policies to conserve crop genetic resources on-farms', *Genetic Resources and Crop Evolution*, vol 51, pp. 121–135

Soemarwoto, O. (1987) 'Home gardens: A traditional agroforestry system with a promising future', in H.A Steppler and P.K.R. Nair (eds), *Agroforestry: A Decade of Development*, ICRAF (International Center for Rearch in Agroforestry), Nairobi, Kenya, pp. 157–170

Sthapit, B.R., Subedi, A., Rijal, D., Rana, R.B., and Jarvis, D.I. (2003) 'Strengthening community-based on-farm conservation of agricultural biodiversity: Experiences from Nepal', in CIP-UPWARD (eds), *Conservation and Sustainable Use of Agricultural Biodiversity: A Source Book. Vol II: Strengthening Local Management of Agricultural Biodiversity*, International Potato Center (CIP) and Users' Perspectives With Agricultural Research and Development (UPWARD), Los Banos, Philippines, pp. 344–354

Sthapit, B.R., Lamers, H., and Ramanatha Rao, V. (eds) (2013) *Custodian Farmers of Agricultural Biodiversity: Selected Profiles from South and South East Asia*, Proceedings of the Workshop on Custodian Farmers of Agricultural Biodiversity, 11–12 February, New Delhi, India; Bioversity International, New Delhi, India

2 Key concepts

Bhuwon Sthapit, V. Ramanatha Rao and Hugo A.H. Lamers

This book uses some key terms that need to be defined at the outset for better understanding and discussion. Before we discuss '*good practices in the management and use of tropical fruit tree diversity*', it is important to understand the wider context in which such practices take place, as some of them are relatively new concepts. Some well-accepted concepts are also included for clarity with respect to tropical fruit trees.

Agricultural biodiversity

Agricultural biodiversity is a subset of the wider term biodiversity and can be distinguished through its functional use for food and agriculture and the need of human intervention for its survival and future use. Just like the general term biodiversity, it entails diversity at genetic, species and ecosystem levels. As defined by the CBD and FAO, agricultural biodiversity, sometimes called 'agrobiodiversity', encompasses the variety within and between animals, plants and micro-organisms that is necessary to sustain key functions of the agroecosystem, its structure and processes for, and in support of, food production and food security (CBD, 1992; FAO, 1998). Thus it includes crops, trees, other associated plants, animals and fish, and interacting species of pollinators, symbionts, pests, parasites, predators and competitors. It comprises domesticated crops or breeds and their wild relatives as found in fields, rangelands or forests, which are used for food and agriculture. Agricultural biodiversity is an important asset for rural households, particularly for the poor in marginal areas of the developing world, as it helps them adapt their crops to particular environmental conditions, lower harvest risks and diversify their production systems for sales or for diversified diets and home use. It often facilitates ecosystem services such as lower pest and disease pressure or the conservation of water and organic matter in soils. As well as this, agricultural biodiversity is the fundamental source for farmers and breeders to identify potential beneficial gene combinations for the development of new or improved cultivars.

Social scientists emphasize the human, social and cultural aspects of agricultural biodiversity. Agricultural biodiversity results from the interaction between the environment, genetic resources and the management systems and practices used by culturally diverse people. Agricultural biodiversity is, therefore, embedded in traditional land use systems, many of which are rapidly changing. Based on Brookfield's definition of 2001, de Boef *et al.* (2012) define the term agrobiodiversity as 'a dynamic and constantly changing patchwork of relations between people, plants, animals, other organisms and the environment, always coping with new problems, always finding new ways' (2012, p. 1). The terms 'dynamic' and 'relations' are of critical importance and very much linked with on-farm management of genetic resources in their own habitats and cultural settings. Agricultural biodiversity is vital for structuring the relationship between people and biodiversity within agroecosystems. Because agricultural biodiversity is closely linked to cultural aspects of people dependent on that diversity in traditional production systems, researchers have developed the concept of 'biocultural diversity'. Biocultural diversity emerges conceptually from an anthropological consideration of the manner in which human societies adapt to the varied biological circumstances in which they live. Biocultural diversity is concerned with the relationships between traditional knowledge, biological diversity and cultural diversity (Johns and Sthapit, 2004). In this book we consider local knowledge and culture to be an integral part of agricultural biodiversity, because it is the human activity of agriculture that conserves and maintains this agricultural biodiversity through sustainable use.

Community Biodiversity Management (CBM)

Community Biodiversity Management (CBM) is a community-driven participatory approach that empowers farmers and communities to organize themselves and to develop livelihood strategies that support the on-farm management of agricultural biodiversity. The CBM approach enhances local knowledge and practices with new sets of skills and knowledge and allows local innovation that matches current needs and social systems and supports conservation and development goals set by participating communities (Sthapit *et al.*, 2008a; 2008b). CBM builds on the capacities and social structures of farming communities to enable them to make inclusive and self-directed decisions in the management and sustainable use of local biodiversity and crop genetic resources. A committed local organization, either a government agency for rural development and extension services or an NGO, builds capacity, facilitates and mentors decision-making processes, and facilitates and reinforces community action plans related to biodiversity-based livelihoods and conservation interventions and reinforces agricultural biodiversity. CBM has been used as a method to realize the on-farm management of agricultural biodiversity. Details about the method are explained in de Boef *et al.* (2012; 2013). With respect to tropical fruit trees in particular, the CBM approach is explored in Chapter 3 of this book.

Community empowerment

Community empowerment refers to the process of enabling communities to increase control over their lives (Chambers, 1993). 'Community' can be considered a specific group of people living in a region, who are organized in a social structure and exhibit some awareness of their identity as a group. 'Empowerment' refers to the process by which people gain control over the factors and decisions that shape their lives (Israel *et al.*, 1994) and also the resources that are available to them. It is the process by which they increase their assets and attributes and build capacities to gain access, partners, networks and a voice in order to gain control. Community empowerment, therefore, is more than the simple involvement, participation or engagement of communities.

Community empowerment as a process is best considered as a continuum representing progressively more organized and broad-based forms of social and collective action. Labonte (1989) developed a five-step continuum model consisting of the following developmental stages: personal action, small mutual groups, community organizations, partnership organization and social and political actions. Further elaborating on the concept, Bartlett (2008) described community empowerment as a transformative change that can be distinguished by three elements of transformation: (i) Means (rights, resources, capabilities and opportunities); (ii) Process (self-directed analysis, decision making and action by agency); and (iii) Ends (greater control of assets and their lives in relative terms). The concept of empowerment has been well developed in the health sector, but is relatively new in the biodiversity conservation sector.

Conservation

The Convention on Biological Diversity (CBD) within its broader framework recognizes two ways of conserving genetic resources: *in situ*, in the place of origin, and *ex situ*, outside the place of origin (CBD, 1992). *In situ* conservation is often further divided into '*in situ* conservation' of genetic resources in their native habitats in the wild, and 'on-farm management' of genetic resources (or *in situ* conservation of traditional varieties in farmers' fields) in agricultural systems.

Ex situ *conservation*

During the last five decades, *ex situ* conservation has been addressed by several nations and considerable numbers of plant genetic resource accessions have been conserved in national genebanks. For example, the total number of accessions conserved *ex situ* worldwide increased by approximately 20 per cent (1.4 million) from 1996 to 2010, reaching 7.4 million (FAO, 2010). Part of the global *ex situ* crop diversity collection is conserved in the Svalbard Global Seed Vault (Westengen *et al.*, 2013).

In situ *conservation*

The concept of *in situ* conservation, especially for agricultural biodiversity, is relatively recent. It is partly an effort by the scientific community to honour its debt to the legacy of the farming communities who created the biological basis of crop production. UNEP (1992) extended the CBD's definition of *in situ* conservation as follows: 'the conservation of ecosystems and natural habitats and the maintenance and recovery of viable populations of species in their natural surroundings and, in the case of domesticated and cultivated species, in the surroundings where they have developed their distinctive properties'.

On-farm conservation or on-farm management

Further elaborating with specific reference to crop genetic resources, Altieri and Merrick (1987) describe on-farm conservation of agricultural biodiversity as the maintenance of traditional crop varieties (landraces) or cropping systems by farmers within the natural habitats where they occur – in farmers' fields and uncultivated plant communities. The goal of on-farm conservation is to encourage farmers to continue to select and manage local crop populations (Brush, 1995). The first report of the *State of the World's Plant Genetic Resources for Food and Agriculture* (*SoW PGRFA*) replaced the term '*in situ* conservation on farm' with 'on-farm management', while maintaining '*in situ* conservation' as the overarching term (FAO, 1998). On-farm management is a highly dynamic form of landrace management, which allows the processes of both natural and human selection to continue to act in the production system (Frankel *et al.*, 1975; Bellon, 2010). The objective of on-farm conservation is to maintain crop evolution in farmers' fields, farms, home gardens and landscapes (Bellon *et al.*, 2014). Farmers' efforts to search for new diversity, select new traits and exchange selected materials with friends and relatives are the processes that allow the genetic material to evolve and change over time. Thus, the conservation of landrace *per se* is secondary to the continuation of the processes that allow material to evolve and change over time (Jarvis and Hodgkin, 2000). This conservation method is increasingly valued for evolving new adaptive diversity and, therefore, enhances farmers' capacity to cope with adversity resulting from the consequences of socio-economic and market forces and climate change (Sthapit *et al.*, 2012).

The second *SoW PGRFA* report notes that, over the last decade, promoting and supporting the on-farm management of genetic resources, whether in farmers' fields, home gardens, orchards or other cultivated areas of high diversity, has become firmly established as a key component of crop conservation strategies, as methodologies and approaches have been scientifically documented and their effects monitored (FAO, 2010). On-farm and *in situ* conservation are considered important approaches to sustain the evolutionary process of creating new diversity, to conserve the associated knowledge about its traits and to sustain important ecosystem services connected to agricultural biodiversity, all factors which will be lost or heavily compromised when pursuing a solely *ex situ* conservation strategy.

Ecosystem services

Ecosystem services are the direct and indirect contributions of ecosystems to human well-being, survival and quality of life. The concept of an ecosystem provides a valuable framework for analyzing and acting on the links between people and their environment. Ecosystem services can be categorized into five main types (MEA, 2005):

- **Provisioning services** are the products obtained from ecosystems such as food, fresh water, wood, fibre, spices and medicines.
- **Regulating services** are defined as the benefits obtained from the regulation of ecosystem processes such as climate regulation, natural hazard regulation, water purification and waste management, pollination or pest control.
- **Habitat services** highlight the importance of ecosystems in providing habitat for migratory species and in maintaining the viability of gene pools.
- **Cultural services** include non-material benefits that people obtain from ecosystems such as spiritual enrichment, intellectual development, recreation and aesthetic values.
- **Evolutionary services** include benefits such as genetic resources that evolve due to selection pressure exerted by humans and nature.

Biodiversity is the source of many ecosystem goods, such as food and genetic resources, and changes in biodiversity can influence the supply of ecosystem services.

Food and nutritional security

Food security exists when all people, at all times, have physical and economic access to enough safe and nutritious food to meet their dietary needs and food preferences for an active and healthy lifestyle (World Food Summit, 1996).

Sustainable diets

Sustainable diets are diets with low environmental impacts that contribute to food and nutrition security and a healthy life for current and future generations (Burlingame and Dernini, 2012). A key feature of a healthy diet is dietary diversity – consuming a variety of foods across and within food groups to ensure a balanced intake of carbohydrates and proteins and sufficient intake of essential vitamins and micronutrients. Minor crops and landraces often contain high levels of important vitamins or micronutrients and thus play an important role in dietary diversity. This crop diversity is grown on farm or *in situ* for home consumption or is purchased through markets. Agricultural biodiversity offers household members a diversified diet that can lead to improved nutrition and family well-being and provides opportunities for the commercialization of traditional recipes or products that make use of nutritious minor crops and landraces.

Sustainable livelihoods

A livelihood comprises the capabilities, assets and activities required for securing a means of living and the necessities of life. A livelihood is sustainable when it can cope with and recover from stresses and shocks and maintain or enhance its capabilities and assets both now and in the future, while not undermining the natural resource base (DFID, 1998). The concept of sustainable livelihoods has been used widely in the development sector and is used in this book to understand good practices for agricultural biodiversity management.

The sustainable livelihood model (DFID, 1998) entails understanding how a rural household achieves or determines its livelihood goals (outcomes) through decisions about livelihood options (strategies and activities) that are based on its resources (capital). The model recognizes five types of capital: physical, human, social, financial and natural. Human capital refers to personal skills, knowledge and information, the ability to work and health. Natural capital includes land, water, livestock, seeds, crops, biodiversity, environment and air. Social capital consists of social networks and connections, relations of trust and mutual support, access to wider institutions and collective action. Financial capital consists of savings, credit, remittances and pensions. Physical capital may consist of post-harvest equipment, shelter, transport, energy, storage, communications and other infrastructure and technology. These assets and their deployment are affected by the vulnerability context of potential shocks, seasonality, trends and changes and by processes, institutions and policies (DFID, 1998). A household makes decisions regarding the use of assets based on their livelihood goals and the wider environmental context.

In this book we use the sustainable livelihoods framework to understand good practices for diversity management. On-farm conservation of agricultural biodiversity is primarily determined by those who manage the farms, i.e. the women and men farmers with their complementary roles. The framework helps us analyze and understand why and how people do what they do on their farms. Farmers primarily have the objective of engaging in farm activities. Conservation is a by-product of their use of genetic resources to obtain outputs of a sustainable livelihood whether they be income, food and nutrition security, cultural and aesthetic or environmental values. If genetic resources are going to be conserved on farm, it must happen as a spinoff of farmers' production activities and livelihood strategies. This means conservation efforts must be carried out within the framework of farmers' livelihoods, income and cultural values. Lessons from previous research show that several interventions are needed on the five basic capitals of the farmers, the farm households and the community for them to adopt a livelihood that conserves and enhances agricultural biodiversity on farm (Sajise and Sthapit, 2005).

References

Altieri, M.A. and Merrick, L.C. (1987) '*In situ* conservation of crop genetic resources through maintenance of traditional farming systems', *Economic Botany*, vol 41, pp. 86–96

Bartlett, A. (2008) 'No more adoption rates! Looking for empowerment in agricultural development programmes', *Development in Practice*, vol 18, no 4, pp. 524–538

Bellon, M. (2010) 'Do we need crop landraces for the future? Realizing the global option value of *in situ* conservation', in A. Kontoleon, U. Pascual and M. Smale (eds), *Agrobiodiversity and Economic Development*, Routledge, London and New York

Bellon, M.R., Gotor, E., and Caracciolo, F. (2014) 'Conserving landraces and improving livelihoods: How to assess the success of on conservation projects?' *International Journal of Agricultural Sustainability*; http://dx.doi.org/10.1080/14735903.2014.986363

Brookfield, H. (2001) *Exploring Agrobiodiversity*, Columbia University Press, New York

Brush, S.B. (1995) '*In situ* conservation of landraces in centers of crop diversity', *Crop Science*, vol 35, pp. 346–354

Burlingame, B. and Dernini, S. (eds) (2012) *Sustainable Diets and Biodiversity: Direction and Solution for Policy, Research and Action*, FAO, Rome, Italy

CBD (1992) 'The Convention on Biological Diversity', Convention on Biological Diversity; available at: http://www.cbd.int/convention/text/, accessed 16 February 2012

Chambers, R. (1993) *Challenging the Professions: Frontiers for Rural Development*, ITDG Publishing, London

de Boef, W.S., Thijssen, M.H., Shrestha, P., Subedi, A., Feyissa, R., Gezu, G., Canci, A., Ferreira, M.A.J.F., Dias, T., Swain, S., and Sthapit, B.R. (2012) 'Moving beyond the dilemma: Practices that contribute to the on-farm management of agro biodiversity', *Journal of Sustainable Agriculture*, vol 36, pp. 788–809. DOI: 10.1080/10440046.2012.695329; available at: http://www.tandfonline.com/eprint/hVuvwUH9HdrZnSQWBiSA/full, accessed 21 August 2012

de Boef, W.S., Subedi, A., Peroni, N., and Thijssen, M. (eds) (2013) *Community Biodiversity Management: Promoting Resilience and the Conservation of Plant Genetic Resources*, Routledge/Earthscan, Abingdon, UK and New York

DFID (1998) DFID sustainable livelihoods guidance sheets, Department For International Development; available at: www.ennonline.net/resources/667

FAO (1998) *The State of the World's Plant Genetic Resources for Food and Agriculture*, FAO, Rome, Italy

FAO (2010) *Status of World Report II of Plant Genetic Resources for Food and Agriculture*, FAO, Rome, Italy

Frankel, O.H., Brown, A.H.D., and Burdon, J.J. (1975) *The Conservation of Plant Biodiversity*, Cambridge University Press, Cambridge, UK

Israel, B.A., Checkoway, B., Schultz, A., and Zimmerman, M. (1994) 'Health education and community empowerment: Conceptualizing and measuring perceptions of individual, organizational and community control', *Health Education Quarterly*, vol 21, no 2, pp. 149–170

Jarvis, D.I. and Hodgkin, T. (2000) 'Farmer decision making and genetic diversity: Linking multidisciplinary research to implementation on farm', in S.B. Brush (ed.), *Genes in the Field: On-Farm Conservation of Crop Diversity*, Lewis Publishers, Boca Raton, FL, pp. 261–278

Johns, T. and Sthapit, B.R. (2004) 'Biocultural diversity in the sustainability of developing country food systems', *Food and Nutrition Bulletin*, vol 2, no 2, pp. 143–155

Labonte, R. (1989) 'Community empowerments: The need for political analysis', *Canadian Journal of Public Health*, vol 80, pp. 87–88

MEA (2005) *Ecosystems and Human Well-being: Biodiversity Synthesis*, Millennium Ecosystem Assessment, World Resources Institute, Washington, DC, no 5, pp. 1–10; available at: http://www.unep.org/maweb/documents/document.354.aspx.pdf

Sajise, P.E. and Sthapit, B.R. (2005) 'Farmers, farms and agrobiodiversity: On-farm conservation experiences of IPGRI', in N.Q. Ng, J.T. Williams and D.A Vaughan (eds), *Proceedings of a Workshop on* In Situ *Conservation of Plant Genetic Resources, held at FAO, Bangkok, Thailand, 29–31 August 2005*; FAO, Bangkok, Thailand, pp. 30–46

Sthapit B.R., Shrestha, P.K., Subedi, A., Shrestha, P., Upadhyay, M.P., and Eyzaguirre, P.E. (2008a) 'Mobilizing and empowering community in biodiversity management', in M.H. Thijssen, Z. Bishaw, A. Beshir and W.S. de Boef (eds), *Farmers, Seeds and Varieties: Supporting Informal Seed Supply in Ethiopia*, Wageningen International, Wageningen, The Netherlands, pp. 160–166

Sthapit, B.R., Subedi, A., Shrestha, P., Shrestha, P.K., Chaudhary, P., and Upadhyay, M.P. (2008b) 'Practices supporting community management of farmers' varieties', in M.H. Thijssen, Z. Bishaw, A. Beshir and W.S. de Boef (eds), *Farmers, Seeds and Varieties: Supporting Informal Seed Supply in Ethiopia*, Wageningen International, Wageningen, The Netherlands, pp. 166–171

Sthapit, B.R., Subedi, A., Lamers, H., Jarvis, D.I., Ramanatha Rao, V., and Reddy, B.M.C. (2012) 'Community based approach to on-farm conservation and sustainable use of agricultural biodiversity in Asia', *Indian Journal of Plant Genetic Resources*, vol 25, no 1, pp 88–104

UNEP (1992) Convention on Biological Diversity, Nairobi, United Nations Environment Programme

Westengen, O.T., Jeppson, S., and Guarino, L. (2013) 'Global *ex-situ* crop diversity conservation and the Svalbard Global Seed Vault: Assessing the current status', *PLOS One*, vol 8; http://www.ncbi.nlm.nih.gov/pmc/articles/PMC3650076/pdf/pone.0064146.pdf

World Food Summit (1996) *Rome Declaration on World Food Security*, FAO, Rome, Italy; http://www.fao.org/docrep/003/w3613e/w3613e00.HTM

3 Community Biodiversity Management as an approach for realizing on-farm management of agricultural biodiversity

Bhuwon Sthapit, Hugo A.H. Lamers, V. Ramanatha Rao and Arwen Bailey

Introduction

The Convention on Biological Diversity (CBD) defined *in situ* conservation as 'the conservation of ecosystems and natural habitats and the maintenance and recovery of viable populations of species in their natural surroundings and, in the case of domesticated and cultivated species, in the surroundings where they have developed their distinctive properties' (CBD, 1992a). Defining *in situ* conservation of agricultural biodiversity was an important step; the main dilemma over the past two decades has been its practical implementation in conservation practices that fit into the context of the sustainable livelihoods of smallholder and poor farmers (Jarvis *et al.*, 2011; Sthapit *et al.*, 2012; Bellon *et al.*, 2014; 2015). While *ex situ* conservation poses largely technical challenges, *in situ* conservation needs additionally to consider several social parameters involving farming communities and the knowledge they hold (de Boef *et al.*, 2012).

Despite the threat of rapidly shrinking biodiversity in farmers' fields with associated loss of evolutionary options for the future, appropriate methods for the on-farm conservation of agricultural biodiversity continue to be meagre. Research carried out to date has remained quite academic and descriptive, so government organizations have difficulty translating the theory into practice on the ground and garnering support from policymakers and communities. How to consolidate the roles of farmers in decision making concerning agricultural biodiversity management has been much debated. Paudel (2015) suggests that conservation of crop genetic diversity in subsistence farming is governed by the economic and environmental constraints of the farmers in the community. Consequently, the biodiversity remaining *in situ* could *de facto* become extinct due to economic and technological development that creates higher economic returns from farming with a narrow range of newly developed varieties. Modern farming systems rely heavily on a small number of improved varieties and crops with a narrow genetic base, and neglect or externalize the costs of maintaining the fundamental genetic base required for the survival of crop species in the long term. Besides, conventional monocropping systems, which have a role in large-scale commercialized industrial agriculture, often are inappropriate to manage the risks that farmers face, especially smallholders living in remote, harsh and marginal farming areas with limited market infrastructure and high transaction costs.

In this book, we have documented practices that showcase the roles of community members and their rural organizations in successful biodiversity management for livelihood benefits. Many conceptual frameworks of on-farm conservation of agricultural biodiversity have been put forward in literature (Bellon, 1996; Bellon *et al.*, 1997; Brush, 2000; Bellon, 2004; Smale, 2006; de Boef *et al.*, 2013) but few studies have conducted long-term research to assess the impact of on-farm conservation projects in developing countries (Bellon *et al.*, 2014). We asked key research questions to assess whether the Community Biodiversity Management (CBM) approach does in fact: (i) empower communities of women and men farmers to make self-directed decisions and secure access to and control over the natural resources required for their well-being; (ii) contribute to improved livelihood strategies and have wider impacts; and (iii) safeguard the evolutionary process of on-farm conservation of biodiversity. The CBM approach takes the three mutually supportive goals to its core and supports the process of dynamic conservation and use of local crop diversity for today and tomorrow (Sthapit *et al.*, 2012; de Boef and Thijssen, 2013). This approach addresses global challenges of maintaining agricultural biodiversity, which is essential to meet current and future needs related to nutrition, food security and ecosystem function and services.

Definition

CBM is a community-driven participatory approach that empowers farmers and communities to organize themselves and develop strategies that support the on-farm management of agricultural biodiversity for the improvement of their livelihoods. The CBM approach integrates knowledge and practices with social systems, institutions and regulations that support conservation and development goals set by participating communities (Subedi *et al.*, 2006b; Subedi *et al.*, 2013). It is a process of implementing a set of good practices that empower local farmer organizations to manage genetic resources for sustainable livelihoods through collective action (Shrestha, 2013) (Figure 3.1). Such local organizations represent the diverse needs of farmers and help them drive their own development and conservation agendas (Dutty and Bolo-Dutty, 2003; Smith, 2009).

Evolution and development of CBM conceptual framework

The CBM approach originally emerged from the experience of on-farm conservation projects carried out in Nepal from 1998 to 2004 during a period of civil conflict (Sthapit *et al.*, 2005; Subedi *et al.*, 2013). In Nepal, the approach has been continued through a variety of different projects managed by a non-governmental organization (NGO) called LI-BIRD (Local Initiatives for Biodiversity, Research and Development) with funding support from the Development Fund, Norway and the International Fund for Agricultural Development (IFAD). It was further tested in the wider context of different seed-system projects as part of a research theme under the Diversity for Livelihoods Programme, Bioversity International, from 2001 to 2006 (DLP-CCER report to External Panel, 2006) and further evolved by diverse actors in several other countries where biodiversity assets are high but other resources are low (see de Boef *et al.*, 2013).

Over the last 10 to 15 years the CBM approach has been tested, demonstrated and further refined or adapted through projects in Nepal (Subedi *et al.*, 2005; Sthapit and Jarvis, 2005; Subedi *et al.*, 2006b; Sthapit *et al.*, 2008a; 2008b); India (Vasudeva *et al.*, 2013); Indonesia, Malaysia and Thailand (Sthapit *et al.*, 2012); Bhutan, Philippines, Thailand and Vietnam (Santos Doctor, 2013); Brazil (Canci *et al.*, 2013); Ethiopia (Feyissa *et al.*, 2013; de Boef *et al.*, 2013); and France (Kendall and Gras, 2013). In 2009, the Centre for Development Innovation (CDI), in collaboration with LI-BIRD, Bioversity International and partners in Brazil, Ethiopia, India and Nepal, launched a global study on CBM and empowerment to compare different practices and realities at 15 selected sites. During the same period Bioversity International also began the UNEP-GEF-funded project 'Conservation and Sustainable Use of Wild and Cultivated Tropical Fruit Tree Diversity: Promoting food security, sustainable livelihoods and ecosystem services' in partnership with India, Indonesia, Malaysia and Thailand in 36 communities for a period of six years (http://tft.atbioversity.net/). The CBM approach was used as the implementation framework of the project. These projects addressed challenges around how to recognize and strengthen the role of rural women and men farmers and forest dwellers in continuing the millennia-old practice of selecting, breeding and maintaining crops and sustainably harvesting from wild trees adapted to different purposes. The case studies carried out in Nepal and other countries demonstrated the value of strengthening a community's capacity to enhance their human, social, natural, financial and physical capital assets (Sthapit and Jarvis, 2005). These assets can be enhanced through the successful implementation of one or many good practices (Sthapit *et al.*, 2006c). A similar approach also independently evolved in Latin America as CIALs[1] (Braun, 2003; Porter-Bolland *et al.*, 2013) and in Asia as the Community Based Natural Resource Management (CBNRM) approach (Vernooy and McDougall, 2003). CBNRM emerged from practical experiences on the ground, built on reflections about past failures of non-community-based natural resource management approaches. In the context of CBNRM, Tyler (2006a; 2006b) featured 11 case studies that illustrated how local innovations in participatory natural resource management can strengthen livelihoods, build capacity for local governance and spark policy change.

Conceptual framework

The goal of the CBM approach is to realize conservation and sustainable use of biodiversity (in this case tropical fruit tree diversity) for present and future generations in farmers' fields, home gardens and orchards, and also in the wild. If crop genetic diversity is going to be conserved on farm, it must happen as an integral part of farmers' production and livelihood strategies (Berg, 1997). This means conservation efforts must be carried out within the framework of farmers' livelihood and income-generating systems (Sthapit *et al.*, 2005; Bellon *et al.*, 2014). Through field experience, a conceptual framework for a CBM approach evolved with the following three specific outcomes to achieve the overarching goal (Figure 3.1):

1. Community empowerment
2. Livelihood development
3. Biodiversity conservation

These outcomes can be achieved if an enabling environment is created for the men and women of the community to enhance their: (i) knowledge, (ii) practices and (iii) institutions.[2] These three types of output influence each of the three outcomes in an individual as well as collective manner and need to be mutually supportive for their sustainability. Communities, after initial community awareness, usually easily agree that the three outcomes are mutually supportive and needed for sustainable agriculture. However, they often lack the confidence, skills and strategies to achieve these objectives in a synergistic manner. The success of the CBM approach hinges on its ability to build the community's commitment, confidence and problem-solving skills to achieve these widely accepted outcomes.

CBM is, therefore, a holistic, inclusive and process-led research-for-development concept that facilitates the identification, improvement, adoption and dissemination of good practices in biodiversity management. The approach helps empower farming communities to combine strategies for the conservation of biodiversity and its related traditional knowledge or ecosystem services with the improvement of health, income and livelihood outcomes. Central to the CBM approach is that sets of interventions enable communities to claim ownership over their natural resources and empower them to make self-directed decisions regarding the sustainable use, management and protection of the species and varieties found within their community. Since the approach essentially deals with empowerment of farmers, there is a need on the part of external agencies and individuals to give up a certain amount of their 'power', for example regarding who actually decides what activities are carried out. This can require radical changes in the mind-set of researchers and officials. This deep transformation process is a long-term effort that requires several years of capacity building, technical assistance, financial resources and institutional support to be successful.

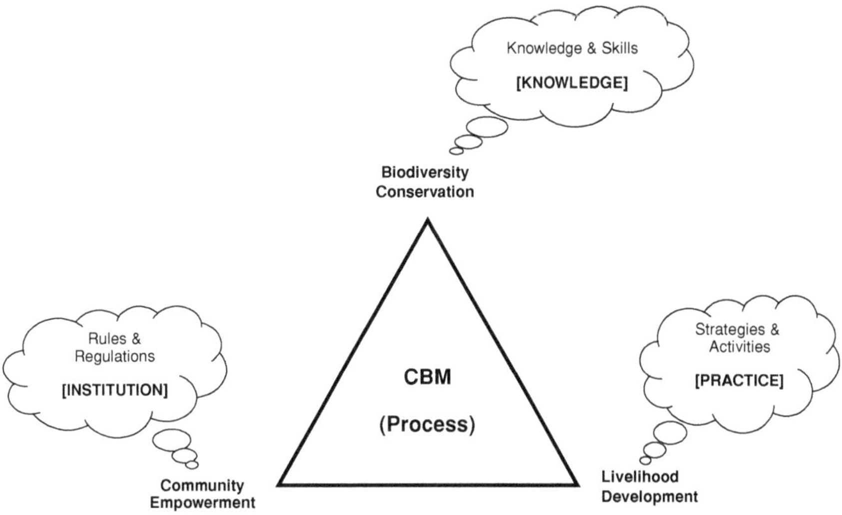

Figure 3.1 Community Biodiversity Management. The cornerstones of CBM, which are shown in the three corners of Figure 3.1, are the three outcomes of: (i) community empowerment, (ii) livelihood development and (iii) biodiversity conservation. The text in the cloud illustrates the means to achieve these outcomes by: (i) strengthening, agreeing and adhering to formal or informal rules and regulations to foster collective action concerning biodiversity; (ii) joint strategies and activities that use local biodiversity to create livelihood benefits such as income, improved diets or reduced risks; and (iii) strengthening the knowledge, skills and values connected with locally available biodiversity to ensure its conservation. The key to the success of CBM is that the community, representatives of local institutions and practitioners see the three outcomes as necessary and mutually reinforcing of each other. Engagement in the CBM process leads to a transformation in the community when individuals see these outcomes not only as important but also feasible and desirable from their perspective. By employing and adapting good practices, community institutions become empowered and community members become active problem solvers and strategists in terms of realizing the three outcomes.

Source: Modified from Sthapit *et al.* (2008a).

Principles of CBM

Each individual farmer is assumed to have some level of traditional knowledge and skill sets to manage the biodiversity that underpins their farming practices (e.g. variety selection, propagation of planting materials, crop management) and to choose livelihood strategies. These practices are deeply embedded in the local culture. These traditional knowledge systems and farming practices are not static in nature. Production practices change as the farmer acquires new sets of scientific knowledge, skills and technologies, and blends them with traditional practices for further livelihood improvement. Farmers tend to unpack promising new farming practices and re-package them to suit their local context. Such local innovation can be scaled up rapidly if local institutions support the creation of an enabling environment (see Figure 3.2) for local innovation and serve as a platform for social learning (Dutty and Bolo-Dutty, 2003; Mburu and Wale, 2006; Smith, 2009).

The CBM approach strengthens farmers' capabilities to mobilize the five types of livelihood assets (natural, social, human, physical and financial) to improve livelihood options without negatively affecting the environmental resource base. A focus on adapting good practices allows the community to develop skills to analyze the household situation and access to the five capitals, and make self-directed decisions that aid community well-being. From the experiences of CBM implementation globally, four guiding principles of CBM stand out for realizing conservation and development goals:

1. Empower local communities to take leadership in planning and decision making
2. Build on local innovations, practices and resources
3. Explore and safeguard biodiversity-based livelihood options for sustainable use
4. Provide a platform for collective action and social learning.

It is important to focus not only on developing formal scientific understanding of on-farm management of agricultural biodiversity, but also to develop the institutions, incentive mechanisms and capacities needed to run internally driven on-farm conservation programmes. Local women and men and village-level opinion leaders need to understand how they can use their own local biodiversity and mobilize their social and human capitals to generate financial resources to develop livelihood options and conservation actions. Once awareness about the potential of community biodiversity and other community resources is increased, conservation efforts have proven much easier to tackle and implement successfully. In light of this, this book presents, based on research into on-farm conservation, a set of good practices that encourage farmers and communities to capitalize on traditional knowledge and skills and locally adapted genetic diversity, and build social capital such as social seed networks and local institutions.

The CBM approach encourages the custodianship of land and agricultural biodiversity as a means for improving the livelihoods of local communities, which simultaneously maintains locally and globally important genetic resources and supports evolutionary processes. While the sets of practices implemented from site to site will be context-specific, the principles of CBM can be employed in all contexts. This suggests that the CBM approach has potential to function as a global framework to assist on-farm conservation efforts and practices if it is further conceptualized, validated, institutionalized and

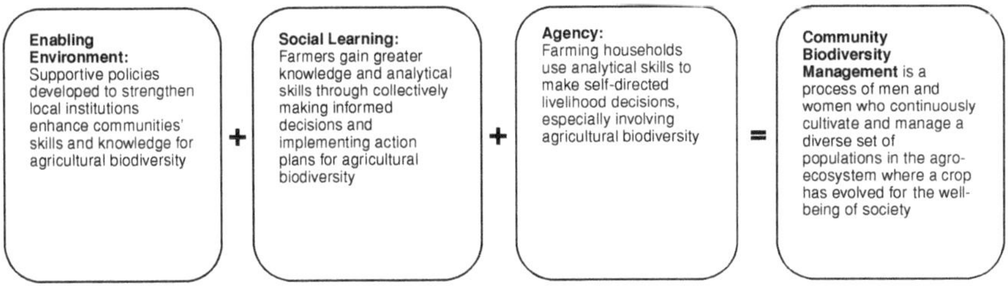

Figure 3.2 On-farm management of agricultural biodiversity is the product of an enabling environment created by institutions to enhance knowledge and skills of its members by providing a platform for social learning all with the end goal of building the agency in terms of informed decision-making to maintain, adapt and exchange local crop diversity for community well-being.

mainstreamed. The approach is considered a pragmatic method to realize on-farm management of local biodiversity and has already been applied by a number of institutions and countries (Sthapit *et al.*, 2012; de Boef *et al.*, 2012; 2013).

Process of CBM

CBM uses participatory methods and builds upon local institutional arrangements to maintain, enhance and create crop genetic diversity over time and space, and to ensure seed and knowledge flow from farmer to farmer and community to community for improved livelihoods on a sustainable basis. There are eight generic steps (Figure 3.3) in the CBM process that enrich farmers' knowledge with scientific knowledge. It is assumed that the process will facilitate local innovation at each step and thereby current farmer practice will evolve and improve with changing challenges and contexts. Each of the steps is guided by the principles of the CBM approach, as described above. A range of methods and tools are available for each step (Table 3.1) and can be customized in the CBM process to suit practitioners' preferences and specific purposes.

Step 1: Selection of site and community3

Sites and communities suitable for the CBM approach, with a goal of supporting on-farm conservation should be selected primarily from centres of diversity. Interspecific and intraspecific diversity are unequally distributed around the world and are usually concentrated in centres of diversity, which often coincide with the crop's centre of domestication (Gepts, 2006). Farming communities living in and around such biodiversity-rich areas should be selected as the main impact group. For greater chances of success with CBM, other site selection criteria must be considered, including ecosystem diversity, intraspecific crop

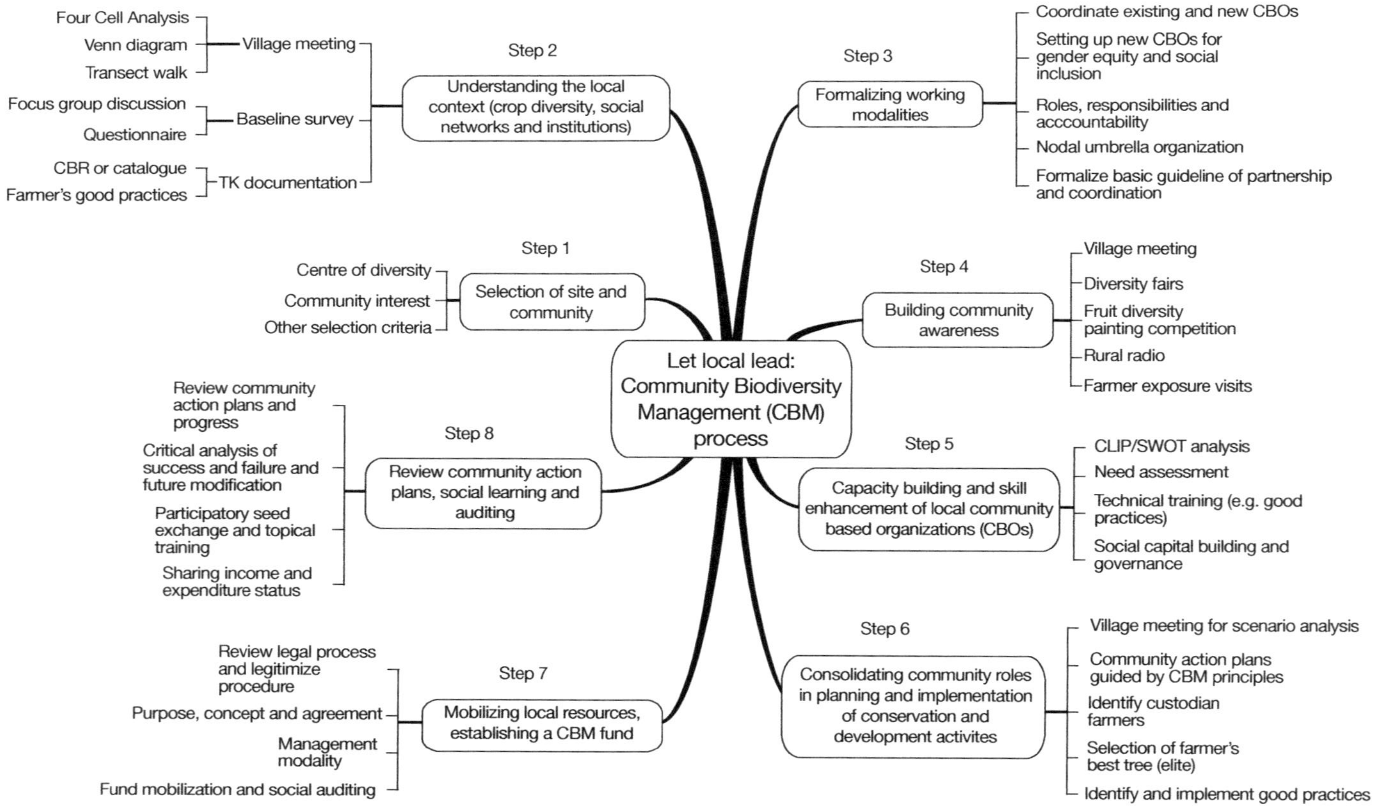

Figure 3.3 Process of CBM.

Source: Modified from Sthapit *et al.*(2012).

diversity within target species, history of cultivation or gathering from the wild, level of genetic erosion, specific adaptations, community interest in native diversity, associated traditional knowledge and multiple use value, collective action, partners' availability and logistics (Jarvis *et al.*, 2000; Ramanatha Rao and Sthapit, 2013).

Step 2: Understanding the local context

Understanding the local context is crucial for planning interventions for all three objectives of the CBM approach. Various staggered events can be used, such as participatory rural appraisal (PRA), on-farm diversity assessment, baseline surveys and participatory documentation of traditional knowledge. There are several participatory tools available for diagnostic purposes to get a quick understanding of the local context (de Boef and Thijssen, 2007). Venn diagrams, Four Cell Analysis (FCA) and timeline analysis help to understand the use and distribution of tropical fruit tree diversity, social networks and active institutional service providers (refer to list and purpose in Table 3.1). A Venn diagram can be used to map out key stakeholders, institutions and organizations to evaluate their relationship with community members. FCA is used to understand the status of local diversity by assessing its abundance (richness) and distribution (evenness) and the main reasons for its current status (Sthapit *et al.*, 2006d). Timeline analysis gives a historical narrative of major social, economic and environmental impacts on the community and its diversity. Such preliminary situation analyses help to identify the major drivers of and threats to diversity, along with an idea about the major players in a community. A baseline survey is not essential at the outset, however, baseline information is usually gathered using participatory methods at each step of CBM approach (Table 3.1) in order to facilitate informed decision making. These initial research-for-development activities help identify custodian households (Sthapit *et al.*, 2013, Chapter 4) and act as an entry point to conduct snowball sampling to map out social seed networks (Paudel *et al.*, 2015; Subedi *et al.* 2013).

Documentation of local diversity and good practices in a catalogue or community biodiversity register helps to monitor changes and facilitates social learning to make local efforts and proposed interventions more efficient, effective and sustainable.

Step 3: Formalizing working modality

In order to meet the diverse needs of community members, local organizations are set up to cultivate innovative partnerships with various local level service providers and develop local leadership to leverage resources and connections. One main feature of the CBM approach is that it builds upon existing organizations. Existing organizations, such as community-based organizations (CBOs)[4], and nodal, umbrella or facilitating organizations, play a pivotal role.

Table 3.1 Steps and core purposes of each CBM (Community Biodiversity Management) process and the availability of various complementary tools and methods for achieving specific objectives. Tools/methods used in the project are shown in bold. The references are listed for each tool/method if a practitioner wishes to gain insights of the detailed procedures and advantages and/or limitations of the tool

Steps	*Core purpose*	*Tools/methods*	*Specific objective of tools/methods*
1. Site and community selection	To identify potential site and communities at the centre of diversity (Jarvis *et al.*, 2000)	**Secondary resources/stakeholder consultations/social mapping**	Shortlist potential sites
		Transect walk (de Boef and Thijssen, 2007)	Rapid site assessment
		Resource and land use mapping (King, 2000)	Rapid assessment of five livelihood capitals (social, financial, physical, human, nature)
		Timeline (de Boef and Thijssen, 2007)	Major historical changes in a given locality
2. Understanding local context (baseline study)	To assess status of local biodiversity, social networks and institutions and identify intervention opportunities (Sthapit *et al.*, 2008a; 2008b) To collect benchmark indicators to measure impact (questionnaire survey)	**Four Cell Analysis** (Sthapit *et al.*, 2006d)	Quick intraspecific diversity assessment and reasons and identify common, rare and unique fruit species and varieties for setting agenda for conservation and development
		Diversity measurement (Jarvis *et al.*, 2008)	Estimate richness and evenness
		Matrix ranking (de Boef and Thijssen, 2007)	Identify key varietal traits and compare
		Pair-wise ranking (de Boef and Thijssen, 2007)	Compare trade-off between varieties
		Venn diagram (de Boef and Thijssen, 2007)	Identify key stakeholders and relationship
		Well-being ranking by card sorting (Rana *et al.*, 2007)	Categorize all the households into one of the three distinct categories; namely, 'resource-rich', 'resource-medium', and 'resource-poor' impact groups
		SWOT analysis (de Boef and Thijssen, 2007)	Assess strengths, weaknesses, opportunities and threats of the community/system
		Social network analysis (Subedi *et al.*, 2003)	Identify nodal farmers and their network for information and materials flow
		Identification of **Custodian farmers** (Sthapit *et al.*, 2013)	Identify diversity minded farmers who maintain, adapt and promote unique and rare diversity over generations

		Sustainable livelihood framework (DFID, 1999) **5-Capital**	Identify key entry point (GPD) amongst five capitals for enhancing other assets and livelihood strategies. Use this SLF framework to encourage social capital building to enhance other assets
		Participatory Market Chain Approach (PMCA) (Bernet *et al*., 2006) with additional tools tailored to the marketing of agricultural biodiversity	Understand and identify value chains for native fruit species and varieties that have unique characteristics and sales potential by making use of tools such as impact filter, rapid market appraisal, value chain map and street theatre play, among others
3. Formalizing working modality	To set up local institutional arrangement for implementing activities; To cultivate innovative partnership and develop local leadership for leveraging local resources; To identify the clear roles and responsibilities of multi-stakeholder partners; To form nodal local organizations for facilitating the smooth implementation of CBM approach (Sthapit *et al.*, 2008)	**Village meeting**	To ensure free prior informed consent (FPIC)
		Venn diagram (de Boef and Thijssen, 2007)	Identify key stakeholders and relationship
		Social analysis using CLIP for power, interest and legitimacy of key stakeholders is essential to mitigate any potential conflict and translate that into collaborative actions (Chevalier, 2007). (CLIP = Conflict/ Collaboration, Legitimacy, Interest and Power)	Identify vulnerable, marginalized and excluded stakeholders and rectify in the project activities or team; identify key gatekeepers
		SWOT analysis (de Boef and Thijssen, 2007)	Assess strengths, weaknesses, opportunities and threats of multi-stakeholder partners
		Four-R analysis (Rights, Responsibilities, Revenues (benefits) of stakeholders, and the Relationships between stakeholder groups)(Mayers, 2005)[1]	Clarify the roles played by different stakeholders and the nature of relationships between them
		Social network analysis (Paudel *et al*., 2015; Friis-Hansen and Sthapit, 2000)	Identify nodal and custodian farmers to consolidate their roles in the project

Table 3.1 continued

Steps	*Core purpose*	*Tools/methods*	*Specific objective of tools/methods*
4. Building community awareness of the importance and values of biodiversity	To sensitize the farming communities in the importance and use value of local biodiversity[2]	**Diversity/seed/food fairs** (Tapia and Rosa, 1993; Sthapit *et al.*, 2006b)	Stimulate community and people's awareness on biodiversity
		Community Biodiversity Register (Subedi *et al.*, 2005)	Traditional knowledge (TK) documentation of genetic resources and good practices and consolidating community role in conservation
		School art competition related to biodiversity	Raise awareness on value of biodiversity among school children
		FM radio/Rural radio (Baral *et al.*, 2006)	Raise public awareness on value of biodiversity in rural communities
		Diversity block (Sthapit *et al.*, 2006b)	Display and compare local crop diversity in farmers' fields for public awareness and education
		Diversity kits (Sthapit *et al.*,2006a)	Deploying new crop diversity to farmers for informal research
		Drama/songs/poetry competition (Dewan *et al.*, 2006)	Raise awareness on value of biodiversity in rural communities
		Community fruit catalogues (Dinesh *et al.*, 2014)	TK documentation and stimulate public awareness of farmer's innovation and further use
5. Capacity building and skills enhancement of community-based organizations	To strengthen a nodal CBO/institution for organizing networks of self-help, women, commodity groups, and other user groups (Carloni and Crowley, 2005; Sthapit and Mijatovic, 2014)	**Diversity/seed fairs** (Sthapit *et al.*, 2006c)	To strengthen farmers' and CBOs' capacity in collective action of organizing diversity fairs
		Community Biodiversity Register (Subedi *et al.*, 2005)	To strengthen farmers' capacity to document TK and local crop diversity and its value
		Four Cell Analysis of agricultural biodiversity (ABD) (Sthapit *et al.*, 2006d)	To build local capacity for on-farm diversity assessment, understanding reasons of changes in availability and distribution of ABD and use information for conservation and development agenda
	To build capacity of agency and institution in situation analysis, critical and analytical thinking	**Diversity Block** (Sthapit *et al.*, 2008a; 2008b)	To demonstrate diversity of village for public awareness, seed multiplication for seed sources for conservation in seedbank and improve access of seed

	and use information for self-directed decision making process (Sthapit *et al.*, 2008a)	**Community seedbank** (Shrestha *et al.*, 2006; Vernooy *et al.*, 2015)	To safeguard local diversity, improves access and availability of seed and empowerment of farmers through collectively managing their own ABD (farmers' rights)
	To strengthen capacity of the local community and its institution with the ability to decide how to structure their organization, develop and implement local policies, prioritize community needs and manage natural and biodiversity resources, and to ensure that the community's interest and participation continue beyond the project's lifetime (Sthapit *et al.*, 2006c).	**Diversity kits** (Sthapit *et al.*, 2006a; Canci *et al.*, 2013) Crowdsourcing (van Etten, 2011)	To promote informal research and development (IRD) capacity by deploying new diversity to farmers for evolutionary breeding and use crowdsourcing to detect portfolio of good varieties
		Grassroots breeding (Sthapit and Ramanatha Rao, 2009)	Build capacity of farmers for selecting best tree from existing diversity and register for wider use/elite material selection
		Participatory Variety Selection (Joshi and Witcombe, 1996)	Build local capacity for selection of fixed lines under target environment by farmer participation
		Participatory Plant Breeding (Sthapit *et al.*, 1996; Witcombe *et al.*, 1996)	Build local capacity for selection of segregating materials under target environment jointly by farmers and breeders
	To create platform for local innovation	**Knowledge share fair**/exchange visits, orientation/training/cross-learning field visit	Share knowledge by cross country or region visit of farmers, researchers and policy makers, such as at the global expo
6. Consolidating the community role in planning and implementation of conservation and development activities	To empower communities to plan, implement, monitor and evaluate both conservation and development activities themselves	**Community action plans (CAP)**	Develop a bottom up plan for managing ABD at the village level driven by local resources
		Community seedbank (Shrestha *et al.*, 2006)	Maintain seeds for local use and provide platform for seed transaction and learning
		Community fruit tree nurseries	Improve easy access of seed to strengthen local seed system and increase options to cope with adversity
	To implement conservation and development plan smoothly in community	**Diversity kits** (Sthapit *et al.*, 2006c)	To provide composite seed packets of multiple species and varieties
		Farmer's Field School (Braun and Duveskog, 2008)	Promote Integrated Pest Management (IPM) or integrated pest and soil fertility management through farmers group learning

Table 3.1 continued

Steps	*Core purpose*	*Tools/methods*	*Specific objective of tools/methods*
		Diversity Field School (DFS)[3]	Provide a platform for farmers to share knowledge and practices to use portfolio of varieties or mixtures for managing disease and pests in the context of farmers' field school
		Participatory Diagnosis guidelines (Jarvis and Campilan, 2006)	Use crop genetic diversity to reduce pests and diseases
		Participatory Variety Selection (PVS)(Joshi and Witcombe, 1996)	Locally select sources of new diversity and seed multiplication
		Participatory Market Chain Approach (PMCA) (Bernet *et al.*, 2006) and value chain stakeholder platform meetings (Thiele *et al.*, 2011)	Improve local entrepreneurs to identify and promote value chains for unique native fruit species and varieties that have sales potential
7. Mobilizing local resources, establishing a CBM fund	To develop process of testing out using the CBM fund as one form of PACS for expanding conservation of a larger amount of agricultural biodiversity in a sustainable manner	**CBM funds** (Shrestha *et al.*, 2012) Group lending Micro-finance Saving and credit scheme Revolving fund	Stimulate local conservation efforts by financial empowerment, livelihood enhancement and sustaining local institutions
		Crowdfunding initiative (http://www.globalgiving.org)	Raise money from informed crowd for specific projects in communities all over the world
	To provide incentive to farming community for sustainable management of biodiversity and empower community for collective actions for meeting both conservation and development goals	**Local institutions** (Carloni and Crowley, 2005) Cooperative Self-help groups Community Based Organizations (CBOs)	Strengthen umbrella local institution to promote collective action and support CBM initiatives

	without direct payment (Shrestha *et al.*, 2013a; Narloch *et al.*, 2011)		
8. Review community action plans, social learning and auditing	To ensure effective, transparent and efficient implementation of CBM process	**Social learning and change** (Aanoudse *et al.*, 2011)	Guide to learning practices in local organizations and social change in diverse community in the adult learning environment
		CBM calendar display (Subedi *et al.*, 2006b)	Display community action plans in a transparent manner to community members
	To build trust legitimizes the local organization and provides a foundation for leveraging resources from local government (Sthapit *et al.*, 2008a)[4]	Travelling seminar (Gyawali *et al.*, 2006)	To facilitate the mutual sharing and learning from direct observation, farmer and professional interactions; participatory monitoring and evaluations and enlighten policy makers by demonstrating field impacts and influencing policy reforms
	To refine community action plans, social learning and auditing by improving social and ethical performance of organizations[5]	Annual general assembly and social auditing	Review progress and report expenditure status of CBM
		Participatory seed and knowledge exchange (Shrestha *et al.*, 2013c)	Foster **knowledge and seed exchange** at the time of village meetings

Note:

[1] **The four Rs** is a tool that can be used to clarify the roles played by different stakeholders and the nature of relationships between them. Assessment and negotiation amongst stakeholders of these roles and relationships is necessary before developing the capacities needed for improving policies and institutions. The four Rs tool aims to operationalize the concept of 'roles' by unpacking these into Rights, Responsibilities, Revenues (benefits) of stakeholders and the Relationships between stakeholder groups. This is particularly useful in contexts where roles need re-thinking, negotiating and developing (Mayers, 2005).

[2] The **diversity fair** is a stepping stone to CBM for two reasons: (i) because it enables the community to begin to develop a local biodiversity database and realization of potential threats, and (ii) because the fair encourages the community to assume responsibility in managing its biodiversity locally.

[3] **Diversity Field School** (DFS) is a platform of social learning – either in a farmer's field or indoor setting – where one can learn to appreciate crop genetic diversity and use for managing pest and diseases but also other abiotic and other adversity for the benefits of the community.

[4] **Social learning theory** suggests that people can learn though observation, including direct instruction, modelling and imitation in order to build capacity of second generation members (Bandura, 1977).

[5]Employing a **social audit** practice by local institutions is a transparent way of measuring, understanding, reporting and ultimately improving an organization's social and ethical performance.

They need to: provide strategic and technical guidance; build collaborative partnerships with other service providers and stakeholders; oversee activities at community level; and reinforce cooperation across the community. Such a facilitating agency helps local groups and organizations to have: (i) knowledge and skills to appreciate and manage agricultural biodiversity (knowledge); (ii) rules and regulations to facilitate collective action and govern the process (institutions) concerning biodiversity; and (iii) strategies and activities to improve biodiversity-based livelihoods and income generation (practices) (Sthapit *et al.*, 2008a; 2008b) (Figure 3.1).

Based upon the analyses of the previous step and experiences, key actors and existing local organizations can be identified for the coordination of certain activities with clear roles and responsibilities in a transparent manner. Village meetings with community men and women leaders and representatives of key institutions are conducted to assess interest, define roles and seek free, prior informed consent (FPIC) from the community to agree upon basic rules regarding the ownership, recognition and sharing of knowledge and germplasm of natural and agricultural biodiversity (Table 3.1). Such FPIC agreements need to follow access and benefit-sharing standards as laid out in the Nagoya Protocol of the CBD and by national governments. A committee of representatives of self-help groups or local institutions is established to discuss and develop community action plans. Villages where no CBOs exist are encouraged to form cooperatives or self-help groups (SHGs) to build social capital. For example, the successful Society for Conservation of Mango Diversity (SCMD) was established by farmers and researchers in Malihabad in India as there was none at the outset of the project. Because of social taboos, SHGs can be organized by gender, which can also foster the empowerment process of women and other socially excluded smallholder farmers. Social tools exist, such as CLIP[5] (Table 3.1), which can be used to identify socially excluded or marginalized men and women groups within the umbrella organization of CBOs in terms of their power, interest and legitimacy criteria (Chevalier, 2007; de Boef and Thijssen, 2007) and rectify them.

Step 4: Building community awareness of the importance and values of biodiversity

Community awareness and education about the values, traits, benefits, uses and costs of agricultural biodiversity are an important foundation for long lasting and sustainable CBM (Shrestha *et al.*, 2013a). Public awareness is needed at different levels, from school children and young men and women in farming communities to urban consumers and policymakers. While the avenues for strengthening awareness vary with place and culture, every society has numerous culturally embedded communication tools at its disposal (Friis-Hansen and Sthapit, 2000; Dewan *et al.*, 2006; Baral *et al.*, 2006). Of the many tools used in the TFT project for creating public awareness and education about agricultural biodiversity (Table 3.1), diversity fairs have been the most popular among country partners, who conducted 61 diversity fairs with various purposes over five years, while formally committed to doing only one each. Diversity fairs are powerful and effective at both community and country levels. Rural drama, FM radio, songs, poetry and art competitions are other useful tools that raise awareness about agricultural biodiversity and reach community members with different educational backgrounds and mobility patterns. Travelling seminars are most appropriate for policymakers.

Step 5: Capacity building and skills enhancement of community-based organizations

One of the worst forms of poverty and disempowerment for farmers and rural communities is the lack of capacities to be able to make self-directed decisions for the betterment of their livelihoods. Step 5 focuses on building basic capacities and skills that help to increase the self-confidence, efficiency and social mobilization capacity of local institutions (Sthapit *et al.*, 2008a; Jarvis *et al.*, 2011). Training focuses at first on the formation of self-help groups, starting savings and loan cycles, financial administration, leadership, setting up institutional objectives, and rules and regulations (see Chapter 29; Shrestha *et al.*, 2013b). In a later stage this involves assistance in forming more complex organizations such as cooperatives, producer companies or SHG federations. Based on needs identified by the CBOs themselves, training can focus on technical aspects related to propagation, nursery or seed management, farming, post-harvest processing and value addition, marketing and conservation. This needs-based capacity building can be made a part of regular activities and can become more sophisticated as capacities of CBO members are enhanced. During the capacity building process the identification of local leaders, to act as change agents and to mobilize local groups, is essential.

Step 6: Consolidating the community role in planning and implementation of conservation and development activities

In line with the concept of community empowerment (Bartlett, 2008), the CBM approach strives to empower communities to plan, implement, monitor and evaluate activities themselves. Once community members and their institutions have participated in the process of understanding the local context (step 2) and have received the opportunity to access new information, skills and knowledge (steps 3 to 5), they are encouraged to make their own decisions through a bottom-up planning process for development and conservation activities (Figure 3.3). To respect traditional cultural norms, give a voice to different community members and also to empower women, it is recommended that this planning process be held in separate groups of men and women, who can subsequently be brought together to discuss important issues and foster mutual understanding in plenary. Activities for conservation and development should preferably be developed simultaneously to avoid unsustainable practices (when conservation is left out) or the loss of interest (when development is left out).

Conservation activities

A wealth of knowledge about species, varieties and their particular traits and uses is often found with custodian farmers and can be captured through the development of community biodiversity registers (CBR) (Subedi *et al.*, 2005), pictorial fruit diversity catalogues (Kiloes *et al.*, 2014) and the identification and documentation of good practices and related traditional knowledge in papers and videos. Nurseries can be established to multiply and disseminate elite materials selected for their potential for markets, particular domestic uses or unique traits that need to be safeguarded for the future. For perennials, diversity blocks with a few trees of all local species and varieties can be established in school gardens, community forests or the private land of custodians. For annuals, community seedbanks or seed networks can be established. Local nurseries, seedbanks, registries and fruit catalogues help ensure access to germplasm for all farmers within the community and also link to niche markets.

Development activities

Limited value chain studies and methods exist that are tailored towards the promotion of agricultural biodiversity for home use and for sales, so most CBM projects have experimented with a mixture of existing and new methods (see Chapters 22 and 24). Rapid market appraisal techniques (participatory market appraisal and participatory value chain mapping) can be used to assess market opportunities, followed by thorough value chain assessments to ascertain major constraints and opportunities. The wealth of local diversity can be screened for traditional uses and marketable traits using participatory methods such as an impact filter to evaluate products and species on their social, economic and environmental impact (Bernet *et al.*, 2006). A range of species for home use, local markets and distant markets can be selected to explore further for commercialization in close collaboration with value chain stakeholders, who jointly develop market or product innovations based on local biodiversity through novel products, processing techniques or equipment, improved labels and innovative branding strategies (Thiele *et al.*, 2011) or agrotourism (see Chapters 23 and 24).

Involvement of service providers, extension agents, researchers, conservation agencies, banks and the private sector are crucial to enrich the participatory planning process of local organizations.

Step 7: Mobilizing local resources, establishing a CBM fund

The purpose of establishing a CBM fund is to enable community organizations to take control over their decision-making on management of biodiversity and development interventions (Table 3.1). A CBM fund can be set up by organizing ongoing savings and credit schemes for members. It is an important driver contributing to the sustainability of the CBM approach. Its operational modality is similar to other microfinance schemes (Basargekar, 2010). In Nepal, where this approach was developed (Shrestha *et al.*, 2012; 2013b), seed money from an international project contributed to the establishment of the fund and matching funds are collected within the community. Every household within the village is eligible to apply for loans from the CBM fund, on condition that they abide by some local codes of conduct such as multiplying seed of rare varieties, or paying a locally determined interest rate in cash or in seeds. Indirectly, the community can mobilize funds and human resources for: (i) livelihood improvements for poor smallholder and women farmers (e.g. a small loan for goat or chicken rearing); (ii) sustaining local institutions (e.g. funds to cover local office costs); (iii) payment of conservation services (e.g. reimburse 1.5kg of seed for each 1kg of seed used or use 20 per cent of the interest earned from the CBM fund loan repayment to finance the following year's local seed production and distribution plans at the local level); and (iv) local benefit sharing of genetic resources (e.g. seed/sapling multiplication). Shrestha *et al.* (2012) documented a similar step-by-step procedure in Nepal. In India, the TFTGR project started a large-scale social capital building exercise by setting up SHGs and CBM funds for sustainability (Chapter 29).

In the context of conservation, the CBM fund creates a local incentive structure that supports on-farm management of rare and useful agricultural biodiversity. This presents a framework for integrating external motives (payments) and internal motives (custodianship) in a way that can lead to more durable behaviours and successful conservation incentive programmes (Sorice and Donlan, 2015). In India, such funds can support diversity-rich gardens of custodian farmers *in situ*. Shrestha *et al.* (2013b) reported that the CBM fund in Nepal has been found an effective means to organize community members, motivating the community and their institutions to implement CBM action plans and developing community ownership for funds' sustainability. Governments and donors may wish to support local communities in establishing CBM funds as a strategy to reach poor people and also as a catalyst for mobilizing local financial capital for the on-farm management of agricultural biodiversity. There is scope to use this CBM fund practice to organize and link a variety of existing groups in a production landscape and coordinate a range of niche-specific conservation, production and livelihood related activities (Sthapit and Mijatovic, 2014). A CBM fund can also serve as one kind of Payment for Agrobiodiversity Conservation Services (PACS)[6] for expanding conservation at the watershed or landscape levels (Narloch *et al.*, 2011). In some cultural contexts, due to perceived negative connotations associated with the concept of 'payments' (despite it being widely used in the literature), rewording payments as 'rewards' can make the schemes more culturally acceptable (Adam Drucker, personal communication).

Experiences with CBM funds show that it is critical to formulate CBM fund guidelines in a participatory manner to ensure proper use of funds and to reach marginalized and resource-poor farmers. The process requires committed local institutions to drive it. Other funds allocated for on-farm conservation in the community can also flow through the CBM fund, thus avoiding different funding agencies doing different things and leaving intact the collective decision of the community.

Step 8: Review community action plans, social learning and social auditing

One important – if not the most important – feature of a powerful CBM approach is to let local men and women lead the CBM process (Figure 3.3). It is important that the executive members of a CBO or SHG review the progress of community action plans and learn from success and failure in a systematic manner. One key link is the nexus between local institutions, farmer–farmer and farmer–researcher knowledge exchange. A clearly defined mechanism needs to be developed so that new knowledge and understanding is shared with farmer institutions and does not remain with the project team. Hence, pedagogical approaches to linking CBM and local institutions are crucial for realizing on-farm conservation goals. Linking different actors also allows transparent sharing of information about activities carried out and expenditures incurred and planned for next year. The CBM process is transformed into a community learning platform to avoid the notion that this is a one-way process whereby learning comes to the participants from outsiders or researchers. Instead, the approach provides a space whereby experiences and lessons are shared by all participants, through an appropriate and participatory approach. The farmers and service providers themselves are a 'pool of knowledge resource' that can be tapped in the learning platform.

One key principle of the CBM approach is that through participation of diverse groups of stakeholders, women and men farmers will learn from each other and improve their knowledge, skills and practices to be able to cope with new challenges, threats, adversity and also new opportunities and connections. An empowered community tends to demonstrate good capability in: (i) situation analysis, (ii) critical thinking about what is good for the environment rather than focusing on short-term gains or external resources, and (iii) appropriate decision making that considers community well-being. This is only possible if local government creates an enabling environment and institutional platforms for social learning and promotes community-level action plans. Most of the participatory tools listed in Table 3.1 and the following case studies (e.g. diversity and product fairs, participatory seed exchange, on-the-spot grafting training) facilitate social learning and scaling up of new ideas by sharing knowledge and encouraging discussion. Although social learning happens at all steps, a final review and audit legitimizes the local organization and provides a foundation from which to leverage resources from local government and build capacity of second-generation members.

Essential CBM activities and tools

Community participation and bottom-up planning alone are not enough to pull together the CBM process. The community must be organized with trained leaders and heightened social and environmental consciousness and ownership by its members. This happens slowly when we empower local communities by providing opportunities, building trust and seeking accountability to decide how to structure their organizations, develop and implement policies, prioritize community needs and manage community resources. To build the capacity and capability of leaders, custodian farmers, change agents and local organizations to develop deeper understanding, a few essential CBM activities, tools and methods might be required (Table 3.1). It is assumed that through using these tools, methods and practices in the different steps for various purposes, community and local organizations will blend the new knowledge, skills and connections with their traditional or current practices, thereby increasing their capability for better productivity, sustainability and ecosystem services.

We recognize, however, that practitioners and project managers can conduct these activities as part of project implementation work while still failing to give true decision making responsibilities to the farmers and achieve the desired change. Hence, equally important to applying the tools is the practice of facilitating understanding among the participating farmers and wider community about why each tool is being used and how it fits within the broader framework of their livelihood and conservation goals. Though these tools are useful on their own, it is the combination of tools over time and participatory selection of them that contribute to the desirable changes brought about through the CBM process (Table 3.1). The tight planning and reporting requirements of development and research projects and the tendency of a value-for-money monitoring approach that prioritizes quantitative numbers while ignoring the qualitative evaluation of these numbers exacerbates the difficulties faced by frontline workers. However, the fact that communities are involved in these good practices provides a good starting point to improve upon how the principles of the CBM approach are integrated in all aspects.

Participatory Four Cell Analysis (FCA) for on-farm diversity assessment

On-farm intraspecific diversity is measured by scientists in different ways (Magurran, 2003; Jarvis *et al.*, 2008) but the methods are difficult to communicate with farming communities. Richness and evenness are two key measurements of biodiversity (Magurran, 2003; Frankel *et al.*, 1975). Richness refers to the number of varieties regardless of their frequencies. Evenness refers to the proportion of area covered by each individual variety (Jarvis *et al.*, 2008). Farmers can easily articulate richness by counting variety names and traits, and evenness by referring to the area under the variety. FCA is a technique to assess the abundance (richness) and distribution (evenness) of local crop diversity within farming communities (Sthapit *et al.*, 2006d). It takes into account the richness and evenness of inter- or intraspecific diversity. FCA was first developed in Nepal (Sthapit *et al.*, 2006d) and has since been used elsewhere for annual crops, but the tool was adapted to perennial fruit tree crops during the TFTGR project. In this context, the purposes of the tool are to: identify the most important fruit tree varieties that play a role in the livelihoods of local people; to facilitate systematic analysis of farmers' logic of extent and distribution of tropical fruit tree diversity; understand farmers' rationale for choosing varieties and area (number of trees); and identify common, unique and rare fruit varieties so that the community and professionals can develop diversified livelihood options and conservation plans.

A focus group of 8 to 12 people from a particular village sharing rich traditional knowledge and a similar environment are selected to conduct FCA. The group could be mixed women and men or disaggregated by gender and should include people whose knowledge and interests in biodiversity may differ, such as those from different ethnic, socio-economic or age groups. The method starts with a focus group discussion among farmers from a specific village. The farmers list all the varieties that thrive within their village boundaries. They

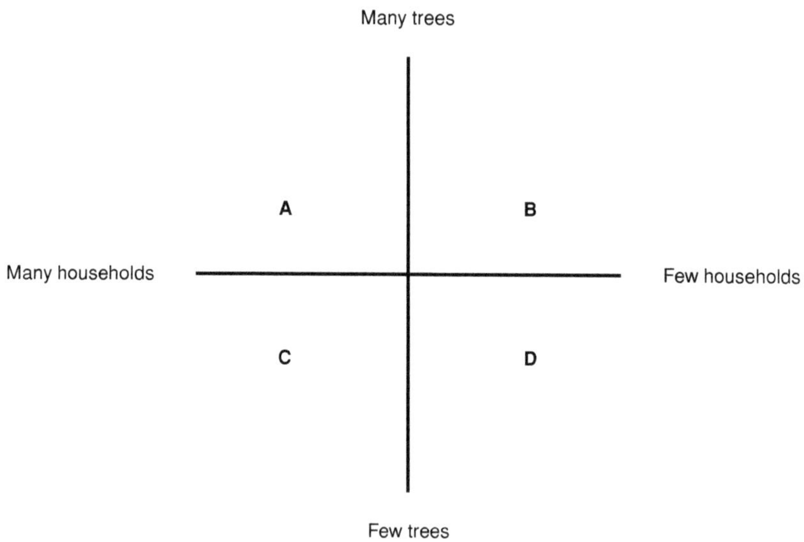

Figure 3.4 Four Cell Analysis for understanding abundance (richness) and spread (evenness) of tropical fruit tree diversity at the community level. Cell A refers to many households cultivating many fruit trees; Cell B refers to few households growing many trees; Cell C refers to many households keeping few trees and Cell D refers to few households cultivating few fruit trees. The cut-off point of each axis is based upon the judgement of the focus group discussion members.

Source: Modified from Sthapit *et al.* (2006d).

then categorize the target fruit trees into four groups using two indicators: abundance (count of varieties) and spread (many or few trees). The group identifies whether a certain variety is abundant or rare and maintained across many or few households by using a chart, placing each variety in one of the four cells (Figure 3.4).

Some practitioners also add a separate step to the exercise, a so-called 'fifth cell' by asking participants, after completing the four-cell exercise, to recall varieties that used to be grown in the area but now are no longer grown. If the varieties were only recently lost, this exercise helps farmers to understand why and helps them make plans to re-establish the varieties if they choose to do so.

The analysis is complemented by visiting four or five gardens to validate findings. Analysis of the results provides useful insights into common patterns between land allocation and crop diversity. This kind of analysis with men and women of the community reveals challenges, threats and opportunities and provides sufficient local-level information to support decision making for conservation and development efforts (Jarvis *et al.*, 2011; Sthapit *et al.*, 2012). This visual process enhances the knowledge of both farmers and researchers in a participatory manner. When the exercise is repeated periodically (after three to five years), it can give valuable insights into trends of biodiversity maintenance in that specific area. Communities and conservation agencies can both use this tool for monitoring crop diversity resulting from interventions, policy changes or stochastic catastrophes.

Diversity fairs

Diversity and seed fairs are used internationally to contribute to increasing conservation efforts and food security and have been cited as a successful and useful tool in different cultural contexts (Rusike *et al.*, 2003; Sthapit *et al.*, 2003; May *et al.*, 2014; Gajanana *et al.*, 2015). Originally used in the Andean countries (Tapia and Rosa, 1993), they are a popular practice employed by various organizations to sensitize communities to the value and importance of biodiversity and traditional knowledge (Rijal *et al.*, 2000; Gajanana *et al.*, 2015). Diversity fairs combine multiple functions: (i) locating rare and unique tropical fruit diversity, (ii) showcasing and providing a quick overview of community-level diversity, (iii) facilitating access to and the exchange of germplasm and knowledge for social learning, (iv) facilitating the generation of income through the sales of saplings and biodiversity-based products, and (v) promoting collective action and recognition for custodian farmers (Sthapit *et al.*, 2003).

A diversity fair is an event to which local people can come and display the diversity from their home gardens, orchards, fields or forests including their rare, unique and preferred species or varieties to compare and share their plant materials, related knowledge or derived products with each other. The main steps in organizing diversity fairs include: (i) agreeing the purpose, (ii) participatory planning with key stakeholders, (iii) setting up norms and procedures for the diversity fair, (iv) planning for implementation of the event, and (v) actual display and evaluation or documentation of participants' materials. Initially external agencies often organize the event, but later this event can be handed over to local organizations. The organizer can give a competitive character to the event to stimulate and increase participation among farmers or communities. Diversity fairs organized by local organizations create ownership and develop local capacity to coordinate events involving various stakeholders. This tool can be considered a first stepping stone to the CBM approach because it enables the community to develop a community fruit catalogue or database, and to start assuming responsibility for managing its own biodiversity (Shrestha *et al.*, 2013a). Later on in the process, the diversity fair can facilitate the evaluation of novel or improved products or saplings and generate income for participants.

The fairs can be embedded within recurring events such as harvest festivals, farmer or trade fairs or cultural–religious celebrations. During a five-year period, as part of the TFTGR project, a total of 61 diversity fairs were carried out in India, Indonesia, Malaysia and Thailand (Gajanana *et al.*, 2015), ranging from a mango diversity festival organized by the Ministry of Tourism in New Delhi, India, to a pomelo diversity fair in a village in Magetan district in East Java, Indonesia. The fair is seen as an excellent platform for small-scale farmer innovation and to showcase locally who has the capacity to innovate, experiment and adapt in the context of biodiversity management and use.

Community fruit catalogue

'Traditional knowledge' refers to the body of wisdom, innovations and practices of indigenous peoples and local communities (CBD, 1992b). With current global trends of formal schooling, which mean that children participate less in their parents' fields, and out-migration from rural areas, this knowledge is being lost at an alarming rate. In order to protect loss of traditional knowledge and consolidate communities' roles in management of agricultural biodiversity, the community fruit catalogue is considered one good practice. It documents traditional knowledge and is a means to transfer genetic materials and information to the next generation or interested outsiders. Table 3.2 emphasizes why the transfer of knowledge is as important as transfer of materials (Sthapit and Quek, 2005; Quek, 2005).

This practice has been successfully adapted from the 'Community Biodiversity Register' (CBR) practice employed in India and in Nepal by the NGO LI-BIRD (Subedi *et al.*, 2005; 2006a; Sthapit and Quek, 2005). In recent years, CBRs have been discussed, proposed and set up in a variety of institutional settings, and for a variety of reasons. Two distinct types of method are emerging: first, listing farmer-named varieties at the community level, assisted by university scholars and government professionals (Hegde *et al.*, 2014) and, second, empowering the local community to document important genetic resources and traditional knowledge by keeping an audio-visual record of them

Table 3.2 Missing components of traditional knowledge and importance of knowledge transfer (Quek, 2005). Knowledge holders are those elderly and experienced farmers who hold traditional knowledge related to biodiversity; recipients are those people who are willing to learn and practise this traditional knowledge; the material refers to genetic resources, seed or planting materials and the situation refers to the local context where the materials are used. For example, the easy availability of paracetamol for fever has eroded the need for herbal cures that require time and effort to procure and to prepare.

Missing component	*Results*			
Knowledge holder	✓	✓	✓	✗
Recipient	✓	✓	✗	✓
Situation	✓	✗	✓	✓
Material	✗	✓	✓	✓
Outcome				
Knowledge is	a story	not needed	threatened	lost

(Quek, 2005). In order to take up the CBR initiative, communities need to be fully empowered and social and human capitals need to be in place. Until such a situation is developed, modified community fruit catalogues can be developed with leadership of a researcher in consultation with farming communities and custodian farmers. In the case of the TFT project, communities were not yet convinced of the importance of documenting traditional knowledge on genetic resources, so researchers helped them to document their local fruit in catalogues. This pictorial and audio database answers the following key questions from the perspective of the community:

- What do we have?
- What do we value most?
- How do we distinguish them?
- What are high value traits?
- Why do we need to maintain them?
- How do we use them?
- Who are the custodians of these genetic resources?

Our experience has been that documenting this process created a platform of social learning among experienced custodian farmers and young farmers and raised community-level awareness and sharing of information and germplasm. This is the foundation of community biodiversity management. These fruit catalogues were printed in local languages and distributed to the community. The process resulted in five kinds of community benefits: (i) high value, rare, unique fruit tree varieties identified, documented and registered as elite materials; (ii) custodian households of rich diversity officially recognized and social status enhanced; (iii) private nurseries have access to elite materials and can conduct large-scale multiplication for income; (iv) farmers have ready-made information to market local fruit diversity locally and for export; and (v) the value of local crop diversity recognized and the access to and control of this information and materials for future use is ensured. These documents were published nationally and locally for future monitoring in India (Dinesh *et al.*, 2014), Indonesia (Kiloes *et al.*, 2014), Malaysia (Idris and Shafie Md Sah, 2014) and Thailand (Somsri *et al.*, 2015). Some of the best materials were selected as elite materials for multiplication as described in the following section.

Elite materials

Farmers' fruit trees in gardens/orchards usually contain many diverse high-value traits and have undergone centuries of observations, selection, exchange and breeding by farmers and communities (Zeven, 1998). Fruit crop improvement is a lifetime of work for any plant breeder because of the perennial nature of fruit species like mangosteen, rambutan, mango and citrus. Elite material is the best planting material from selected trees, which have a set of farmer-preferred traits or a set of high-value traits preferred in the markets, and which are superior to current commercial cultivars. One low-cost, efficient strategy to strengthen community biodiversity management is to work with custodian farmers (see Chapter 4; Sthapit *et al.*, 2013) to identify elite materials, which are the best trees ('plus trees') available in the community, characterize and evaluate them and further multiply them for community benefits. In this way selected unique elite material in the name of the farmer or community can be registered with the appropriate government authority, according to Farmers' Rights (http://www.planttreaty.org/content/farmers-rights) and making the information available for public and private nurseries. Once propagated, the elite trees can be integrated into farming systems.

In five years, 95 elite varieties of *Mangifera*, 32 of *Citrus*, 5 of *Garcinia* and 2 of *Nephelium* were identified, characterized and documented by means of farmer fruit tree catalogues from four countries. Of these, a total of 75 farmer varieties of *Mangifera*, 16 of *Citrus*, 5 of *Garcinia*, and 2 of *Nephelium* were registered by the respective competent government authority. These elite farmer materials are potentially valuable natural assets developed by farmer innovation that help income and livelihoods of farmers. They can be multiplied through networks of fruit tree nurseries and directly used as rootstocks or scions for grafting or budding. Selected elite trees were collected and multiplied in 126 fruit tree nurseries and made available to farming communities for general cultivation.

Custodian farmers

Whilst searching for good practices of tropical fruit management, researchers identified what they have called 'custodian farmers' (Sthapit *et al.*, 2013; 2015). These are farmers who maintain portfolios of diverse crop species and varietal diversity of agricultural biodiversity. Three features characterize these farmers: (i) they maintain high richness of agricultural biodiversity, (ii) they adapt or select available diversity, and (iii) they disseminate materials and knowledge on a wider scale in the community. These farmers select varieties adapted to local conditions and promote the use and conservation of local diversity among their friends and neighbours, even in the absence of any extrinsic incentives. Despite commercialization and global biodiversity loss, a few such farmers can be found in every country. They can be considered local role models for change; hence, it is useful to identify custodian farmers from different gender, age, ethnic or socio-economic groups so they can inspire people from similar backgrounds. These individuals can be an entry point for any community biodiversity management initiative as they are often knowledge and material holders without whom knowledge is lost (Table 3.2). The enabling environment of these custodian farmers in society needs to be strengthened so that their numbers and roles do not shrink and their knowledge and genetic resources are maintained.

During the process of GPD identification in the TFTGR project, the research team developed a methodology to identify custodian farmers and document their unique genetic materials (Sthapit *et al.*, 2013). Chapter 4 describes the roles, functions, responsibilities and rights of these farmers and highlights how custodian farmers play a dynamic role in on-farm and *in situ* conservation of perennial tropical fruit trees. Some national research institutions, like the Indian Council of Agricultural Research (ICAR), have officially taken up the responsibility of identifying custodian farmers across different crops in India (Gajanana *et al.*, 2015).[7] Similar interest has been reported from Bolivia (Gruberg *et al.*, 2013), Nepal (Sthapit *et al.*, 2015), Malaysia and Indonesia (Lamers *et al.*, 2015).

Concluding remarks

On-farm conservation efforts are not sustainable without local efforts and there are considerable knowledge gaps globally about how to do this on the ground. The CBM approach engages men and women, rich and poor, in a collective planning and learning process for conservation and development. Community empowerment is a key driver to achieve these dual goals. Farming communities and their organizations can enhance knowledge of local intraspecific diversity and improve traditional practices through platforms designed for social learning and progressive improvement in horticultural practices. A set of CBM principles and practices facilitates the blending of traditional knowledge and wisdom with scientific knowledge and skills to improve the productivity and resilience of tropical fruit trees. The CBM process builds over time the adaptive capacity of a farming community to synthesize diverse sources of knowledge on fruit tree diversity through cross learning, integration, co-production of knowledge and practices – often termed small-scale farmer innovation – that have major impacts in their lives and society in general.

Such local-level platforms of social learning and change are possible with continued interactions between farmers, extension agents, researchers, private sector and other key stakeholders. In the case of the TFTGR project, changes have been observed in practices of propagation techniques, production and crop management and commercialization and home use, which are documented in Part 3 of this book.

References

Aarnoudse, A., Reeler, D., and Martin, T. (2011) *The Barefoot Guide 2: Learning Practices in Organizations and Social Change*, The Barefoot Guide Resource Centre, www.barefootguide.org

Bandura, A. (1977) *Social Learning Theory*, Prentice Hall, Englewood Cliffs, NJ

Baral, K.P., Sapkota, T., Bhandari, B., Adhikari, A., Dewan, S., Pageni, S., Subedi, A., and Upadhyay, M.P. (2006) 'Rural radio programme: An effective tool for reaching unreachable on biodiversity conservation issues', in B.R. Sthapit, P.K. Shrestha and M.P. Upadhyay (eds), *Good Practices: On-farm Management of Agricultural Biodiversity in Nepal*, NARC, LI-BIRD, IPGRI and IDRC, Nepal

Basargekar, P. (2010) 'Measuring effectiveness of social capital in microfinance: A case study of urban microfinance programme in India', *International Journal of Social Inq*uiry, vol 3, no 2, pp. 25–43

Bartlett, A. (2008) 'No more adoption rates! Looking for empowerment in agricultural development programmes', *Development in Practice*, vol 18, no 4–5, pp. 524–538. DOI:10.1080/09614520802181269

Bellon, M.R. (1996) 'The dynamics of crop intraspecific diversity: A conceptual framework at the farmer level', *Economic Botany*, vol 50, pp. 26–39

Bellon, M.R., Pham, J.L., and Jackson, M.T. (1997) 'Genetic conservation: A role for rice farmers', in N. Maxted, B.V. Ford-Lloyd and J.G. Hawkes (eds), *Plant Genetic Conservation: The In Situ Approach*, Chapman & Hall, London, pp. 263–289

Bellon, M.R. (2004) 'Conceptualizing interventions to support on-farm genetic resource conservation', *World Development*, vol 32, pp. 159–172

Bellon, M.R., Gotor, E., and Caracciolo, F. (2014) 'Conserving landraces and improving livelihoods: How to assess the success of on-farm conservation projects?' *International Journal of Agricultural Sustainability*, http://dx.doi.org/10.1080/14735903.2014.986363

Bellon, M.R., Gotor, E., and Caracciolo, F. (2015) 'Assessing the effectiveness of projects supporting on-farm conservation of native crops: Evidence from the High Andes of South America', *World Development*, vol 70, pp. 162–176, http:dx.doi:10.1016/j.worlddev.2015.01.014

Berg, T. (1997) 'The role of *in situ* genetic resources management', a summary of conclusions of the workshop "Towards a synthesis between crop conservation and development" held in Baarlo, the Netherlands, 29 June to 2 July 1997

Bernet, T., Thiele, G., and Zschocke, T. (2006) *Participatory Market Chain Approach (PMCA) User Guide*, CIP, Lima, Peru, p. 184; available at: http://tools4valuechains.org/tool/participatory-market-chain-approach-0

Braun, A. (2003) 'The farmer research group (CIAL) as a community-based natural resource management organization', in B. Pound, S. Snapp, C. McDougall, and A. Braun (eds), *Managing Natural Resources for Sustainable Livelihoods: Uniting Science and Participation*, Earthscan, London

Braun, A. and Duveskog, D. (2008) *The Farmer Field School Approach – History, Global Assessment and Success Stories*, background paper for the IFAD Rural Poverty Report 2010

Brush, S.B. (2000) *Genes in the Field: On-farm Conservation of Crop Diversity*, Lewis Publishers, Washington, DC

Canci, A., Gaudagnin, C.A., Henke, J.P., and Lazzari, L. (2013) 'The diversity kit: Restoring farmers' sovereignty over food, seed and genetic resources in Guaraciaba, Brazil', in W.S. de Boef, A. Subedi, N. Peroni, M. Thijssen, and E. O'Keeffe (eds), *Community Biodiversity Management: Promoting Resilience and the Conservation of Plant Genetic Resources*, Earthscan/Routledge, Abingdon, UK and New York, pp. 32–36

CBD (1992a) 'The Convention on Biological Diversity'; available at: http://www.cbd.int/convention/text/, accessed 16 February 2012

CBD (1992b) 'Traditional knowledge, innovations and practices', Secretariat of the Convention on Biological Diversity, Montreal, Canada, Article 8(j); available at: http://www.cbd.int/traditional/, accessed 11 January 2012

Carloni, A.S. and Crowley, E. (2005) *Rapid Guide for Missions: Analysing Local Institutions and Livelihoods*, FAO Guideline, Food and Agriculture Organization of the United Nations, Rome, Italy, p. 34

Chevalier, J.M. (2007) 'SAS2 1.0: Social Analysis CLIP,' in Social Analysis Systems 2 1.0, http://www-sas-pm.com/

de Boef, W.S. and Thijssen, M. (2007) *Participatory Tools Working with Crops, Varieties and Seeds: A Guide for Professionals Applying Participatory Approaches in Agrobiodiversity Management, Crop Improvement and Seed Sector Development*, research report, Wageningen International, Wageningen, The Netherlands

de Boef, W.S. and Thijssen, M. (2013) 'Community biodiversity management and in situ conservation of plant genetic resources', in W.S. de Boef, A. Subedi, N. Peroni, M. Thijssen, and E. O'Keeffe (eds), *Community Biodiversity Management: Promoting Resilience and the Conservation of Plant Genetic Resources*, Earthscan/Routledge, Abingdon, UK and New York, pp. 44–51

de Boef, W.S., Thijssen, M.H., Shrestha, P., Subedi, A., Feyissa, R., Gezu, G., Canci, A., Ferreira, M.A.J.F., Dias, T., Swain, S., and Sthapit, B.R. (2012) 'Moving beyond the dilemma: Practices that contribute to the on-farm management of agro biodiversity', *Journal of Sustainability Agriculture*, vol 36, pp. 788–809. DOI: 10.1080/10440046.2012.695329; available at: http://www.tandfonline.com/doi/abs/10.1080/10440046.2012.695329, accessed 21 August 2012

de Boef, W.S., Subedi, A., Peroni, N., and Thijssen, M. (eds) (2013) *Community Biodiversity Management: Promoting Resilience and the Conservation of Plant Genetic Resources*, Earthscan/Routledge, Abingdon, UK and New York

Dewan, S., Tiwari, R.K., Pageni, S., Rijal, D., Baral, K.P., Chaudhary, P., Rana, R.B., Subedi, A., and Sthapit, B. (2006) 'Rural poetry journey: An effective approach to sensitize farming community', in B.R. Sthapit, P.K. Shrestha, and M.P. Upadhyay (eds), *Good Practices: On-farm Management of Agricultural Biodiversity in Nepal*, NARC, LI-BIRD, IPGRI and IDRC, Nepal

Dinesh, M.R., Rajan, S., Singh, I.P., Singh, A., Singh, S.K., Vasudeva R, Vinoth, S., Patil, P., Parthasarathy, V.A., Reddy, B.M.C., and Sthapit, B.R. (2014) *National Fruit Catalogue of Tropical Fruit Tree Diversity* (Mangifera, Citrus *and* Garcinia), GEF, UNEP, Bioversity International and ICAR, India

DFID (1999) DFID sustainable livelihoods guidance sheets, www.ennonline.net/resources/667

DLP CCER report to External Panel (2006) Diversity for Livelihood Programme CCER Report to External Panel, Bioversity, Rome

Dutty, S. and Bolo-Dutty, B. (2003) 'Empowering people's organizations in community based forest management in the Philippines: The community organizing role of NGOs', *Annals of Tropical Research*, vol 25, no 2, pp. 13–27

Feyissa, R., Gezu, G., Tsegaye, B., and Alegn, T.D. (2013) 'On-farm management of plant genetic resources through community seed banks in Ethiopia', in W.S de Boef, A. Subedi, N. Peroni, M. Thijssen, and E. O'Keeffe, E. (eds), *Community Biodiversity Management: Promoting Resilience and the Conservation of Plant Genetic Resources*, Earthscan/Routledge, Abingdon, UK and New York, pp. 26–31

Frankel, O.H., Brown, A.H.D., and Burdon, J.J. (1975) *The Conservation of Plant Biodiversity*, Cambridge University Press, Cambridge, UK

Friis-Hansen, E. and Sthapit, B.R. (eds) (2000) *Participatory Approaches to the Conservation and Use of Plant Genetic Resources*, IPGRI (International Plant Genetic Resources Institute), Rome

Gajanana, T.M., Dinesh, M.R., Rajan, S., Vasudeva R, Singh, S.K., Parthasarathy, V.A., Sthapit, B.R., and Ramanatha Rao, V. (2015) 'Motivation for on-farm conservation of mango (India – a case study)', *Indian Journal of Plant Genetic Resources*, vol 28, no 1, pp. 1–6. DOI 10.5958/0976-1926.2015.00001.7

Gepts, P. (2006) 'Plant genetic resources conservation and utilization: The accomplishments and future of a societal insurance policy', *Crop Science*, vol 46, pp. 2278–2292

Gruberg, H., Meldrum, G., Padulosi, S., Rojas, W., Pinto, M., and Crane, T. (2013) 'Towards a better understanding of custodian farmers and their roles: Insights from a case study in Cachilaya, Bolivia', Bioversity International, PRONIPA, Wageningen UR, CCAFS and IFAD, Rome

Gyawali, S., Gautam, R., Upadhyay, M., and Gauchan, D. (2006) 'Traveling seminar: Influencing policy', in B.R. Sthapit, P.K. Shrestha, and M.P. Upadhyay (eds), *Good Practices: On-farm Management of Agricultural Biodiversity in Nepal*, NARC, LI-BIRD, IPGRI and IDRC, Nepal

Hegde, V.M., Kamtekar, S., Vasudeva R, Parthasarathy, V.A., and Sthapit, B.R. (2014) 'Community Biodiversity Register (CBR) for tropical fruit tree (TFT) species of Salkani village in Uttara Kannada district', technical report of the UNEP/GEF project – Sirsi Site, College of Forestry, Sirsi; University of Agricultural Science, Dharwad, India

Idris, S. and Shafie Md Sah, M. (2014) *'Fruit Catalogue' Conservation and Sustainable use of Cultivated And Wild Tropical Fruit Diversity: Promoting Sustainable Livelihood, Food Security and Ecosystem Services*, GEF, UNEP, Bioversity International, MARDI, Malaysia

Jarvis, D.I., Myer, L., Klemick, H., Guarino, L., Smale, M., Brown, A.H.D., Sadiki, M., Sthapit, B.R., and Hodgkin, T. (2000) *A Training Guide for In Situ Conservation On-farm*, version I, IPGRI, Rome

Jarvis, D.I., Brown, A.H.D., Cuong, P.H., Collado-Panduro, L., Latourniere-Moreno, L., Gyawali, S., Tanto, T., Sawadogo, M., Mar, I., Sadiki, M., Hue, N.T.H., Arias-Reyes, L., Balma, D., Bajracharya, J., Castillo, F., Rijal, D., Belqadi, L., Rana, R., Saidi, S., Ouedraogo, J., Zangre, R., Keltoum, R.O., Chavez, J.L., Schoen, D., Sthapit, B.R., Santis, P.D., Fadda, C., and Hodgkin, T. (2008) 'A global perspective of the richness and evenness of traditional crop genetic diversity maintained by farming communities', in *Proceedings of the National Academy of Sciences PNAS (USA)*, vol 105, no 14, pp. 5326–5551

Jarvis, D.I., Hodgkin, T., Sthapit, B.R., Fadda, C., and Lopez-Noriega, I. (2011) 'A heuristic framework for identifying multiple ways of supporting the conservation and use of traditional crop varieties within the agricultural production system', *Critical Reviews in Plant Sciences*, vol 30, no 1–2, pp. 1–49

Jarvis, D.I. and Campilan, M. (2006) *Crop Genetic Diversity to Reduce Pests and Disease On-farm: Participatory Diagnosis Guidelines*, Version I, Bioversity Technical Bulletin No. 12, Bioversity International, Rome, Italy

Joshi, A. and Witcombe, J.R. (1996) 'Farmer participatory crop improvement II. Participatory varietal selection, a case study in India', *Experimental Agriculture*, vol 32, pp. 461–477

Kendall, J. and Gras, E. (2013) 'The Maison de la Semence Paysanne and diversity platform: Promoting agrobiodiversity in France', in W.S. de Boef, A. Subedi, N. Peroni, M. Thijssen, and E. O'Keeffe (eds), *Community Biodiversity Management: Promoting Resilience and the Conservation of Plant Genetic Resources*, Earthscan/Routledge, Abingdon, UK and New York, pp. 43–51

Kiloes, A.M., Winarno, Kuntoro, B.A., Rafieq, A., and Arsanti, I.W. (2014) *Community Fruit Catalogue: Mango and Citrus in East Java and South Kalimantan, Indonesia*, Office of National Project Management Unit of GEF/UNEP Project 'Conservation and Sustainable Use of Wild and Cultivated Tropical Fruit Diversity: Promoting Sustainable Livelihoods, Food Security, and Ecosystem Services', Jakarta

King, A. (2000) 'A brief review of participatory tools and techniques for the conservation and use of plant genetic resources', in E. Friis-Hansen and B.R. Sthapit (eds), *Participatory Approaches to the Conservation and Use of Plant Genetic Resources*, IPGRI (International Plant Genetic Resources Institute), Rome

Lamers, H.A.H, King, E.D.I.O, Sthapit, S., Bernhart, A., Rafieq, A., Boga Andri, K., Gajanana, T.M., Shafie Md. Sah, M., Umar, S., Brooke, P., Rajan, S., Sripinta, P., Nimkingrat, T., Singh, S.K., Singh, T., and Sthapit, B. (2015) 'Characteristics and motivations of custodian farmers in South and South East Asia: A preliminary reflection', in S. Sthapit, G. Meldrum, S. Padulosi, and N. Bergamini (eds), *Strengthening the Role of Custodian Farmers in the National Conservation Programme of Nepal*, Proceedings from the National Workshop, 31 July to 2 August 2013, Pokhara, Nepal, pp. 15–20

Magurran, A.E. (2003) *Measuring Biological Diversity*, Blackwell, Oxford, UK

May, C., McClelland, S., Scrafford, H., and Saarli, K. (2014) 'Seed and Diversity Fairs – A Literature Review', IARD 4850, Course, Cornell University, USA, p. 25

Mayers, J. (2005) *The Four Rs. From Power Tools*, IIED, London

Mburu, J. and Wale, E. (2006) 'Local organizations involved in the conservation of crop genetic resources: Conditions for their emergence and success in Ethiopia and Kenya', *Genetic Resources and Crop Evolution*, vol 53, pp. 613–629. DOI 10.1007/s10722-004-2683-1

Narloch, U., Drucker, A.G., and Pascual, U. (2011) 'Exploring the potential of payments for ecosystem services for *in-situ* agrobiodiversity conservation', Ecosystem Services Economics (ESE) working paper series, no. 14, Division of Environmental Policy Implementation, UNEP

Paudel, D. (2015) 'On farm conservation of crop genetic resource: Declining de facto diversity and optimal funding strategy', *Natural Resources*, vol 6, pp. 196–207, http://www.scirp.org/journal/nr, http://dx.doi.org/10.4236/nr.2015.63018

Paudel, D., Sthapit, B. R., and Shrestha, P. (2015) 'An analysis of Social Seed Network and its Contribution to On-Farm Conservation of Crop Genetic Diversity in Nepal',

International Journal of Biodiversity, vol. 2015, Article ID 312621, 13 pages. DOI: 10.1155/2015/31262

Porter-Bolland, L., Ruiz-Mallen, I., Camacho-Benavides, C., and McCandless, S.R. (eds) (2013) *Community Action for Conservation Mexican Experiences*, Springer, London, p. 173

Quek, P. (2005) 'The IK Journal: A concept in TK documentation, IPGRI APO, Serdang, Malaysia', IPGRI unpublished document

Ramanatha Rao, V., and Sthapit, B.R. (2013) 'Conservation of tropical plant genetic resources – *in situ* approach', in H.F. Chin, B.M. Reed, and N.M. Noor (eds), *Conservation of Tropical Plant Species*, Springer Science and Business Media, New York

Rana, R.B., Garforth, C., Sthapit, B.R., and Jarvis, D.I. (2007) 'Influence of socio-economic and cultural factors in rice varietal diversity management on-farm in Nepal', *Agriculture and Human Values*, vol 24, no 4, pp. 461–472

Rijal, D., Rana, R., Subedi, A., and Sthapit, B. (2000) 'Adding value to landraces: Community-based approaches for *in-situ* conservation of plant genetic resources in Nepal', in E. Friis-Hansen and B. Sthapit (eds), *Participatory Approaches to the Conservation and Use of Plant Genetic Resources*, IPGRI (International Plant Genetic Resources Institute), Rome, pp. 166–172

Rusike, E. Mujaju, C., Mafa, A., and Zinhanga, F. (2003) 'Increasing seed diversity through community seed fairs in Zimbabwe', in CIP-UPWARD (eds), *Conservation and Sustainable Use of Agricultural Biodiversity: A Source Book, Vol II: Strengthening Local Management of Agricultural Biodiversity*, International Potato Center (CIP) and Users' Perspectives With Agricultural Research and Development (UPWARD), Los Banos, Philippines, pp. 277–283

Santos Doctor, J.A.P. (2013) 'Practices that contribute to the empowerment of farming communities for managing their agro biodiversity in Asia', in W.S. de Boef, A. Subedi, N. Peroni, M. Thijssen, and E. O'Keeffe (eds), *Community Biodiversity Management: Promoting Resilience and the Conservation of Plant Genetic Resources*, Earthscan/Routledge, Abingdon, UK and New York, pp. 37–43

Shrestha, P., Subedi, A., Sthapit, S., Rijal, D., Gupta, S.K., and Sthapit, B. (2006) 'Community Seed Bank: Reliable and effective option for agricultural biodiversity conservation', in B.R. Sthapit, P.K. Shrestha, and M.P. Upadhyay (eds), *Good Practices: On-farm Management of Agricultural Biodiversity in Nepal,* NARC, LI-BIRD, IPGRI and IDRC, Nepal

Shrestha, P., Sthapit, S., Paudel, I., Subedi, S., Subedi, A., and Sthapit, B.R. (2012) 'A guide to establishing a community biodiversity management fund for enhancing agricultural biodiversity conservation and rural livelihoods', LI-BIRD, Nepal; http://www.libird.org/app/publication/view.aspx?record_id=68&origin=results&QS=QS&f1_4417=Pitambar&union=AND&top_parent=221

Shrestha, P. (2013) 'An assessment of community-based biodiversity management approach in improving livelihoods: A case study of Shivagunj, Jhapa', master's thesis, Indira Gandhi National Open University (IGNOU), Maidan Garhi, New Delhi

Shrestha, P, Subedi, A., and Sthapit, B. (2013a) 'Enhancing awareness on the value of local biodiversity in Nepal', in W.S. de Boef, A. Subedi, N. Peroni, and M. Thijssen (eds), *Community Biodiversity Management: Promoting Resilience and the Conservation of Plant Genetic Resources*, Earthscan/Routledge, Abingdon, UK and New York

Shrestha, P., Sthapit, S., Subedi, A., and Sthapit, B.R. (2013b) 'Community biodiversity management fund: Promoting conservation through livelihood development in Nepal', in W.S. de Boef, A. Subedi, N. Peroni, and M. Thijssen (eds), *Community*

Biodiversity Management: Promoting Resilience and the Conservation of Plant Genetic Resources, Earthscan/Routledge, Abingdon, UK and New York

Shrestha, P., Sthapit S., and Paudel, I. (2013c) 'Participatory seed exchange for enhancing access to seeds of local varieties', LI-BIRD, Pokhara, Nepal (in Nepali)

Smale, M. (2006) 'Valuing crop biodiversity: On-farm genetic resources and economic change', CABI (Centre for Agriculture and Biosciences International), Wallingford, UK

Smith, R.B. (2009) 'Let the local lead', *Nature*, vol 462, pp. 280–281

Somsri, S., Noppornphan, C., and Yoovatana, M. (eds) (2015) 'Fruit catalogue: *Citrus, Garcinia, Mangifera* and *Nephelium* in Chanthaburi, Chiang Mai, Nakhon Si Thammarat and Trang, Thailand', Department of Agriculture, Bangkok, p. 146

Sorice, M.G. and Josh Donlan, C. (2015) 'A human-centred framework for innovation in conservation incentive programs', *AMBIO*, doi 10.1007/s13280-015-0650-z, published online, 28 April 2015

Sthapit, B.R., Joshi, K.D., and Witcombe, J.R. (1996) 'Farmer participatory crop improvement III. Participatory plant breeding: A case study for rice in Nepal', *Experimental Agriculture*, vol 32, pp. 479–496

Sthapit, B.R., Rijal, D., De, N.N., and Jarvis, D.I. (2003) 'A role for diversity fairs: Experiences from Nepal and Vietnam', in CIP-UPWARD (eds), *Conservation and Sustainable Use of Agricultural Biodiversity: A Source Book. Volume II: Strengthening Local Management of Agricultural Biodiversity*, International Potato Center (CIP) and Users' Perspectives With Agricultural Research and Development (UPWARD), Los Banos, Philippines, pp. 271–276

Sthapit, B.R., Upadhyay, M.P., Shrestha, P.K., and Jarvis, D.I. (eds) (2005) *On-farm Conservation of Agricultural Biodiversity in Nepal: Assessing the Amount and Distribution of Genetic Diversity on Farm*, vol I, Proceedings of the Second National Workshop, 25–27 August 2004, Nagarkot, Nepal; IPGRI, NARC, LI-BIRD, Nepal

Sthapit, B. and Quek, P. (2005) 'Community biodiversity register: Overview, concepts and some systematic approaches', in A. Subedi, B. Sthapit, M. Upadhyay, and D. Gauchan (eds), *Learning from Community Biodiversity Register in Nepal*, Proceedings of the National Workshop, 27–28 October 2005, Khumaltar, Nepal, pp. 14–27

Sthapit, B.R. and Jarvis, D.I., (2005) 'Good practices of community based on-farm management of agricultural biodiversity in Nepal: Lessons learned', in B.R. Sthapit, M.P. Upadhyay, P.K. Shrestha, and D.I. Jarvis (eds), *On-farm Conservation of Agricultural Biodiversity in Nepal: Managing Diversity and Promoting its Benefits*, vol II, Proceedings of the Second National Workshop, 25–27 August 2004, Nagarkot, Nepal; IPGRI, NARC, LI-BIRD, Nepal, pp. 91–111

Sthapit, B., Gautam, R., Gyawali, S., Joshi, B.K., Subedi, A., Yadav, R.B., Chaudhary, P., and Rijal, D. (2006a) 'Diversity kits: Deploying new diversity to farmers', in B.R. Sthapit, P.K. Shrestha, and M.P. Upadhyay (eds), *Good Practices: On-farm Management of Agricultural Biodiversity in Nepal*, NARC, LI-BIRD, IPGRI and IDRC, Nepal

Sthapit, B., Shrestha, P., Baral, K.P., Subedi, A., Bajracharya, J., and Yadav, R.B. (2006b) 'Diversity block: Assessing and demonstrating local crop diversity', in B.R. Sthapit, P.K. Shrestha, and M.P. Upadhyay (eds), *Good Practices: On-farm Management of Agricultural Biodiversity in Nepal*, NARC, LI-BIRD, IPGRI and IDRC, Nepal

Sthapit, B.R., Shrestha, P.K., and Upadhyay, M.P. (eds) (2006c) *Good Practices: On-farm Management of Agricultural Biodiversity in Nepal*, NARC, LI-BIRD, IPGRI and IDRC, Nepal

Sthapit, B.R., Rana, R.B., Subedi, A., Gyawali, S., Bajracharya, J., Chaudhary, P., Joshi, B.K., Sthapit S., Joshi, K.D., and Upadhyay, M.P. (2006d) 'Participatory four cell analysis (FCA) for local crop diversity', in B.R. Sthapit, P.K. Shrestha, and M.P. Upadhyay (eds), *Good Practices: On-farm Management of Agricultural Biodiversity in Nepal*, NARC, LI-BIRD, IPGRI and IDRC, Nepal

Sthapit, B.R., Shrestha, P., Subedi, A., Shrestha, P., Upadhyay, M.P., and Eyzaguirre, P. (2008a) 'Mobilizing and empowering community in biodiversity management', in M.H. Thijssen, Z. Bishaw, A. Beshir, and W.S. de Boef (eds), *Farmers, Seeds and Varieties: Supporting informal Seed Supply in Ethiopia*, Wageningen International, Wageningen, The Netherlands, pp. 160–166

Sthapit, B.R., Subedi, A., Shrestha, P., Shrestha, P., and Upadhyay, M.P. (2008b) 'Practices supporting community management of farmers' varieties', in M.H. Thijssen, Z. Bishaw, A. Beshir and W.S. de Boef (eds), *Farmers, Seeds and Varieties: Supporting Informal Seed Supply in Ethiopia*, Wageningen International, Wageningen, The Netherlands, pp. 166–171

Sthapit, B.R. and Ramanatha Rao, V. (2009) 'Consolidating community's role in local crop development by promoting farmer innovation to maximise the use of local crop diversity for the well being of people', in H. Jaenicke, J. Ganry, I. Höschle-Zeledon and R. Kahane (eds), *Proceedings of the International Symposium on Underutilized Plants for Food, Nutrition, Income and Sustainable Development*, Arusha, Tanzania, 3–7 March 2008. ISHS Acta Horticulturae 806, International Society for Horticultural Science, Leuven, Belgium, p. 739

Sthapit, B.R., Subedi, A., Lamers, H., Jarvis, D.I., Ramanatha Rao, V., and Reddy, B.M.C. (2012) 'Community based approach to on-farm conservation and sustainable use of agricultural biodiversity in Asia', *Indian Journal of Plant Genetic Resources*, vol 25, no 1, pp. 88–104

Sthapit, B.R., Lamers, H., and Ramanatha Rao, V. (2013) *Custodian Farmers of Agricultural Biodiversity: Selected Profiles from South and South East Asia*, Proceedings of the Workshop on Custodian Farmers of Agricultural Biodiversity, 11–12 February, New Delhi, India; Bioversity International, New Delhi, India

Sthapit, S., and Mijtavic, D. (2014) 'The evolving landscape of agriculture biodiversity conservation, community based biodiversity management (CBM): A landscape approach to the conservation of agriculture biodiversity cultivated on fifteen years of experiences in Begnas, Nepal', LI-BIRD, Bioversity International, Nepal

Sthapit, S., Meldrum, G., Padulosi, S., and Bergamini, N. (2015) *Strengthening the Role of Custodian Farmers in the National Conservation Programme of Nepal*, Proceedings of the National Workshop 31 July to 2 August 2013, Pokhara, Nepal; Bioversity International, Rome, Italy, and LI-BIRD, Pokhara, Nepal

Subedi, A., Chaudhary, P., Baniya, B., Rana, R., Tiwari, R.K., Rijal, D.K., Jarvis, D., and Sthapit, B. (2003) 'Who maintains crop genetic diversity and how: Implications for on-farm conservation and utilization', *Culture & Agriculture*, vol 25, no 2, pp. 41–50

Subedi, A., Sthapit, B., Upadhyay, M.P., and Gauchan, D. (eds) (2005) *Learning from Community Biodiversity Register in Nepal*, Proceedings of the National Workshop, 27–28 October 2005, Khumaltar, Nepal

Subedi, A., Sthapit, B., Rijal, D., Gauchan, D., Upadhyay, M.P., and Shrestha, P.K. (2006a) 'Community Biodiversity Register: Consolidating community role in management of agricultural biodiversity', in B.R. Sthapit, P.K. Shrestha, and M.P.

Upadhyay (eds), *Good Practices: On-farm Management of Agricultural Biodiversity in Nepal*, NARC, LI-BIRD, IPGRI and IDRC, Nepal

Subedi, A., Shrestha, P., Shrestha, P.K., Gautam, R., Upadhyay, M.P., Rana, R.B., Eyzaguirre, P., and Sthapit, B.R. (2006b) 'Community Biodiversity Management: Empowering community to manage agricultural biodiversity', in B.R. Sthapit, P.K. Shrestha, and M.P. Upadhyay (eds), *Good Practices: On-farm Management of Agricultural Biodiversity in Nepal*, NARC, LI-BIRD, IPGRI and IDRC, Nepal

Subedi, A., Shrestha, P., Upadhyay, M.P., and Sthapit, B.R. (2013) 'The evolution of community biodiversity management as a methodology for implementing *in situ* conservation of agrobiodiversity in Nepal', in W.S. de Boef, A. Subedi, N. Peroni, and M. Thijssen (eds), *Community Biodiversity Management: Promoting Resilience and the Conservation of Plant Genetic Resources*, Earthscan/Routledge, Abingdon, UK and New York

Tapia, M.E. and Rosa, A. (1993) 'Seed fairs in the Andes: A strategy for local conservation of plant genetic resources', in W. de Boef, K. Amanor, K. Wellard, and A. Bebbington (eds), *Cultivating Knowledge Genetic Diversity, Farmer Experimentation and Crop Research*, Intermediate Technology Publications, London, pp. 111–118

Thiele, G., Devaux, A., Reinoso, I., Pico, H., Montesdeoca, F., Pumisacho, M., Andrade-Piedra, J., Velasco, C., Flores, P., Esprella, R., Thomann, A., Manrique, K., and Horton, D. (2011) 'Multi-stakeholder platforms for linking small farmers to value chains: Evidence from the Andes', *International Journal of Agricultural Sustainability*, vol 9, no 3, pp.423–433

Tyler, S.R. (2006a) *Communities, Livelihoods and Natural Resources: Action Research and Policy Change in Asia*, Intermediate Technology Publications, IDRC, London

Tyler, S.R. (2006b) 'Comanagement of Natural Resources: Local Learning for Poverty Reduction', IDRC (International Development Research Centre), Canada

van Etten, J. (2011) 'Crowdsourcing crop improvement in sub-Saharan Africa: A proposal for a scalable and inclusive approach to food security', *IDS Bulletin*, vol 42, no 4, July, Institute of Development Studies, UK

Vasudeva R, Hegde, N., and Sthapit, B.R. (2013) 'Promoting community management of *Garcinia* resources in the central Western Ghats, India: Initiatives of Tropical Fruits Diversity Project', in W.S. de Boef, A. Subedi, N. Peroni, and M. Thijssen (eds), *Community Biodiversity Management: Promoting Resilience and the Conservation of Plant Genetic Resources*, Earthscan/Routledge, Abingdon, UK and New York

Vernooy, R. and McDougall, C. (2003) 'Principles for good practice in participatory research: Reflecting on lessons from the field', in C. McDougall and A. Braun (eds), *Managing Natural Resources for Sustainable Livelihoods: Uniting Science and Participation*, Earthscan, London and Sterling, VA, pp. 113–137

Vernooy, R., Shrestha, P., and Sthapit, B. (eds) (2015) *Community seed banks: Origins, evolution and prospects*, Routledge, London; http://www.biodiversityinternational.org/e-library/publications/detail/community-seed-banks-origins-evolution-and-prospects

Witcombe, J.R., Joshi, A., Joshi, K.D., and Sthapit, B.R. (1996) 'Farmer participatory crop improvement I: Varietal selection and breeding methods and their impact on biodiversity', *Experimental Agriculture*, vol 32, pp. 445–460

Zeven, A.C. (1998) 'Landraces: A review of definitions and classification', *Euphytica*, vol 104, pp. 127–139

Notes

1 CIAL refers to the Spanish acronym (Comité de Investigación Agrícola Local) of Local Agricultural Research Committees in Latin America.

2 Rural institutions mean the rules that govern intangible institutions like kinship and marriage, and organizations that operate at community level and are controlled by their members. Community rural institutions include both informal and formal local institutions.
3 A community refers to the locus where all members of a group of people having some form of collective claim over a territory and recognizing some form of collective governance can be given the opportunity to influence decisions in matters of public choice that affect their livelihood. We used the term 'community' as roughly interchangeable with the term village with a population of around 500 households (Carloni and Crowley, 2005).
4 Community-based organizations (CBOs) include self-help groups (SHGs), farmer cooperatives, village forest committees or any local group or association active at community level. Supportive organizations can be government offices, agricultural extension services, universities, NGOs, private companies, etc.
5 Collaboration/conflict, Legitimacy, Interests and Power (CLIP) is the social analysis of ongoing relations of collaboration/conflict, legitimacy, interests (gains or loss) and power (Chevalier, 2007; http://www-sas-pm.com/).
6 All the factsheets are available at http://www.bioversityinternational.org/pacs-related-publications/
7 http://andamanchronicle.net/index.php?option=com_content&view=article&id=4553:custodian-farmers-meet-on-agricultural-biodiversity-conservation-held-at-ciari&catid=37&Itemid=142

4 Custodians of tropical fruit tree diversity

Identifying and strengthening the roles and rights of custodian farmers

Bhuwon Sthapit, Hugo A.H. Lamers and V. Ramanatha Rao

Background

The global economy's heavy reliance on a narrow diversity of crops puts future food and nutrition security at risk. Over the past century, more than 75 per cent of plant genetic resources have been lost and one third of today's diversity could disappear by 2050 (FAO, 2011). Despite this trend, there are some farmers who continue to actively maintain and employ agricultural biodiversity on their farms, and who possess specialized knowledge about its use and cultivation. We recognize them as the 'custodians' of diverse crop species and varieties. These custodian farmers select crop varieties adapted to local conditions and preferences and promote their use and conservation in family and local networks. We as researchers came to discover the role of custodian farmers in rural communities through the process of seeking good practices for diversity management. We were interested in why some farmers were keen to grow and save seeds of a number of crops and varieties, whereas other farmers were either uninterested or have opted to engage in specialized commercial farming using a limited number of varieties. We collected and described 20 case studies of farmers in India, Indonesia, Malaysia and Thailand to understand what motivates custodian farmers to conserve, innovate and disseminate tropical fruit tree diversity, find ways to formally recognize such farmers and create mechanisms to support and expand their management of local crop biodiversity *in situ* and on farm (Sthapit *et al.*, 2013).

Our research collaboration[1] carried out field studies in 36 communities across the four countries. The studies aimed to: (1) develop a deeper understanding of the roles of custodian farmers in the conservation, use and dissemination of tropical fruit tree diversity; (2) highlight their contribution to the national plant genetic resource system and sustainable agricultural development in general; and (3) raise their visibility and contribute to the longer term development of institutional and policy support for their ongoing contributions to genetic resource management.

Defining custodian farmers

The term 'custodian' literally means a guardian, caretaker, protector or warden (www.thefreedictionary.com). A custodian is usually defined as someone who is responsible for looking after something important or valuable. The term 'custodian' does not necessarily refer to an individual or to either gender or mean that the people it describes always act solely in the area of conservation (van Oudenhoven, 2011). For the purposes of biodiversity conservation, Sthapit *et al.* (2013) defined custodian farmers as 'those farmers (men and women) who actively maintain, adapt and promote agricultural biodiversity and related knowledge at farm and community levels over an extended period of time, and are recognized by community members for doing so'. Often, custodian farmers do not act alone, but rather are actively supported in their efforts by family or household members.

Methodology

Our first attempts to identify custodian farmers found that the concept was not always evident to villagers, researchers or genetic resource specialists of on-farm conservation. Often it was confused with the more widely known terms 'progressive farmer' or 'innovative farmer'. Rather than identifying custodian farmers on the basis of their key functions (such as maintenance, selection and adaptation, promotion), there was a tendency to identify a village leader or farmer who uses modern varieties and technologies, or to pinpoint a wealthier large-holder farmer with strong institutional connections. In order to avoid this inherent bias, the following simple guide for selection of custodian farmers was provided to national partners:

1. Discuss the definition and characteristics of custodian farmers among the implementing partners
2. Identify potential custodian farmers using secondary sources of information such as local records showing farmers who grow unique or a higher number of crops or varieties in a village
3. Conduct focus group discussions with men and women in the community
4. Consult key informants to further clarify the definition and characteristics of custodian farmers before gathering information on potential candidates
5. Depending on the size of the community, shortlist three to ten potential custodian farmers in each community based on the focus group opinions and consultations
6. Validate candidates with personal field visits and informal interviews to assess the profile and characteristics of custodian farmers
7. Use Four Cell Analysis – a participatory method used to assess the richness level of diversity [modified from Sthapit *et al.* (2006b) and described in Chapter 3 of this book] – for individual farmers to identify the unique traits or characteristics of the genetic resources they maintain, adapt or promote

8. Most importantly, explore and document the rationale and motivations of the shortlisted farmers by identifying triggers or driving forces that prompt them to assume their conservation practices.

Using this method the national partners identified and interviewed 20 custodian farmers. The information they obtained was used to develop custodian farmer profiles using the following structure (Sthapit *et al.*, 2013):

- Introduction: Household, landscape, farm, livelihood activities.
- Maintain: Which crops and landraces, how many?
- Promote: Share knowledge and seeds – which and how?
- Adapt: Improve, evaluate or select seeds – which and how?
- Motivations: Anecdotal stories showcasing why they maintain.
- Unique features: Why is this custodian special or different from the others?
- Continuation: Involvement of younger generation.
- Support: Response to needs and requests.

In February 2013, a workshop on 'Custodian farmers of agricultural biodiversity: Policy support for their roles in use and conservation' was organized to bring together global experts on agricultural biodiversity conservation and the 20 custodian farmers from South and Southeast Asia to share expertise and experiences.[2]

Through the workshop discussions, we refined our understanding of custodian farmers and identified four broad types of custodian farmers in agricultural communities (see Figure 4.1):

1. Farmers who *maintain* a rich and unique portfolio of species and varieties
2. Farmers who *maintain* and *promote* a portfolio of species and varieties
3. Farmers who *maintain* and *adapt* a portfolio of species and varieties
4. Farmers who actively *maintain, adapt* and *promote* their portfolio of species and varieties.

'Maintain' refers to the number of species or varieties the farmer has. 'Promote' refers to the sharing of material (seeds, saplings) or related knowledge about traits. 'Adapt' refers to simple selection, breeding involving crossing, or experiments such as trait identification or adaptation to local conditions. Discussions and case studies suggest that the boundaries distinguishing the different types of custodian farmer may be blurred depending on local factors such as crop type, local culture, exposure to new knowledge and settings and environmental conditions. Therefore, the purpose of this categorization is simply to shed light on the diversity of custodian farmer types one may expect to encounter in the field. The custodian farmer role is dynamic; as farmers acquire more knowledge, skills, social connections and recognition, they may choose to take on more functions (Figure 4.1). This change in behaviours has already been noticed with some farmers following the regional workshop,

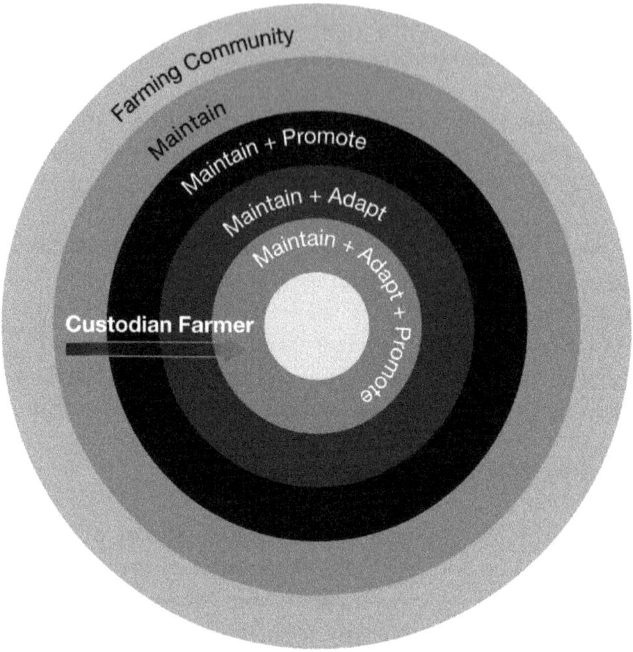

Figure 4.1 Typology of custodian farmers. The arrow illustrates that, with increased awareness and capability, custodian farmers are able to 'take on' additional functions, which make them increasingly proactive agents of change for on-farm conservation of agricultural biodiversity.

Source: Sthapit *et al.* (2013).

who have expedited the exchange of germplasm and knowledge among fellow farmers.

Characteristics of custodian farmers

Both the term and concept of 'custodian farmer' are relatively new in the field of *in situ* and on-farm conservation of agricultural biodiversity (Negri, 2003). Subedi *et al.* (2003) originally called them 'nodal farmers' in their studies of social seed networks. The general characteristics of custodian farmers are presented in Table 4.1. As can be seen from the table, custodian farmers may not be progressive or innovative farmers as in the context of modern agriculture, but they are an important component of the rural agricultural scene and command the respect of people in the region for their role in the informal seed system.

The 20 case study custodian farmers and selected profiles of these farmers are illustrated in Plate 1 and Table 4.2. Selected profiles are included because they stand out amongst examples from each country.

Table 4.1 Characteristics of custodian farmers of local crop diversity

Characteristic	*Description*	*Indicators*
Driven by conservation ideology	Maintains rich diversity of tropical fruit species (richness in terms of inter- and/or intraspecific diversity) over and above the average farmer.	Diversity richness figure and guardian of at least one unique and rare or very valuable variety that may be difficult to propagate.
Knowledge holder	Holds knowledge on the usefulness of variety traits. Holds knowledge about propagation techniques, production, storage, processing or use of the local varieties and seeds.	Cited by multiple community members (men and women) as a source of knowledge of diversity, traits and techniques.
Community recognition	Recognized by community members as someone who conserves local seeds and/or knowledge.	Community members cite his/her contribution in management of unique local crop diversity.
Highly motivated and self-directed	Has strong personal motivation to conserve local varieties without depending on external support for continued conservation and use.	Empowered individual, self-motivated and self-directed, often revealed by the capacity to provide anecdotal stories about his or her motivations.
Consistent commitment	Grows diverse varieties (even on small plot of land) over a long period even without immediate use or income generation from it. Uses varieties herself/himself and encourages others to do the same.	Personal orchards or home gardens contain relatively high number of crops and varieties compared to the average community member. Evidence of experimentation, comparison, crossing or selections made from existing or new germplasm.

Source: Sthapit *et al.* (2013).

Sources of motivation

The 20 case studies revealed that sources of motivation for the custodian farmer role can be diverse: personal, social, economic, cultural, environmental and policy/legal factors all, in varying degrees, drive their approach to farming (Sthapit *et al.*, 2013). Similar findings were also reported by van Oudenhoven (2011), Gruberg *et al.* (2013) and Sthapit *et al.* (2015b). A survey of 66 custodian farmers also reported multiple motivations for maintenance of crop and fruit tree diversity. They include personal home use (91 per cent) followed by conservation (86 per cent), heritage (85 per cent), adaptation (74 per cent), income (62 per cent), hobby (52 per cent) and culture (36 per cent). Such a wide array of motivating factors is to be expected given the diversity of local conditions, customs and pre-existing practices. The challenge is how to use

Table 4.2 Selected profiles of custodian farmers from south and southeast Asia

Region, country	*Custodian farmer**	*Richness (varieties/species)*	*Description*
Sirsi, India	Dattatreya Hedge	52 mango varieties and 4 *Garcinia indica*	Cultural and personal hobby driven by passion of local diversity. He is a custodian of 12 unique 'appe' mango of the Western Ghats, India. Major source of income is commercial orchard of arecanut, banana, cardamom and pepper but also fruit crop nursery or fruit tree nursery.
Sirsi, India	Vishweshwar Ganapati Hedge "Eshanna"	25 mango varieties including 14 'appe' mango	A master grafting expert and barefoot breeder of local 'appe' mango varieties. Personal zeal to look for best scions of good pickle making varieties from wild as his wife is known expert for pickle making. He also maintains unique *Varate Giduga* mango variety (see Chapter 28).
Malihabad, India	Chhote Lal Kashyap	135 mango varieties including seedling types	Limited resources and poor sandy soils led farmer to search and test portfolio of seedling and grafted mango varieties, resulting in highest number of mango varieties in his 2 ha orchard. *Tukmi Heera*, *Deshi Lambui* and *Tukmi Surkha* are unique types.
South Kalimantan, Indonesia	Ahmad Kusasi	6 *Mangifera* spp. with *Kasturi*, *Rawa-rawa*, *Kuini* and *Hambawang* unique types	Custodian of six species of *Mangifera*: *casturi*, *griffithi*, *odorata*, *applantat*, *foetida* and *indica* and three varieties of *M. foetida* (Hambawang) in his orchards.
Papar, Malaysia	Palin Along	16 species of tropical fruits; 2 varieties of aroi aroi	Despite small size of orchard, he is custodian of underutilized tropical fruit species as personal hobby. He learned to appreciate diversity from his father. He maintains unique aroi aroi (*Garcinia forbesii*) that has a thick rind used in local cuisine.
Bukit Gantang, Malaysia	Razali Yahya	4 Cultivars *Garcinia atroviridis* (Asam gelugor), 1 mangosteen and 3 rambutan with 6 other species	He maintains nine species of tropical fruits in his orchards and is known guardian of Asam gelugor. He domesticated planting of wild Asam gelugor in agroforestry systems and promoted in the village.
Chiang Mai, Thailand	Suradech Tapuan	21 varieties of mango and 4 wild relatives	A champion of side-grafting and custodian of unique mango diversity. He developed unique side-grafting technique suitable for rain-fed conditions. He has a passion for grafting and cultivating multiple varieties of mango in a single tree.

Source: Sthapit *et al.* (2013).

*The project worked specifically with the head of the household, who in these cases was always male. It should be noted, however, that many custodian farmers reported that their custodian role is carried out jointly with their wife and other family members.

knowledge about the motivating factors to create conducive environments for fostering such local innovations. The UNEP/GEF project for the on-farm conservation of tropical fruit trees has shown that custodian farmers are a useful entry point within a community when planning interventions that strengthen on-farm conservation and use practices. Starting with a thorough understanding of their motivations, characteristics and knowledge will enhance the effective, efficient and sustainable adoption of such practices, and thus the implementation of community biodiversity management projects.

Responsibilities and rights of custodians

De facto responsibilities assumed by custodian farmers

The study confirmed that custodian farmers exist in all of the countries examined and play a distinct and important role in those countries' agricultural systems (Sthapit *et al.*, 2013; 2015a; Dinesh *et al.*, 2014; Rajan *et al.*, 2014; Gautam *et al.*, 2014). They maintain and conserve a wide range of tropical fruit tree species and varieties based on their own interests and the local context. They often are the nodal points for the informal exchange of seed and plant material among farmers, and they are also important providers of materials and related knowledge to breeders and seed distribution programmes. Custodian farmers play a key role in linking traditional and modern seed systems, and thus may contribute to the evolutionary process of crop adaptation in a dynamic and competitive arena. However, though their roles as conserver, innovator and promoter are often well appreciated in local communities, their contributions often go unnoticed at national and global levels. Policies and institutions that could be developed to support their further efforts – for example, granting property rights for the varieties they develop, or guaranteeing the right of facilitated access to quality reproductive materials, and the right to save, exchange and sell such materials – are often underdeveloped (or non-existent) at national and global levels.

All the custodian farmers are self-motivated and have taken on the responsibility of conservation, adaptation and to some extent promotion of plant genetic diversity. There is no clear indication whether the costs involved are actually and fully recovered in a financial sense from the benefits they derive by maintaining such crop diversity on their own. Many of the benefits take the form of goods and services for the family well-being. Yet these custodian farmers also generate important spillover benefits beyond their immediate context, for other farmers in their own communities, and for their countries (or even the global community) considered more broadly. For example, custodian farmer Vishweshwar Ganapati Hedge from the Western Ghats of India has a personal zeal to select the best scions from the wild 'appe' mango and distribute them to more than 1,000 farmers and students. Similarly, Suradet Tapuan from Thailand has developed his own side-grafting technique for mango that is now being used by mango farmers from dry and rain-fed areas.

Both of these farmers inadvertently contribute to the global public good of fruit tree diversity as they maintain rich diversity of tropical fruits.

Policy support needed to strengthen the custodian farmer role

From a purely utilitarian point of view, it makes sense for governments to invest and to develop institutional and policy support to encourage these farmers in their continued efforts and to encourage other farmers to assume similar practices. One could also argue that there is a moral obligation on the part of society in general and the public plant genetic resource community in particular to give formal recognition to custodian farmers' roles and mainstream their activities, providing whatever support is required. In Table 4.3 we provide a list of the key activities and responsibilities that have voluntarily been assumed by the custodian farmers that were the subject of this study. And, for each set of activities, we list the concomitant forms of institutional or policy support (including rights, privileges, freedoms, rewards and incentives) that are necessary for their current efforts and those of custodian farmers in future generations.

The provisions of the two international agreements that together provide the international legal framework for access and benefit-sharing – the Convention on Biological Diversity (CBD; with its Nagoya Protocol) and the International Treaty on Plant Genetic Resources for Food and Agriculture (ITPGRFA) – are presented and analyzed in Table 4.3. Many of the policy supports listed in the third column constitute means by which member states can implement their commitments under the ITPGRFA concerning the sustainable use and conservation of plant genetic resources for food and agriculture, and the promotion of Farmers' Rights (under articles 6, 5 and 9, respectively; FAO, 2002). Farmers' Rights are basically about: enabling farmers to continue their work as stewards and innovators of agricultural biodiversity; recognizing and rewarding them for their contribution to the global pool of genetic resources; and elevating the level of their participation in national agricultural biodiversity-related planning. The sections of the Treaty on conservation and sustainable use include undertakings to support on-farm conservation, increased use of crop diversity in farming systems, and increased participation of farmers in plant breeding. Our research with custodian farmers confirms that the recognition of their rights is inextricably linked to the promotion of their efforts to conserve and sustainably use genetic resources for food and agriculture and, by extension, their contributions to food security – today and in the future. Countries do not need to be members of the ITPGRFA to develop these kinds of policy support nationally.

It is beyond the scope of this chapter to provide an analysis of the extent to which the list of desirable policy measures in Table 4.3 are being addressed in the four countries that were involved in this research project. Instead, our intention is that Table 4.3, with its list of voluntarily assumed responsibilities and desirable supportive policies, will be useful in itself, not only for the four research partner countries but also for all countries where custodian farmers are, or could be, playing an important role.

Table 4.3 Activities and responsibilities assumed by custodian farmers and the policy support needed to strengthen them

Role	*Activities/responsibilities assumed*	*Necessary policy supports*
Maintain	Save seed or planting materials of a richness of species, variety or trait diversity conserved at household level and document associated traditional knowledge Take care of at least one unique, rare, special or difficult-to-propagate variety Maintain sets of traits when old varieties are replaced	Recognition in national policies of the value of their efforts Protection of associated traditional knowledge (individual or collective) to ensure that benefits of use of specialist knowledge are shared in ways that are acceptable to the farmers and lead to community benefits Right to facilitated access to materials they need from the national or international genebanks Financial support, subsidies or rewards to compensate for custodian-related costs, and to support voluntary cooperation with formal sector actors where appropriate (e.g. genebanks and breeders in dynamic forms of *in situ* and *ex situ* conservation)
Adapt	Identify, domesticate, select or improve traits of interest Blend and use ecological indigenous knowledge of diversity, heritability and selection with scientific knowledge	Property rights (farmer plant breeders' rights) for the varieties they improve, tailored to accommodate the conditions under which the farmers work and for the diversity of the materials they improve Right to save, exchange and sell propagating materials Public policy to promote participation of custodian farmers in research and development activities Financial support, subsidies or rewards for voluntary cooperation with formal sector actors in participatory plant breeding, multiplication of quality reproductive materials for use by others, and other aspects of the fruit tree value chain based on uses of custodian farmer varieties Right to facilitated access to information and materials from genebanks and public sector plant breeders Protection of farmers' traditional knowledge and innovation in a secure way that leads to community benefits
Promote	Share materials and knowledge with other farmers High frequency of exchange of seed and associated knowledge Try to ensure family continues to harbour portfolio of species and varieties Transfer of knowledge and practices to next generation or kin Ensure alternative options for crops and varieties under threat	Right to save, exchange and sell reproductive materials Right to participate in decision making and benefit-sharing through community-based approaches Right to multiply and sell seed as Community Based Seed Producers (CBSP) groups Share in collective benefits through mechanisms such as: community biodiversity register (CBR), community seedbank (CSB), participatory plant breeding (PPB), farmer field school (FFS), community development, community biodiversity management (CBM) fund (Sthapit *et al.*, 2006a). Empowered to participate in a network of custodian farmers Recognition of shared custodianship within households and whole communities Access to new materials Land tenure security (women farmers for intergenerational transfer of farms)

Challenges

Continuity

Although in some farm families custodianship is passed from generation to generation, this method of transmission is by no means guaranteed. In light of increasing rural migration and higher education rates, the new generation of potential custodian farmers is not unanimously keen on continuing their parents' work. Inviting custodian farmers to participatory seed or planting material exchanges or diversity fair events provides alternative options for transferring knowledge, germplasm and roles to others. One proposed mechanism to maintain custodianship in this context is the establishment of a network in which the 'tenure' of one custodian farmer can be taken over or shared by other farmers when they are no longer willing or able to continue their efforts. Potential manifestations of this mechanism include community gardens or community seedbanks, which will help to preserve current information while linking it to young and future farmers.

Recognition

A second challenge is that the important role played by custodian farmers in conservation, innovation and development is often underestimated, undervalued and unrecognized (Sthapit *et al.*, 2013; Gruberg *et al.*, 2013). This can be attributed to the relative rarity of such farmers combined with their lack of connection to mainstream research and development institutes or networks. Mechanisms that establish connections between custodian farmers, the wider network of regional farmers and both national and international genetic resource systems would address this challenge. The decision to facilitate the process of mainstreaming the efforts by custodian farmers into national R&D systems should be taken up by the responsible authorities in the countries.

Identification and selection of the best fruit trees from farmer-managed genetic resources can provide immediate benefits to communities. Sthapit and Ramanatha Rao (2009) proposed a simple process by which custodian farmers' unique, rare or elite varieties could be formally registered and thus enter the commercial multiplication and distribution system. A number of such elite materials in India have been identified and registered under the Protection of Plant Variety and Farmers' Rights Authority (PPV&FRA; http://plantauthority.gov.in/). Similar actions could be used with farmers' best varieties from Indonesia, Malaysia, Thailand and other countries. On the other hand, national seed laws that require farmers' varieties to satisfy minimum standards of distinctness, uniformity and stability, or have exceedingly high standards for seed producers, can have the effect of preventing those materials, and the farmers who have developed them, from being recognized and from entering the market. Similarly, national access and benefit-sharing laws can present impediments for farmer exchange of materials if permission from national authorities is required and minimum processing fees have to be paid. Such policies can, perhaps inadvertently, present challenges to the wider recognition of materials developed by custodian farmers. One potential method for addressing this obstacle is to require immediate registration into a type of 'national seed board' authority (e.g. NBPGR and PPV&FRA in India). It is also important to ensure free prior informed consent from custodian farmers, following clear, easy-to-follow procedures, in order to make sure that access to materials the farmers develop is subject to benefit-sharing conditions (Ruiz and Vernooy, 2012). Research into mechanisms for such issues is still in an infant stage. Recognition of local crop innovations by government authorities and the NGO sector has started in a few countries, for example in India, Bolivia and Nepal, but the identification and selection procedure may require more scientific rigour. It may also require investment on the part of national competent authorities to work with farmers, to help them develop materials and present them in ways that meet national registration standards. The Indian PVP&FRA has provided considerable community-level assistance to farmers in this regard, to register their farmer varieties under the Act. There are many research opportunities and significant excitement in this field as an entry point for on-farm conservation of agricultural biodiversity.

Further development of the framework of responsibilities and rights of custodian farmers as set out in Table 4.3 is essential in those countries where a relevant policy is not in place. This includes the right to participate in national decision-making processes, especially those related to plant genetic resources and benefit-sharing policies, as well as in international agreements. This will only become possible if we advocate for the formal recognition of custodian farmers, similar to the special recognition already afforded to outstanding progressive farmers or genebank curators, as stewards of the world's food and nutritional security.

A way forward

1. Assess the importance of custodian farmers for on-farm/*in situ* conservation of local fruit tree diversity and the informal and formal seed system.
2. Establish fruit tree custodian farmers' networks as an integral part of national and international conservation strategies and link them directly to agricultural biodiversity conservation institutions (e.g. genebanks to document diversity and involve custodian farmers in research programmes).
3. Use a community-based approach to improve the capacities of custodian farmers in: (i) protection of traditional knowledge of fruit tree genetic diversity for food and agriculture through documentation, use and conservation of this knowledge (e.g. community fruit catalogue, community biodiversity register); (ii) the right to save, use, exchange and own saved seeds or planting material (e.g. community seedbanks and participatory crop improvement); (iii) the right to participate in decision making at a national level on matters of the conservation and use of plant genetic diversity, as well as overall community development (e.g.

community biodiversity management, establishing CBM fund); and (iv) the right to equitably participate in benefit sharing arising from the use of plant genetic resources by creating economic and nutritional benefits (e.g. product development, marketing and home processing).

4. Support locally driven CBM funds that can directly maintain the multiplication and exchange of rare and unique materials at the local level.
5. Advocate formal recognition of custodian farmers and their roles in conservation of plant genetic resources and promote their participation in national-level decision making.

Acknowledgements

Insightful comments and reviews of Michael Halewood, Rooney Vernooy, M. P. Vasimalai and Arwen Bailey are gratefully acknowledged.

References

Dinesh, M.R., Rajan, S., Singh, I.P., Singh, A., Singh, S.K., Vasudeva R, Vinoth, S., Patil, P., Parthasarathy, V.A., Reddy, B.M.C., and Sthapit, B. (2014) *National Fruit Catalogue of Tropical Fruit Tree Diversity (*Mangifera, Citrus *and* Garcinia*)*, GEF, UNEP, Bioversity and ICAR, India

FAO (2002) *The International Treaty on Plant Genetic Resources for Food and Agriculture: A Global Treaty for Food Security and Sustainable Agriculture*, FAO (Food and Agriculture Organization of the United Nations), Rome

FAO (2011) *Save and Grow: A Policymaker's Guide to the Sustainable Intensification of Smallholder Crop Production*, FAO (Food and Agriculture Organization of the United Nations), Rome

Gautam, R.K., Sankaran, M., Zamir Ahmed, S.K., Sunder, J., Ram N., and Dam Roy, S. (2014) *Custodian Farmers and Communities of Biodiversity Conservation and Utilization in Andaman and Nicobar Islands*, CIARI/01/Technical Bulletin/2014, CIARI/ICAR, India

Gruberg, H., Meldrum, G., Padulosi, S., Rojas, W., Pinto, M., and Crane, T. (2013) *Towards a Better Understanding of Custodian Farmers and Their Roles: Insights from a Case Study in Cachilaya, Bolivia*, Bioversity International, Rome and Fundación PROINPA, La Paz, Bolivia

Negri, V. (2003) 'Landraces in Central Italy: Where and why they are conserved and perspectives for their on-farm conservation', *Genetic Resources and Crop Evolution*, vol 50, pp. 871–885

Rajan, S.M.R., Ravishankar, K.V., Singh, A., Singh, S.K., Singh, I.P., Vasudeva R, Reddy, B.M.C., Parthasarathy, V.A., and Sthapit, B.R. (2014) *Heirloom Varieties of Important Tropical Fruits: A Community Initiative to Conservation*, ICAR/GEF UNEP/Bioversity International, India

Ruiz, M. and Vernooy, R. (2012) *The Custodians of Biodiversity: Sharing Access to and Benefits of Genetic Resources*, Earthscan from Routledge, Abingdon, UK and New York

Sthapit, B.R. and Ramanatha Rao, V. (2009) 'Consolidating community's role in local crop development by promoting farmer innovation to maximise the use of local crop diversity for the well-being of people', in H. Jaenicke, J.Ganry, I. Höschle-Zeledon, and R. Kahane (eds), *Acta Horticulturae 806: Proceedings of the International*

Symposium on Underutilized Plants for Food Security, Nutrition, Income and Sustainable Development, Arusha, Tanzania, 3–7 March 2008, International Society for Horticultural Science, Leuven, Belgium, pp. 659–676

Sthapit, B.R., Shrestha, P.K., and Upadhyay, M.P. (2006a) *Good Practices: On-farm Management of Agricultural Biodiversity in Nepal*, NARC, LI-BIRD, IPGRI and IDRC, Nepal

Sthapit, B., Rana, R.B., Subedi, A., Gyawali, S., Bajracharya, J., Chaudhary, P., Joshi, B.K., Sthapit, S., Joshi, K.D., and Upadhyay, M.P. (2006b) 'Participatory four cell analysis (FCA) for local crop diversity', in B.R. Sthapit, P.K. Shrestha, and M.P. Upadhyay (eds), *Good Practices: On-farm Management of Agricultural Biodiversity in Nepal*, NARC, LI-BIRD, IPGRI and IDRC, Nepal

Sthapit, B.R., Lamers, H., and Ramanatha Rao, V. (eds) (2013) *Custodian Farmers of Agricultural Biodiversity: Selected Profiles from South and South East Asia*, Proceedings of the Workshop on Custodian Farmers of Agricultural Biodiversity, 11–12 February 2013, New Delhi, India; Bioversity International, New Delhi, India

Sthapit, B.R., Vasudeva R, Rajan, S., Sripinta, P., Reddy, B.M.C., Arsanti, I.W., Idris, S., Lamers, H., and Ramanatha Rao, V. (2015a) 'On-farm conservation of tropical fruit tree diversity: Roles and motivations of custodian farmers and emerging threats and challenges', in H. Jaenicke, S.E. Ashmore, M.E. Dulloo, L. Guarino, and M. Taylor (eds), *Acta Horticulturae 1101: XXIX International Horticultural Congress on Horticulture: Sustaining Lives, Livelihoods and Landscapes (IHC2014): IV International Symposium on Plant Genetic Resources*. DOI 10.17660/ActaHortic.2015.1101.11

Sthapit, S., Meldrum, G., Padulosi, S., and Bergamini, N. (eds) (2015b) *Strengthening the Role of Custodian Farmers in the National Conservation Programme of Nepal*, Proceedings of the National Workshop 31 July to 2 August 2013, Pokhara, Nepal; Bioversity International, Rome, Italy and LI-BIRD (Local Initiatives for Biodiversity, Research and Development), Pokhara, Nepal

Subedi, A., Chaudhary, P., Baniya, B., Rana, R., Tiwari, R.K., Rijal, D., Jarvis, D., and Sthapit, B.R. (2003) 'Who maintains crop genetic diversity and how? Implications for on-farm conservation and utilization', *Culture and Agriculture*, vol 25, no 2, pp. 41–50

van Oudenhoven, F. (2011) *Roots of Our People: Fruits Trees and Their Custodians in Kyrgyzstan and Tajikistan*, Bioversity International, Rome, Italy

Notes

1 The research was part of a project on 'Conservation and Sustainable Use of Cultivated and Wild Tropical Fruit Diversity: Promoting Sustainable Livelihoods, Food Security and Ecosystem Services', funded by the Global Environment Facility (GEF) with implementation support from the United Nations Environment Programme (UNEP). The project, from 2009 to 2015, was carried out by Bioversity International, in collaboration with the Indian Council of Agricultural Research (ICAR), the Indonesian Centre for Horticultural Research and Development (ICHORD), the Malaysian Agricultural Research and Development Institute (MARDI) and the Department of Agriculture, Thailand.

2 The workshop was held by the Global Environment Facility (GEF), the United Nations Environment Programme (UNEP) and Bioversity International in collaboration with ICAR, the National Bureau of Plant Genetic Resources (NBPGR) and the Protection of Plant Varieties and Farmers' Rights Authority (PPV&FRA), both in India.

Part 2

Good practices of agricultural biodiversity management and sustainable use

5 Good practices for conservation and sustainable use of crop wild relatives of tropical fruit tree diversity

Danny Hunter and Suchitra Changtragoon

Introduction

Broadly speaking, a crop wild relative (CWR) may be defined as a 'wild plant species more or less closely related to a particular crop and to which it may contribute genetic material, but which unlike the crop species has not been domesticated' (Heywood *et al.*, 2007, cited in Hunter and Heywood, 2011, p. 4). An alternative definition is that CWRs include 'the wild congeners or closely related species of a domesticated crop or plant species, including relatives of species cultivated for medicinal, forestry, forage or ornamental reasons' (Meilleur and Hodgkin, 2004, p. 670).[1] Wild relatives are a critical resource to meet the challenges of providing for food security and nutrition in the context of a rapidly growing world population and accelerated climate change (Hunter and Heywood, 2011). They represent a substantial reservoir of genetic variation that can be employed in plant breeding and tree improvement, and which has also a potentially large economic value. For example, germplasm of the wild and threatened apple species *Malus sieversii*, collected from Kazakhstan in the 1990s, has demonstrated resistance to a range of biotic and abiotic stresses including drought, apple scab and fire blight and is being used to enhance disease resistance in apple cultivars in the US for an industry worth USD 2.7 billion in 2011 (FAO, 2014).

The world's wild fruit trees represent a hugely valuable resource (Sthapit *et al.*, 2015). In tropical zones, wild fruit harvested from forests contributes significantly to the total income and sustainable nutritious diets of many rural households, as well as contributing substantially to important ecosystem services (Vinceti *et al.*, 2013). Wild relatives and wild-growing semi-domesticated species of tropical fruit trees also provide services to domesticated fruit trees in terms of resistance to extreme abiotic and biotic stresses through their high levels of genetic diversity.

Conserving wild relatives of tropical fruit trees in their natural habitats allows populations to evolve and continue to generate new genetic diversity that helps them adapt to changing environments. However, these natural habitats and the genetic diversity of these wild tropical fruit trees are increasingly under threat (Sthapit *et al.*, 2015). Efforts to address these threats to date have included endeavours to conserve wild fruit tree species *ex situ*. This is problematic, however, because most tropical fruit tree species seeds are recalcitrant, making seed storage impossible. Field genebanks offer opportunities for conserving the wild relatives of tropical fruit trees, especially for ongoing evaluation and characterization and the selection of superior types for improvement (Hunter and Heywood, 2011), but they are often expensive because of the costs involved for management and maintenance. Field genebanks are also increasingly vulnerable to the impacts of natural catastrophes and shocks. Possible alternative *ex situ* methods such as *in vitro* and cryopreservation are not yet fully viable, requiring further research and development.

In situ approaches seem feasible for conserving wild relatives of tropical fruit trees, but experiences with targeted species and actions inside and outside protected areas appear to be relatively few. Consequently, wild relatives of tropical fruit trees remain a largely underconserved natural resource, both *ex situ* and *in situ*, and are continuously under threat in their natural habitat from neglect and overharvesting.

Any strategy to conserve the wild relatives of tropical fruit trees should contain elements of both *in situ* and *ex situ* conservation and have a focus on conservation both inside and outside protected areas. It should also ensure coordination of planning and implementation, institutionalize the practice of wild relative conservation, promote public awareness and understanding, create a suitable policy environment and highlight the many benefits derived from their sustainable conservation and use. Unfortunately few countries have developed such strategies for wild relatives generally, despite guidance and good practice being documented and disseminated. For the conservation of wild relatives of tropical fruit trees specifically, few if any national plans of action have been developed (Hunter and Heywood, 2011).

Threats to wild relatives of tropical fruit trees

Wild relatives of tropical fruit trees and the habitats in which they arise, like other crop wild relatives, are increasingly threatened in various ways. For example, eco-geographic surveys of wild mangoes in Borneo and west Malaysia found that a large proportion of the rich genepool (including most notably *Mangifera blommesteinii*, *M. leschenaulitii*, *M. superba* and *M. paludosa*) were highly threatened, on the verge of being lost (Kostermans and Bompard, 1993). What is most alarming in regards to this particular example is that no one really knows for sure the current status of these species, since, like most wild relatives of tropical fruit trees, they have never been inventoried. Many threats are directly or indirectly as a result of human action. The main threats to the existence of wild relatives are habitat loss, fragmentation and degradation, changes in disturbance regimes, invasive species and, increasingly, climate change:

- At population level: small subpopulations caused through fragmentation of habitat, low numbers in a population, narrow or small distributional range
- Changes in disturbance regime: for example, as a result of fragmentation and the consequent effects on dispersal and gene flow between isolated populations
- Fire: changes in components of fire regimes, including season, extent, intensity, or frequency, inhibiting regeneration from seed or by vegetative reproduction
- Threats of biotic origin: pest and disease or predation (e.g. fungal disease); interactions with native species (e.g. allelopathy, competition, parasitism, feral animal grazing and trampling); invasive alien species
- Habitat loss or destruction, degradation (e.g. due to contamination or pollution), modification or simplification as a result of land use change such as clearing for agriculture (for crops and pastures, draining swamps and wetlands), forestry, plantations; housing and urban and coastal development; energy production and mining; agriculture edge effects (including herbicides, pesticides, drainage, etc.)
- Overexploitation for commercial, recreational, scientific or educational purposes; tourism and ecotourism
- Global change (demographic, climatic), such as human population pressure, changes in rainfall and temperature patterns.

(Modified from Hunter and Heywood, 2011)

Conserving wild relatives of tropical fruit trees in protected areas

Protected areas play a vital role in the conservation of forest genetic resources (Thomson *et al.*, 2001a) and awareness is growing of the benefits for food security and well-being of conserving genetic diversity within them (Stolton *et al.*, 2015). Globally significant populations of wild relatives of tropical fruit trees occur in existing protected areas, although detailed information on the extent and distribution of such species is rarely available. While it may be assumed that by dint of being protected, protected areas automatically afford wild relatives of tropical fruit trees some degree of protection, such a 'hands-off' approach does not always translate into effective *in situ* conservation of these genetic resources, as the result of poor representation, or viability, of populations. Some degree of active management of the target species is required, particularly if the species is threatened. Good practices, recommendations and key steps have been described to plan *in situ* conservation and develop species and area management plans that can be employed for wild relatives of tropical fruit trees in conventional protected areas (Thomson *et al.*, 2001b; Thomson and Theilade, 2001; Hunter and Heywood, 2011; Hunter *et al.*, 2012). As yet, for the many wild relatives of tropical fruit trees such efforts and plans rarely exist. Some notable exceptions are a handful of genetic reserves, gene management zones (GMZ), gene parks or sanctuaries, which provide options for *in situ* conservation of wild relatives of tropical fruit trees. For example, the National Citrus Gene Sanctuary in Nokrek National Park in the West Garo Hills, northeast India, was created in 1981 for citrus wild relatives, and a similar genetic reserve for wild relatives, including relatives of lychee, longan and citrus, was established in Vietnam under a project supported by the Global Environmental Facility (Hunter and Heywood, 2011).

Biosphere reserves

Biosphere reserves, with their zonation of core area, transition area and buffer zone, can play an important role in ensuring the *in situ* conservation of wild relatives of tropical fruit trees, where sustainable human resource management and use practices are permissible in the transition zones. Certain wild species of mangoes and their wild relatives are known to occur in biosphere reserves, national parks and other reserves in India, Indonesia, Singapore, the Philippines, Thailand and Sri Lanka, but little targeted *in situ* conservation has been undertaken as mentioned above (Heywood and Dulloo, 2005).

Sacred sites and the conservation of wild relatives of tropical fruit trees

One good diversity practice is seen in sacred groves and forests. These sites represent an important type of nature conservation practised as part of the religion-based conservation beliefs of groups of people in certain parts of the world. Such belief systems often include a range of institutional prescriptions, such as taboos, that help regulate human behaviour and contribute to a more conservative use of natural resources (Heywood and Dulloo, 2005; Hunter and Heywood, 2011). For example, the Mahafaly and Tandroy communities, in collaboration with relevant local authorities in southern Madagascar and the Malagasy government, apply such community biodiversity management practices within sacred forests (Heywood and Dulloo, 2005). It has been estimated that there are between 100,000 and 150,000 sacred groves in India. Their role as local havens of biodiversity and micro-watersheds, which help to meet the water needs of local communities, is of the utmost importance (Warrier, undated). Growing awareness and networks around sacred sites present considerable opportunities for the greater application of community biodiversity management principles (Bhagwat and Rutte, 2006), including for the conservation of wild relatives of the tropical fruit trees contained therein. A similar opportunity is presented in the increasing numbers of participatory reserves, Indigenous and Community Conserved Areas (ICCAs) and areas such as People's Protected Areas in India, which employ models of community-based participatory management, non-destructive harvesting and equitable benefit sharing (Sharma, 2003, cited in Heywood and Dulloo, 2005).

Certain sacred trees in some geographic regions are wild relatives of tropical fruit trees. For example, the bael trees (*Aegle marmelos*) growing in many Shiva temples are conserved because their leaves are used as offerings to the gods. Another typical example is the conservation of pomelo in Bihar, which is central to the religious festival of *Chhat Puja* (Chapter 17; Somashekhar, 2014). Historically, the people of Telaga Langsat in South Kalimantan, Indonesia revered many of the trees found in their villages and surrounding buffer forest (Sthapit *et al.*, 2013). These sacred trees were protected from felling by taboos, and the informal rules protected these globally significant wild relatives including kasturi (*Mangifera casturi*) and kuini (*M. odorata*). However, this belief system and the protection provided to these species have now eroded to the extent that villagers have started felling and selling them to local timber mills. The kasturi, endemic to South Kalimantan, has, as a result, been declared extinct in the wild, with only a few rare remnants found in a few village home gardens. Alarmed by this situation, some community members are collaborating with village authorities and the District Forestry Service to develop informal village regulations enforced by Village Protection Groups that limit the felling of trees.

Community forest management and conservation of wild relatives of tropical fruit trees

The frequent use by local people of either the target species or other elements of the ecosystem means that the maintenance of genetic reserves for tropical fruit trees and their wild relatives will depend on community participation in their management (Heywood and Dulloo, 2005). Local communities in biodiversity-rich countries have been closely linked to their natural environments for millennia and often have intimate ecological knowledge about habitats and their wild plant species, including wild relatives of tropical fruit trees, and time-tested knowledge of their sustainable management. In many instances this intimacy has been disrupted by conventional conservation approaches and requires greater commitment through community biodiversity management approaches (Thomson and Theilade, 2001; Hunter and Heywood, 2011; de Boef *et al.*, 2013; Kothari *et al.*, 2015).

Community forestry, whereby local communities play a major role in decision making around land use and forest management, is a potential good diversity practice for the conservation of wild relatives of tropical fruit trees. In community forests, local people are able to collect forest products to meet their needs and have the right to decide how the forest is managed, as long as it is in a sustainable manner (UNDP, 2004). There are 11,400 villages in Thailand that manage community forests, representing about 15.5 per cent of all villages in Thailand. Of these, 8,331 villages have formally registered their forest with the Royal Forest Department of Thailand. These community forests cover an area of 196,667 ha, as both national forest reserves (112,869 ha) and other forest areas (83,798 ha), accounting for about 1.2 per cent of Thailand's total forest area (Community Forest Management Bureau, 2010; Changtragoon *et al.*, 2012).

Community forests serve as sources of food and medicinal plants as well as natural products (Changtragoon, 2004). Some community forests even act as genebanks to preserve genetic stocks of local plant varieties. There are a number of species of wild fruit trees and medicinal plants found in community forests in Thailand (Srisutham and Kaewjampa, 2010; Inta and Pongmornkul, 2012). Some edible wild tropical fruit species are also commonly known and exploited by Thai local people as food and medicinal products, such as *Phyllanthus embilca* Linn., *Mangifera cochinchinensis* Engl., *M. foetida* Lour., *Antidesma bunius* (Linn.) Spreng,. *Terminalia bellirica* (Gaertn.) Roxb., *Garcinia cowa* Roxb., *Terminalia chebula* Retz., *Aegle marmelos* Linn. and *Baccaurea ramiflora* Lour.

The United Nations Development Programme (UNDP, 2004) identified four examples of best practice of community forest management in Thailand[2] and, from analyzing these, listed eight factors critical for success, which are relevant for community forests elsewhere:

1. A strong sense of community ownership, where trust and relationships among members are strong
2. Strong potential for healthy recovery of the forest
3. Mutual community benefits from conservation and ecosystem goods and services such as for the protection of water sources, water purification, food and medicines, etc.
4. An intense public awareness of forest conservation well beyond immediate usage
5. A strong and wise leader
6. Local organization set up to represent villagers' conservation interests
7. A strong community belief in the concepts of common resources and common rights and a shared sense that the forests belong to the community
8. A set of enforced regulations and conditions for the use of community forests.

These eight key factors play an important role in the conservation and sustainable use of wild plant resources including wild relatives of tropical fruit trees in forest communities in Thailand and could be applied in other tropical countries.

Strengthening community biodiversity management (de Boef *et al.*, 2013) of wild relatives of tropical fruit trees offers many challenges, but the opportunities for sustainable solutions to address many of the threats posed to wild relatives and ensure their enhanced conservation and use would make this worthwhile. Already, as seen in the case of Thailand above, many examples of good diversity practices at a local level are emerging that, if applied at greater geographic scale, could have major impact. Further, opportunities are emerging to link such local initiatives to increasing interest in ecosystem and eco-agriculture landscape approaches (Sthapit and Mijatovic, 2014) and initiatives, such as FAO's Globally Important Agricultural Heritage Systems (GIAHS), the Satoyama Initiative and an expanding network of Indigenous and Community Conserved Areas (ICCAs). These would appear to offer win–win scenarios for wild relative conservation (Hunter and Heywood, 2011; Hunter *et al.*, 2011).

Other good practices for managing the diversity of wild relatives of tropical fruit trees

In addition to using protected areas and supporting community forestry as outlined above, numerous good practices exist that support the conservation and use of wild relatives of tropical fruit trees. In the rest of this chapter, we describe a selection of these.

Custodian farmers and farmer networks

Custodian farmers and their networks provide considerable opportunities for enhancing the conservation and use of wild relatives of tropical fruit trees, and efforts could focus on strengthening and supporting many of the good diversity practices they currently engage in, including safeguarding wild relatives in their natural habitat, community mobilization, local policy and decision making, domestication processes and integration into home gardens and orchards, propagation and grafting, value-adding and income generation, education and awareness raising (Sthapit *et al.*, 2013; 2015; Chapters 3 and 4 in this book). Custodian farms are sites of experimentation and innovation. Sthapit *et al.* (2013) have highlighted how custodian farmers of tropical fruit tree diversity frequently venture into forest areas in order to find rare and indigenous fruit varieties that they bring to their farms and communities for propagation and wider cultivation. Although the uses of such wild relatives in plant breeding and improvement are poorly documented or acknowledged, it is generally believed that many official varieties that have been released are the product of wild saplings and seedlings of unknown origin. This and much more could be supported by greater government recognition, provision of relevant incentives, training and capacity-building, identification of markets and development of a better enabling policy environment, including linking wild relatives of tropical fruit trees to a REDD+ mechanism[3] as the source of carbon sequestration and storage as well as non-carbon benefits. Future work in this area should focus on a wider identification of those custodian farmers actively working to conserve wild relatives of tropical fruit trees and documentation of their full range of good diversity practices and how these might be strengthened or gaps addressed. Formal recognition of this important role by governments, development and scientific agencies should be advocated, and these custodians and their networks should be linked to the wider crop wild relative networks, for further capacity exchanging opportunities (Chapters 3 and 4 in this book).

Integrating tropical fruit tree wild relatives into home gardens

Integrating wild tropical fruit tree species (e.g. *Mangifera* spp., *Garcinia* spp., *Nephelium* spp.) into home gardens is another good practice for conservation of wild relative diversity. Some of these wild species have been incorporated into orchards and are cultivated on the high ridges separating rice paddies. Many custodian farmers grew up with a strong tradition of visiting forested areas with farming parents to identify and collect wild relatives for cultivating back in family home gardens (Sthapit *et al.*, 2013). Not only does this take pressure off the remaining wild types in the forested areas, it can also become a source of income, and farmers can further distribute seedlings as well as raising awareness about the value of wild relatives among communities, especially among young children. Such good diversity practice can be enhanced through the provision of support for training in specialized techniques such as grafting and budding (Chapters 3 and 4 in this book). In one documented case, in the Western Ghats of India, a network of grafting experts provides support to farmers and assists with grafting rare and valuable wild-aromatic pickle mango types. Over a period of years the network has evolved from a social activity to an informal network of expert grafters and in the process hundreds of rare types of pickling mango have been conserved (Chapter 29 in this book).

Value adding and marketing

Value-adding and marketing activities provide incentives for local communities to conserve and sustainably use wild relatives, and thus represent a good diversity practice. Aroi aroi (*Garcinia forbesii* King.), a wild relative of the more common mangosteen (*G. mangostana*), commonly grows along the west coast of Sabah, Malaysia, as a wild species in the lowland forests in the foothills of the Crocker Range and also as a semi-wild or cultivated fruit tree in home gardens and orchards (Wong *et al.*, in Chapter 21 of this book). Once popular in traditional cuisine for the sour flavour of its rind, it has become less popular especially with the younger generations due to outmigration and change in food culture. Recently, however, there has been a surge of interest in the species and so farmers are exploring promotion and value addition options including the introduction of solar dryers for small-scale processing and the testing of techniques to improve quality and reduce investment costs. Increased local demand is leading to increased income and improved livelihoods among fruit farmers. In order to increase incentives, small-scale local farmers are collaborating with researchers at the national Malaysian research institute to explore how to create better value addition from aroi aroi, for example by developing new products such as aroi aroi pickle, cordial, drinks, candied fruit, chutneys and jams. All of this has contributed to the sustainable conservation and use of this valuable wild relative of mangosteen (Vasudeva *et al.*, 2010). Similar market potential is posited for other wild tropical fruit species: *Mangifera* (*M. pajang*, *M. caesia*, *M. odorata*, *M. casturi*), *Citrus* [*C. indica*, *C. macroptera*,

C. assamensis, *C. ichangensis*, *C. megaloxycarpa* (Sour Pummelo)], *Garcinia* (*G. cowa*, *G. atroviridis*, *G. indica*, *G. xanthochymus* and *G. gummi-gutta*) and *Nephelium* (*N. ramboutan-ake*) (Sthapit *et al.*, 2015).

When coupled with value chain development, participatory domestication can help raise awareness of the economic value of preserving genetic diversity of wild relatives of tropical fruit trees and raise interest in more sustainable production practices. Scientists and agricultural extension officers can work with groups of women and men farmers to identify and select preferred traits. The process has been proved useful in documenting traditional management practices (dos Santos *et al.*, 2013)

Awareness raising and education

Good practices for communication and raising awareness about wild relatives can strengthen the approaches outlined above and have been described elsewhere by Hunter and Heywood (2011) and Shrestha *et al.* (2013). They include practices such as rural poetry journeys and rural drama. Education and training can be particularly effective when it targets the intergenerational transmission of traditional knowledge from older people to children and young adults. Paraprofessional training is a further approach that can be used to build the capacity of key individuals and conservation leaders in local communities involved in wild relative conservation programmes. The approach can develop their conservation skills through workshops, training courses, seminars or attachment to conservation professionals through a project or national programme. This can expose individuals to a range of skills and provide local communities with an enhanced capacity to implement, monitor and evaluate actions aimed at conserving wild relatives of tropical fruit trees (Hunter and Heywood, 2011).

The conscious management of wild relatives can be fostered by enhancing awareness of important collective knowledge on the sustainable management of these species and of the value of conserving wild relatives of tropical fruit tree diversity. Additional good diversity practices exist that can be implemented either independently or as part of a long-term strategy to strengthen community capacity and to encourage communities to make informed decisions over these resources. Some of these practices include:

- Village workshops and community platforms can help foster community awareness of wild relatives of tropical fruit tree diversity. Usually different stakeholders are involved (e.g. collectors, processors, cooperatives and women's groups) and participatory tools are used to assess the threat level of a species and to jointly develop strategies for its conservation (Vasudeva *et al.*, 2013).
- Diversity fairs can serve as a platform for knowledge exchange among a diverse set of stakeholders – farmers, scientists, local entrepreneurs and decision makers. The fairs can foster greater appreciation of wild relatives

of tropical fruit tree diversity and of the traditional knowledge linked to management of a particular species. They can boost community pride and identification of biodiversity champions, and lead to innovations in sustainable management practices. The network of grafting experts for *Garcinia* that is described above was formed following one such diversity fair (Vasudeva *et al.*, 2013; 2015).

- Community biodiversity registers (CBRs) are inventories maintained by forest community members who normally form a biodiversity management committee. Registers are used to document biodiversity resources, including wild relatives of tropical fruit trees, and traditional knowledge of their management and use. They represent a valuable tool to raise awareness of biodiversity richness and even more importantly to foster community ownership of these resources (Subedi *et al.*, 2005; Sthapit *et al.*, 2006).

Concluding remarks and a way forward

The world's wild relatives of tropical fruit trees represent a unique globally important natural resource, critical to meet the challenges of enhancing food security and nutrition. They represent a substantial reservoir of genetic variation, although this is generally poorly documented, which can be employed to enhance resistance to extreme abiotic and biotic stresses through improvement programmes. Wild fruit and other products harvested from forests contribute significantly to providing sustainable diets and improving the livelihoods of many rural households, as well as contributing substantially to important ecosystem services. Yet the natural habitats where many wild relatives of tropical fruit tree species are found continue to come under threat, which undermines the genetic diversity they provide. In many instances safeguarding this genetic diversity through conventional *ex situ* approaches is not practical for technical or economic reasons. This means that appropriate *in situ* conservation approaches and strategies must be explored and developed if we wish to secure these resources for future use. Although *in situ* conservation presents challenges, it has the important advantage of conserving wild relatives of tropical fruit trees in their natural habitat, thus allowing populations to continue to evolve and generate new genetic diversity. This is critical in a rapidly changing and uncertain future. This brief chapter has attempted to survey the limited work to date on how to conduct *in situ* conservation of wild relatives of tropical fruit trees, and reviews and explores options for conservation inside and outside protected areas, including strengthening the role of local communities to promote appropriate community biodiversity management approaches to safeguard the significant rich wild fruit tree species diversity that exists in some regions of the world. In addition, it briefly highlights possibilities for value-adding and marketing and income generation as a way of enhancing the benefits that can be derived from wild fruit tree species diversity as a means to safeguard them for the future.

References

Bhagwat, S.A. and Rutte, C. (2006) 'Sacred groves: Potential for biodiversity management', *Frontiers in Ecology and the Environment*, vol 4, pp. 519–524

Changtragoon, S. (2004) 'Opportunities in using the conservation of biodiversity to alleviate poverty in Thailand', in H.C. Sim, S. Appanah, and Y.C. Youn (eds), *Proceedings of the Workshop Forests for Poverty Reduction: Opportunities with Clean Development Mechanism, Environmental Services and Biodiversity*, 27–29 August 2003, Seoul, Korea; RAP, FAO Bangkok, Thailand, pp. 167–172

Changtragoon, S., Ongprasert, P., Tangmitchareon, S., Diloksumpun, S., Luangviriyaseang, V., Sornsathapornkul, P., and Pattanakait, S. (2012) *Country Report on Forest Genetic Resources of Thailand*, Department of National Parks, Wildlife and Plant Conservation, Bangkok, Thailand, p. 128

Community Forest Management Bureau (2010) *Annual Report of Community Forest, Royal Forest*, Royal Forest Department, Bangkok, Thailand (in Thai)

de Boef, W.S., Subedi, A., Peroni, N., Thijssen, M., and O'Keeffe, E. (2013) *Community Biodiversity Management: Promoting Resilience and the Conservation of Plant Genetic Resources*, Earthscan/Routledge, Abingdon, UK and New York

dos Santos, K., Peroni, N., Guries, R.P., Donazzola, J., and Nodari, R.O. (2013) 'Participatory domestication of the fruit tree species feijoa (*Acca sellowiana*) in Brazil', in W.S. de Boef, N. Peroni, A. Subedi, and M.H. Thijssen (eds), *Community Biodiversity Management: Promoting Resilience and the Conservation of Plant Genetic Resources*, Earthscan/Routledge, Abingdon, UK and New York, pp. 240–245

FAO (2014) *The State of the World's Forest Genetic Resources*, Food and Agriculture Organization of the United Nations, Rome, Italy

Heywood, V. and Dulloo, E. (2005) In Situ *Conservation of Wild Plant Species – A Critical Global Review of Good Practices*, IPGRI Technical Bulletin, no 11, FAO and IPGRI, IPGRI, Rome, Italy

Heywood, V., Casas, A., Ford-Lloyd, B., Kell, S., and Maxted, N. (2007) 'Conservation and sustainable use of crop wild relatives', *Agriculture, Ecosystems and Environment*, vol 121, pp. 245–255

Hunter, D. and Heywood, V. (2011) *Crop Wild Relatives: A Manual of In Situ Conservation*, Earthscan, London

Hunter, D., Guarino, L., Khoury, C., and Dempewolf, H. (2011) 'A field divided: lessons from the conservation of crop wild relatives around the world', in N. Maxted, M.E. Dulloo, B.V. Ford-Lloyd, J.M. Iriondo, and M.A.A.P. de Carvalho (eds), *Agrobiodiversity Conservation: Securing the Diversity of Crop Wild Relatives and Landraces*, CABI, UK

Hunter, D., Maxted, N., Heywood, V., Kell, S., and Borelli, T. (2012) 'Protected areas and the challenge of conserving crop wild relatives', *PARKS: The International Journal of Protected Areas and Conservation*, vol 18, pp. 87–97

Inta, A., Siri-Ard, P., and Pongmornkul, W. (2012) 'Medicinal plants in Ban Hua Thung community forest, Chaing Dao wildlife sanctuary, Chiang Dao district, Chiang Mai province', *Thai Journal of Botany*, vol 4, no 2, pp. 213–232

Kostermans, A.J.G.H. and Bompard, J.M. (1993) *The Mangoes, Their Botany, Nomenclature, Horticulture and Utilization*, IBPGR, Rome, Italy and the Linnean Society of London, UK

Kothari, A., Cooney, R., Hunter, D., McKinnon, K., Muller, E., Nelson, F., Oli, K., Pandey, S., Rasheed, T., and Vavrova, L. (2015) 'Managing resource use and development', in G.L. Worboys, M. Lockwood, A. Kothari, S. Feary, and I. Pulsford (eds), *Protected Area Governance and Management*, Australian National University Press, Canberra, Australia, pp. 789–822

Maxted, N., Ford-Lloyd, B.V., Jury, S., Kell, S., and Scholten, M. (2006) 'Towards a definition of a crop wild relative', *Biodiversity and Conservation*, vol 15, pp. 2673–2685

Meilleur, B.A. and Hodgkin, T. (2004) '*In situ* conservation of crop wild relatives: Status and trends', *Biodiversity and Conservation*, vol 13, pp. 663–684

Shrestha P., Subedi A., and Sthapit, B.R. (2013) 'Enhancing awareness on the value of local biodiversity in Nepal', in W.S. de Boef, N. Peroni, A. Subedi, and M.H. Thijssen (eds), *Community Biodiversity Management: Promoting Resilience and the Conservation of Plant Genetic Resources*, Earthscan/Routledge, Abingdon, UK and New York

Somashekhar, B.S. (ed.) (2014) *Tropical Fruit Trees & Farmers' Traditional Knowledge – Community Primer on Good Practices for Diversity Management*, National Project Management Unit – India, Bioversity International, Tropical Fruit Tree Genetic Resources Project, Indian Institute of Horticultural Research, Bangalore, India

Srisutham, M. and Kaewjampa, N. (2010) 'Indigenous community forest management in northeastern Thailand: Biodiversity conservation through rural development', *Journal of Environmental and Rural Development*, vols 1–2, pp. 28–32

Sthapit B.R., Shrestha P.K., and Upadhyay, M.P. (2006) *Good Practices: On-farm Management of Agricultural Biodiversity in Nepal*, NARC, LI-BIRD, IPGRI and IDRC, Nepal

Sthapit, B., Lamers, H., and Ramanatha Rao, V. (2013) *Custodian Farmers of Agricultural Biodiversity: Selected Profiles from South and South East Asia*, Bioversity International, Rome, Italy

Sthapit, B., Vasudeva R, Idris, S., Changtragoon, S., Winarno, M., Roongrattanaku, P., Limpiyaprapan, S., and Ramanatha Rao, V. (2015) 'Harnessing the diversity of wild relatives of tropical fruit tree species for sustainable livelihoods', in H. Jaenicke, S.E. Ashmore, M.E. Dulloo, L. Guarino, and M. Taylor (eds), *Acta Horticulturae 1101: XXIX International Horticultural Congress on Horticulture: Sustaining Lives, Livelihoods and Landscapes (IHC2014): IV International Symposium on Plant Genetic Resources*. DOI 10.17660/ActaHortic.2015.1101.31

Sthapit, S. and Mijatovic, D. (2014) *The Evolving Landscape of Agricultural Biodiversity Conservation. Community-based Biodiversity Management (CBM): A Landscape Approach to the Conservation of Agricultural Biodiversity Cultivated on Fifteen Years of Experiences in Begnas, Nepal*, LI-BIRD and Bioversity, Nepal

Stolton, S., Dudley, N., Avcıoğlu Çokçalışkan, B., Hunter, D., Ivanic, K.-Z., Kanga, E., Kettunen, M., Kumagai, Y., Maxted, N., Senior, J., Wong, M., Keenleyside, K., Mulrooney, D., and Waithaka, J. (2015) 'Values and benefits of protected areas', in G.L. Worboys, M. Lockwood, A. Kothari, S. Feary, and I. Pulsford (eds), *Protected Area Governance and Management*, Australian National University Press, Canberra, Australia, pp. 145–168

Subedi, A., Sthapit, B.R., Upadhyay, M.P., and Gauchan, D. (eds) (2005) *Learning from Community Biodiversity Register in Nepal*, Proceedings of the National Workshop, 27–28 October 2005, Khumaltar, Nepal; NARC, LI-BIRD and IPGRI, Nepal, pp. 28–40

Thomson, L. and Theilade, I. (2001) 'Protected areas and their role in conservation of forest genetic resources', in *Forest Genetic Resources Conservation and Management, Vol. 2: In Managed Natural forests and Protected Areas (*In Situ*)*, International Plant Genetic Resources Institute, Rome, Italy

Thomson, L., Graudal, L., and Kjaer, E. (2001a) 'Conservation of genetic resources in their natural environment', in *Forest Genetic Resources Conservation and Management,*

*Vol 2: In Managed Natural forests and Protected Areas (*In Situ*)*, International Plant Genetic Resources Institute, Rome, Italy

Thomson, L., Graudal, L., and Kjaer, E. (2001b) 'Selection and management of *in situ* gene conservation areas for target species', in *Forest Genetic Resources Conservation and Management, Vol. 2: In Managed Natural forests and Protected Areas (*In Situ*)*, International Plant Genetic Resources Institute, Rome, Italy

UNDP (2004) *Community Forests in Thailand*, United Nation Development Programme, p. 144

Vasudeva R, Janagoudar, B.S., Reddy, B.M.C., Sthapit, B.R., and Singh, H.P. (2010) 'Garcinia genetic resources: Linking diversity, livelihood and management', in Proceedings of National Symposium on Garcinia, 8–9 May 2010, College of Forestry, Sirsi; UNEP GEF, Bioversity, ICAR, University of Agricultural Sciences Dharbad, NABARD, India

Vasudeva R, Hegde, N., and Sthapit, B. (2013) 'Promoting the community management of Garcinia genetic resources in the central Western Ghats, India', in W.S. de Boef, N. Peroni, A. Subedi, and M.H. Thijssen (eds), *Community Biodiversity Management: Promoting Resilience and the Conservation of Plant Genetic Resources*, Earthscan/Routledge, Abingdon, UK and New York, pp. 163–168

Vasudeva R, Reddy, B.M.C., Sthapit, B., and Ramanatha Rao, V. (2015) 'An informal network of grafting experts to help communities to conserve and utilize wild-aromatic pickle-mango (*Mangifera indica*) diversity in the Central Western Ghats', India, in H. Jaenicke, S.E. Ashmore, M.E. Dulloo, L. Guarino, and M. Taylor (eds), *Acta Horticulturae 1101: XXIX International Horticultural Congress on Horticulture: Sustaining Lives, Livelihoods and Landscapes (IHC2014): IV International Symposium on Plant Genetic Resources*. DOI 10.17660/ActaHortic.2015.1101.10

Vinceti, B., Ickowitz, A., Powell, B., Kehlenbeck, K., Termote, C., and Hunter, D. (2013) 'Challenges and opportunities in strengthening the contribution of forests to sustainable diets', *Sustainability*, vol 5, pp. 4797–4824

Warrier, Kannan C.S. (undated) *Sacred Groves and Community Participation in FGR Conservation and Management*, Institute of Forest Genetics and Tree Breeding, Indian Council of Forestry Research and Education, Coimbatore, India

Notes

1 See Hunter and Heywood (2011, Chapter 1) and Maxted *et al.* (2006) for detailed discussions on what constitutes a crop wild relative.

2 The four communities identified in Thailand by UNDP as examples of good community forest management were: the Pakayor communities in Chiangmai and Lampang provinces in the north of Thailand; the In-Paeng Group in Sakon Nakhon, Udon Thani, Kalasin and Mukdahan provinces in the northeast of Thailand; the Dond Na Tam community network in Ubon Ratchathani province in the northeast of Thailand; and the Community Forests Rehabilitation Network in Nong Bua Lan Phu province in the northeast of Thailand.

3 Reducing Emissions from Deforestation and Forest Degradation (REDD) is an effort to create a financial value for the carbon stored in forests, offering incentives for developing countries to reduce emissions from forested lands and invest in low-carbon paths to sustainable development. 'REDD+' goes beyond deforestation and forest degradation and includes the role of conservation, sustainable management of forests and enhancement of forest carbon stocks. http://www.un-redd.org/aboutredd/tabid/102614/default.aspx

6 Exotic and indigenous fruit tree diversity on farm and the conservation of tree genetic resources

Case studies from sub-Saharan Africa

Katja Kehlenbeck, Martin Wiehle, Anne Sennhenn and Jens Gebauer

Introduction

Wild and cultivated fruit trees on farms are important for the food and nutrition security of smallholder farming households in sub-Saharan Africa (SSA) and can contribute significantly to families' incomes. Cultivation of fruit species can bring significant revenues and diversify the crop production options of small-scale farmers (Keatinge *et al.*, 2010). Women often benefit greatly from the local fruit business, as they are usually heavily involved in collection, processing and trade of wild fruits from their farms and forests (Schreckenberg *et al.*, 2006). Despite their importance in local and regional markets, many indigenous fruit tree species (IFTS) occurring in SSA both on farms and in natural habitats are relatively unknown in global markets. Some of these IFTS have high potential for domestication and on-farm production (Akinnifesi *et al.*, 2008), thus contributing to improved family nutrition and increased income generation. One recent example for successfully promoting a wild fruit tree species from Africa is the baobab (*Adansonia digitata* L.), a remarkable, huge, multipurpose food tree of the African savannahs. The approval of baobab fruit pulp as a 'novel food' ingredient in the European Union and the US in 2008 and its subsequent promotion as a 'superfruit'[1] with high antioxidant, mineral and vitamin content substantially increased the demand for baobab products outside of Africa. The still growing export market offers income-generating opportunities for farmers, collectors and small-scale processors of baobab fruit pulp in regions that are well integrated into supply chains, such as South Africa, Zimbabwe, Malawi and Senegal,[2] whereas other parts of Africa, particularly the Eastern African countries, are still on the way to entering the mentioned markets.

Contribution of cultivated and wild fruits to nutrition and livelihoods of rural communities

Fruits of cultivated and wild fruit tree species in SSA often have a very high content of vitamins and sometimes even minerals (Kehlenbeck *et al.*, 2013; Table 6.1). About 10–20 g of baobab pulp or a glass of its juice meets the daily vitamin C requirements of a child under eight years old, while 40–100 g of the berries of white crossberry (*Grewia tenax*) cover this child's daily iron requirement. The high sugar content of some fruits, such as baobab or tamarind (*Tamarindus indica*), make them important sources of energy (Table 6.1), particularly in times of food shortage.

Income generation by rural farming households through fruit cultivation or collection can be of significant importance. For example, a case study on the economic benefits from mango production in 87 farms in drylands in Eastern Kenya performed in 2012 showed that each household generated an annual mean income of almost 30,000 KES (equivalent to about US$350) from

Table 6.1 Nutrient content of selected African indigenous and exotic fruits per 100 g edible portion of their pulp (high values are highlighted in bold)

Species	*Energy (Kcal)*	*Protein (g)*	*Vit C (mg)*	*Vit A (RE) (μg)*	*Iron (mg)*	*Calcium (mg)*
Indigenous fruits:						
Baobab (*Adansonia digitata* L.)	**340**	3.1	150–**500**	0.03–0.06	1.7	360
White crossberry (*Grewia tenax* (Forrsk.) Fiori)	N.A.	3.6	N.A.	N.A.	**7.4–20.8**	**610**
Marula (*Sclerocarya birrea* Hochst.)	225	0.5	68–**200**	0.035	0.1	6
Waterberry (*Syzygium guineense* (Willd.) DC.)	70	1.6	12	N.A.	**7.9**	67
Tamarind (*Tamarindus indica* L.)	270	**4.8**	3–9	0.01–0.06	0.7	260
Jujube (*Ziziphus mauritiana* Lam.)	21	1.2	70–165	0.070	1.0	40
Exotic fruits:						
Guava (*Psidium guajava* L.)	68	2.6	**228**	0.031	0.3	18
Mango (*Mangifera indica* L.)	65	0.5	28	0.038	0.1	10
Orange (*Citrus sinensis* L. Osbeck)	47	0.9	53	0.008	0.1	40
Passion fruit (*Passiflora edulis* L.)	43	0.7	7	**0.054**	0.1	5
Pawpaw (*Carica papaya* L.)	39	0.6	62	**0.135**	0.1	24

RE = retinol equivalents.

Sources: Kehlenbeck *et al.* (2013; compiled from different sources), Lukmanj *et al.* (2008) and Stadlmayr *et al.* (2012; 2013).

mango farming, which corresponds to almost 30 per cent of the total household income (Kehlenbeck and Jamnadass, 2014). However, the benefits are not reaped evenly by men and women. Only few female-headed households grew mango, and these raised lower annual incomes than male-headed households as women had smaller farms and fewer mango trees than men. Similarly, a study performed in the Nuba Mountains, Sudan, assessed the contribution to the income of 170 households (half of them female-headed) of non-timber forest products (NTFPs) collected from the wild, including fruits from 19 IFTS (El Tahir and Gebauer, 2004). NTFPs contributed 70–88 per cent of the total income in female-headed households, but only 25–33 per cent in male-headed households.

Conservation status of wild fruits

Despite their importance for rural livelihoods, wild fruit trees are threatened because of their decreasing abundance. Partly, this decrease is caused by unsustainable fruit harvesting and overexploitation of the resource. Multiple uses of fruit trees and destructive harvesting techniques of species, such as the desert date (*Balanites aegyptiaca*) for vegetables in Uganda (Okia *et al.*, 2011) and the chocolate berry (*Vitex payos*) for fuelwood and timber in Eastern Kenya (Kimondo *et al.*, 2010), lead to the loss of tree individuals and to genetic erosion. Rejuvenation of fruit tree species such as baobab can be lacking in harsh environmental conditions under pressure from grazing livestock in some regions such as northern South Africa (Venter and Witkowski, 2013). Under better conditions, many young baobab trees can be documented (Gebauer and Luedeling, 2013), but intensive commercial fruit harvesting and the increasingly high demand for baobab pulp in regional and international markets could hamper the rejuvenation of future baobab populations (Venter and Witkowski, 2013).

Smallholder agroforestry systems such as multistrata home gardens (i.e. small plots of land around the house, which are often part of a larger farm, fenced and cultivated with useful annual and perennial plants, such as vegetables, spices, fruits and medicinal plants mainly for family consumption) or traditional mixed farms (i.e. whole farm units used mainly for staple crop production, but interplanted with other crops, including pulses, vegetables and cash crops, and with fruit and fuelwood/timber trees) are said to be suitable for *circa situm* conservation (i.e. on-farm conservation of species originally occurring in surrounding natural habitats) of tree genetic resources (Dawson *et al.*, 2013). Farmers often retain valuable trees such as wild fruit trees when clearing forests and woodlands for cultivation. In addition, they may allow regrowth of these valuable species and protect the seedlings and saplings of desired species on their farms. In some cases, transplanting of indigenous fruit tree wildlings from forests to farms and gardens is reported (Dhillion and Gustad, 2004). However, valuable indigenous tree species are often found only in low numbers and on few plots in mixed agroforestry farms (Pinard *et al.*, 2014). This is partly due to the market value of wild fruits, which is often low compared with commercialized fruit species, and to the non-availability of planting material for wild fruit trees in tree nurseries. A low population size, often combined with poor connectivity between isolated remaining wild or on-farm populations in the landscape, may lead to inbreeding depression and low fruit production of these rare species, resulting in a loss of intraspecific genetic diversity (Dawson *et al.*, 2013). Similarly, domestication of a fruit tree species is sometimes based on propagating offspring of only very few initially selected individuals, which in the long run can lead to loss of intraspecific genetic diversity as a result of the 'bottleneck' effect (Dawson *et al.*, 2013).

In the rest of this chapter, selected case studies from different regions and agroclimatic zones of SSA are presented and discussed with the aim of documenting inter- and intraspecific fruit tree diversity and evaluating the suitability of smallholder mixed farms and home gardens for *circa situm* conservation of fruit tree genetic resources. Finally, some recommendations are made for improving the conservation status and use of wild fruit tree species and traditional fruit varieties.

Case study 1: Fruit tree diversity in urban and peri-urban gardens in semi-arid Niamey, Niger

Peri-urban and urban agriculture can supply a large part of the fruit and vegetables offered in urban markets and it contributes to the family nutrition and livelihoods of the producers involved. However, urban agricultural systems might be of low species and varietal diversity as producers are often very market-oriented and focus on producing a few fruit species with high market demand only. Urban producers could also face many challenges such as lack of space and water for irrigation that may have further negative impacts on species diversity. In urban and peri-urban Niamey, where there is a mean annual rainfall of 540 mm, 51 gardens were studied in 2007 by inventorying all plant species grown in these gardens, combined with interviews with the household member most responsible for gardening (Bernholt *et al.*, 2009). Gardens were small (mean size 860 m^2) and mostly (80 per cent) managed for commercial vegetable production (Plate 2). However, many fruit trees were integrated in the gardens, resulting in a total of 29 fruit tree species plus six other tree species with a secondary or potential use as fruit trees (Bernholt *et al.*, 2009). These trees were mainly exotic: out of the 35 fruit tree species, 57 per cent were of exotic origin, and of the 1,247 fruit tree individuals counted, 92 per cent were exotic (H. Bernholt, personal communication, 2008). In 18 of the 51 gardens, particularly in the smaller gardens used mainly for intensive vegetable production, no fruit trees were cultivated. The most frequent fruit tree species was the indigenous species baobab, which occurred in 45 per cent of the gardens surveyed (Bernholt *et al.*, 2009). However, baobab was mainly grown for its leaves, which are used fresh and dried as a very nutritious vegetable or sauce ingredient and are in high demand in urban markets, and not for its fruit. The exotic species mango (*Mangifera indica*), lemon (*Citrus limon*), date palm (*Phoenix dactylifera*) and pawpaw (*Carica papaya*) were also frequently grown for both home consumption and sale at urban markets. Though the exotic fruit tree species were often found in large numbers in the gardens surveyed, the indigenous species, such as wild custard apple (*Annona senegalensis*), gingerbread palm (*Hyphaene thebaica*) or white crossberry, were frequently represented by only one or two individuals each. On average, only two indigenous fruit tree individuals were found per garden (range 0–31), whereas 23 exotic ones per garden were found (range 0–695; H. Bernholt, personal communication).

Urban farmers most probably mainly consider high market value when selecting fruit trees to be integrated into their space-limited, commercialized gardens. Products from exotic species such as mango, lemon and pawpaw have a ready market in urban centres, while there is low demand for indigenous fruits, partly because urban populations may have lost knowledge about these wild fruits and their uses and values for nutrition. In addition, it is easy to find planting material of exotic fruits in local tree nurseries, but not of indigenous fruits, so the trees in the gardens studied were often remnants of the previous vegetation and decreasing in numbers as the result of old age.

In summary, the gardens surveyed were dominated by exotic fruit species and harboured only few indigenous species. Although fruit and vegetable production in the urban gardens studied contributed much to the nutrition and income generation of the families managing them, the value of these gardens for *circa situm* conservation of indigenous fruit tree genetic resources seems to be rather low, as tree numbers may further decrease as the result of the old age of trees and removal as gardeners focus more and more on commercial production.

Case study 2: Fruit tree diversity in rural home gardens in semi-arid Nuba Mountains, Kordofan, Sudan

Rural home gardens contribute to the food and nutrition security of the families managing them as their products often complement and diversify the families' production from staple crop fields. In South Kordofan, Sudan, almost every rural family owns a home garden around their house, locally called 'jubraka', in addition to their 'far fields', where staple crops such as sorghum and millet are produced. During the rainy season, women cultivate vegetables, spices, pulses, oilseeds and early maturing cereal crops in their fenced home gardens, where fruit, fuelwood and timber trees also occur as well as ornamental plants. An inventory of useful plant species, including fruit trees, was performed in 61 rural home gardens in the semi-arid Nuba Mountains, South-Kordofan Province, Sudan, where there is an annual mean rainfall of 500–800 mm (Wiehle *et al.*, 2014a). Ninety percent of the relatively large gardens (mean size 2000 m^2) were managed by female household members, and in 88 per cent of cases production was mainly for family consumption and performed in a very traditional way (only 3 per cent of gardeners used mineral fertilizer, no respondent had ever had contact with an agricultural extension worker and planting material was almost exclusively procured through informal networks). In total, 32 fruit tree species were documented, 18 with a main use as fruit and 14 with a secondary fruit use (Wiehle *et al.*, 2014a). Most of the fruit tree species (73 per cent) were indigenous, whereas only 27 per cent of all the fruit tree individuals counted were exotics. The most frequent fruit tree species included the indigenous Christ's thorn jujube (*Ziziphus spina-christi*; found on 61 per cent of the gardens), baobab (46 percent; Plate 3) and desert date (43 per cent) (Wiehle *et al.*, 2014a). On average, 12 indigenous fruit tree individuals were found per garden (range 0–63), but only 5 (0–28) exotic ones (e.g. mango or pawpaw) were found (M. Wiehle, personal communication, 2013).

In summary, home gardens in the Nuba Mountains were dominated by indigenous fruit trees, whereas exotic species were rare. Both indigenous and exotic fruits were mainly produced for home consumption and played an important role in family nutrition, and only rarely some surplus of fruits was sold. One reason for the poor market integration and the low numbers of exotic fruit trees might be the remoteness of the research area, which was rather isolated during a decade-long civil war until 2005 (Wiehle *et al.*, 2014a). Remnants from previous vegetation and natural regeneration are the most important sources for fruit tree planting material in addition to some informal seed exchange for exotic species. Lack of fruit tree nurseries and poor access to markets may have further hindered the adoption of commercial exotic fruit tree cultivation. Fruit tree diversity in the home gardens surveyed therefore largely represents the species composition of the surrounding woodlands, where the same wild fruit trees often occur in abundance. However, the currently high value for *circa situm* conservation of fruit tree genetic resources of the home gardens surveyed might be threatened because of the rapid transformation processes observed in the research area, including commercialization of production, particularly of light-demanding vegetables (Plate 3), which results in the felling of shady trees and introduction of exotic fruit tree species, for example by NGOs.

Case study 3: Fruit tree diversity in mixed smallholder farms in a high-potential agricultural area in Machakos County, Eastern Kenya

In the densely populated, high-potential agricultural areas of Kenya such as Machakos, farmers usually have rather small areas of farmland located around their houses, similar to large home gardens, where they mainly cultivate the staple crops maize and beans, mixed with some vegetables and fruit trees. It is often surrounded by a living fence consisting of fuelwood, timber and fodder trees. Production is both for family consumption and sale; therefore, often cash crops such as coffee are also cultivated. A tree inventory, including fruit trees, was performed on 90 smallholder mixed farms in Machakos County, Eastern Kenya, in an area with mean annual rainfalls between 700 and

1,200 mm and altitudes from 1,200 to 2,100 masl. (Mutunga *et al.*, unpublished data). The mean farm size was 1.6 ha and a total of 28 fruit tree species were documented, 17 with a primary fruit use and 11 with a potential fruit use. Though almost half of the fruit tree species (46 per cent) were indigenous, almost all (96 per cent) of the 3,505 fruit tree individuals counted were exotics. The most frequent fruit tree species were the exotic species mango (occurring in 79 per cent of farms), avocado (*Persea americana*; 60 per cent) and lemon (46 per cent). Indigenous fruit tree species such as tamarind or waterberry (*Syzygium guineense*) were rare and represented by few individuals. On average, as many as 37 exotic fruit tree individuals were cultivated per farm (range 0–276), but only an average of two indigenous ones (range 0–21). Fruit cultivation in the farms surveyed contributed greatly to family nutrition and livelihoods as most of the farmers produced fruits for both home consumption and income generation, the latter particularly from selling exotic fruits such as mango and avocado.

One possible explanation for the clear dominance of exotic fruits is the relatively good market access of the study area, which is close to Kenya's capital Nairobi. Farmers reacted to the strong demand and high selling prices of exotic fruits such as mango by planting high numbers of these species, for which planting material is easily available from the many tree nurseries in the area. Almost 70 per cent of the planting material for exotic fruit trees was procured from commercial nurseries or raised in the gardener's own nursery. A second explanation is the lack of natural forests and woodlands for seed dispersal of wild fruit tree species, which further contributes to the low abundance of these species on farms. Almost 70 per cent of the indigenous fruit trees originated from natural regeneration or were remnants from the previous vegetation. In addition, the high agricultural potential and long cultivation period in the study region also resulted in rather intensive production systems offering few niches for the survival of indigenous fruits. Third, wild fruits are seen as 'poor people's food' by many Kenyans and there is hardly any market for wild fruit products in urban centres. Finally, the existing extension systems in the area, both governmental and NGO-managed, only promote the cultivation of a few exotic fruit tree species, while indigenous species are completely neglected.

Case study 4: Intraspecific diversity of two fruit tree species in Sudan and Kenya

As an example of intraspecific diversity of an indigenous species, Christ's thorn jujube in the home gardens and adjacent forests of Sudan is used. For exotic species, the diversity of mango landraces on mixed farms in Kenya is presented. In the Nuba Mountains in Sudan, 250 individual *Z. spina-christi* trees were sampled from home gardens and adjacent forests in five locations (Wiehle *et al.*, 2014b). Morphological tree and fruit traits were documented and molecular characterization performed using AFLP markers. Individual trees in home gardens had slightly larger fruit than those in the forests, which may be caused by better environmental conditions in the home gardens or by human selection, the first step towards domestication. Contrary to expectations, the genetic diversity of *Z. spina-christi* was higher in the home gardens than in the forests, possibly caused by the introduction of mixed germplasm from markets originating most probably from a diversity of wild collection locations (Wiehle *et al.*, 2014b). The home gardens studied were therefore seen as a suitable agroforestry system for *circa situm* conservation of this threatened IFTS in Sudan.

In Eastern and Central Kenya, fruits and leaves of 38 mango accessions were collected from farms for morphological and genetic characterization using SSR markers (Sennhenn *et al.*, 2013). The morphological variability of fruits among the 38 accessions was high (Plate 4) and cluster analysis using the 10 most discriminant traits resulted in the identification of six different types or landraces of local mangoes.

Similarly, the genetic diversity among the accessions was high and only few samples were genetically identical. A cluster analysis of the SSR results revealed eight different types or landraces of local mangoes, partly confirming the results of the morphological clusters (Sennhenn *et al.*, 2013). However, the owners of the sampled mango accessions mentioned in interviews that they had removed many local mango trees from their farms and replaced them with improved, introduced mango cultivars despite some advantages of the local landraces such as higher pest and disease tolerance, better taste and higher juice content (A. Sennhenn, personal communication, 2012). This trend will most probably continue as market demand and prices for local mangoes in Kenya are low. Documentation of the advantages of local mango landraces including their nutritional value followed by the development of 'conservation through use' strategies is urgently needed to raise the awareness of farmers and fruit processors about the value of local mangoes to avoid further loss of valuable genetic resources.

Conclusions and recommendations

Fruit tree cultivation is very common in smallholder farming systems and contributes much to family nutrition and livelihoods via income generation. Farmers in areas close to markets and with a commercial production goal may prefer exotic fruit trees over indigenous species and may favour improved varieties over traditional landraces. These farmers mainly consider the ready markets and good profits for species and variety selection. On the other hand, farmers in remote areas may rather focus on subsistence production and still maintain indigenous fruit trees and local landraces on their farms. However, as soon as market access improves and improved species and varieties are available, they may switch to commercial production, leading to a decrease in indigenous fruit tree diversity. Farmers should therefore be better informed about the advantages of indigenous fruit trees such as high nutritional values or drought resistance. Wild fruit tree species should be included in domestication programmes and high-quality planting materials of indigenous species offered in tree nurseries. Traditional knowledge on the use of wild fruits needs to be documented and used for developing 'conservation through use' strategies for threatened species and varieties. Integrated approaches, involving stakeholders from governmental, private and NGO sectors, to raise awareness of the value of indigenous fruits for nutrition and income generation and to increase production, processing and consumption of these fruits are largely lacking in SSA and should be developed and disseminated soonest to avoid further loss of valuable genetic resources.

References

Akinnifesi, F.K., Leakey, R.R.B., Ajayi, O.C., Sileshi, G., Tchoundjeu, Z., Matakala, P., and Kwesiga, F.R. (2008) *Indigenous Fruit Trees in the Tropics: Domestication, Utilization and Commercialization*, CAB International, Wallingford, in association with the World Agroforestry Centre, Nairobi

Bernholt, H., Kehlenbeck, K., Gebauer, J., and Buerkert, A. (2009) 'Plant species diversity in urban and peri-urban gardens of Niamey, Niger', *Agroforestry Systems*, vol 77, pp. 159–179

Dawson, I.K., Guariguata, M.R., Loo, J., Weber, J.C., Lengkeek, A., Bush, D., Cornelius, J., Guarino, L., Kindt, R., Orwa, C., Russell, J., and Jamnadass, R. (2013) 'What is the relevance of smallholders' agroforestry systems for conserving tropical tree species and genetic diversity in *circa situm*, *in situ* and *ex situ* settings? A review', *Biodiversity and Conservation*, vol 22, pp. 301–324

Dhillion, S.S. and Gustad, G. (2004) 'Local management practices influence the viability of the baobab (*Adansonia digitata* L.) in different land use types, Cinzana, Mali', *Agriculture, Ecosystems and Environment*, vol 101, pp. 85–103

El Tahir, B.A. and Gebauer, J. (2004) 'Non-timber forest products: Opportunities and constraints for poverty reduction in the Nuba Mountains, South Kordofan, Sudan', full paper presented at Tropentag Conference on International Agricultural Research, 5–7 October 2004, Berlin, Germany; http://www.tropentag.de/2004/abstracts/full/93.pdf

Gebauer, J. and Luedeling, E. (2013) 'A note on baobab (*Adansonia digitata* L.) in Kordofan, Sudan', *Genetic Resources and Crop Evolution*, vol 60, pp. 1587–1596

Keatinge, J.D.H., Waliyar, F., Jamnadass, R.H., Moustafa, A., Andrade, M., Drechsel, P., Hughes, J.A., Kadirvel, P., and Luther, K. (2010) 'Relearning old lessons for the future of food – by bread alone no longer: Diversifying diets with fruit and vegetables', *Crop Science*, vol 50, pp. S51–S62

Kehlenbeck, K. and Jamnadass, R. (2014) 'Chapter 6.2.1: Food and nutrition – Fruits, nuts, vegetables and staples from trees', in J. De Leeuw, M. Njenga, B. Wagner, and M. Iiyama (eds), *Treesilience: An Assessment of the Resilience Provided by Trees in the Drylands of Eastern Africa*, ICRAF, Nairobi; http://www.worldagroforestry.org/knowfordocs/Treesilience_Book_2014.pdf

Kehlenbeck, K., Asaah, E., and Jamnadass, R. (2013) 'Diversity of indigenous fruit trees and their contribution to nutrition and livelihoods in sub-Saharan Africa: Examples from Kenya and Cameroon', in J. Fanzo, D. Hunter, T. Borelli and F. Mattei (eds), *Diversifying Food and Diets: Using Agricultural Biodiversity to Improve Nutrition and Health*, Earthscan, New York, pp. 257–269

Kimondo, J.M., Agea, J.G., Okia, C.A., Abohassan, R.A.A., Mulatya, J., and Teklehaimanot, Z. (2010) '*Vitex payos* (Lour.) Merr fruit trees in the drylands areas of Eastern Kenya: Use, marketing and management', *Botany Research Journal*, vol 3, pp. 14–21

Lukmanj, Z., Hertzmark, E., Mlingi, N., Assey, V., Ndossi, G., and Fawzi, W. (2008) *Tanzania Food Composition Tables*, MUHAS-TFNC (Muhimbili University of Health and Allied Sciences and Tanzania Food and Nutrition Centre), Dar es Salaam, Tanzania and HSPH (Harvard School of Public Health), Boston, MA; http://www.hsph.harvard.edu/nutritionsource/files/tanzania-food-composition-tables.pdf

Mutunga *et al.*, unpublished data extracted from: Kehlenbeck, K., Mutunga, C.N., Giliba, R.A. (2013) *Tree Species Richness and Diversity on Farms in Machakos, Eastern Kenya, and Mbarali, Tanzania*, Report for the Evergreen Agriculture Project of the World Agroforestry Centre ICRAF, Nairobi, Kenya

Okia, C.A., Agea, J.G., Kimondo, J.M., Abohassan, R.A.A., Okiror, P., Obua, J., and Teklehaimanot, Z. (2011) 'Use and management of *Balanites aegyptiaca* in drylands of Uganda', *Research Journal of Biological Sciences*, vol 6, pp. 15–24

Pinard, F., Joetzjer, E., Kindt, R., and Kehlenbeck, K. (2014) 'Are coffee agroforestry systems suitable for *circa situm* conservation of indigenous trees? A case study from Central Kenya', *Biodiversity and Conservation*, vol 23, pp. 467–495

Schreckenberg, K., Awono, A., Degrande, A., Mbosso, C., Ndoye, O., and Tchoundjeu, Z. (2006) 'Domesticating indigenous fruit trees as a contribution to poverty reduction', *Forests, Trees and Livelihoods*, vol 16, pp. 35–51

Sennhenn, A., Prinz, K., Gebauer, J., Whitbread, A., Jamnadass, R., and Kehlenbeck, K. (2013) 'Identification of mango (*Mangifera indica* L.) landraces from Eastern and Central Kenya using a morphological and molecular approach', *Genetic Resources and Crop Evolution*, vol 61, pp. 7–22

Stadlmayr, B., Charrondiere, R., Eisenwagen, S., Jamnadass, R., and Kehlenbeck, K. (2013) 'Nutrient composition of selected indigenous fruits from sub-Saharan Africa', *Journal of the Science of Food and Agriculture*, vol 93, pp. 2627–2636

Stadlmayr, B., Charrondiere, R., Enujiugha, V.N., and Bayili, R.B. (2012) *West African Food Composition Table*, FAO, Rome; http://www.fao.org/docrep/015/i2698b/i2698b00.pdf

Venter, S.M. and Witkowski, E.T.F. (2013) 'Where are the young baobabs? Factors affecting regeneration of *Adansonia digitata* L. in a communally managed region of southern Africa', *Journal of Arid Environments*, vol 92, pp. 1–13

Wiehle, M., Goenster, S., Gebauer, J., Mohamed, S.A., Buerkert, A., and Kehlenbeck, K. (2014a) 'Effects of transformation processes on plant species richness and diversity in home gardens of the Nuba Mountains, Sudan', *Agroforestry Systems*, vol 88, pp. 539–562

Wiehle, M., Prinz, K., Kehlenbeck, K., Goenster, S., Mohamed, S.A., Buerkert, A., and Gebauer, J. (2014b) 'The role of home gardens and forest ecosystems for domestication and conservation of *Ziziphus spina-christi* (L.) Willd. in the Nuba Mountains, Sudan', *Genetic Resources and Crop Evolution*, vol 61, pp. 1491–1506

Notes

1 http://baobabsuperfruit.com/; http://www.baobabfoods.com/
2 http://phytotrade.com/products/baobab/; http://www.baobab-fruit.com/

7 Promoting community management of underutilized tropical and subtropical fruit genetic resources in Brazil

Josué Francisco da Silva Júnior, Dalva Maria da Mota, Rosa Lia Barbieri and Adriana Alercia

Introduction

In 1992, ten centres of fruit species diversity were defined in Brazil, located from the Amazon to the southern region, which house more than 500 native species of major economic importance (Giacometti, 1993) such as pineapple (*Ananas comosus*), cashew (*Anacardium occidentale*) and passion fruit (*Passiflora* spp.). Besides these, a large number of native fruit species not only play an important role in the economic well-being but are also essential to the survival of small-scale farmers.

In 2003, in order to promote research, conservation and use of fruit species, the Ministry of the Environment in Brazil prioritized native species of current and potential economic value through a project called 'Plants for the Future' involving research institutions, educational and non-governmental organizations. Research covered various scientific disciplines, but did not include *in situ* conservation or the role of traditional communities in the management of biodiversity in natural habitats.

Studies on good practices for managing Brazilian native fruits are scarce and the information available focuses mainly on post-harvest and processing activities. Research efforts on sustainable management of resources, especially in natural habitats, are rare. However, a few of them, such as good management practices for pequi (*C. brasiliense*) (Oliveira and Scariot, 2010), mangaba (*Hancornia speciosa*) (Lima and Scariot, 2010) and other native fruits, have been published by EMBRAPA (the Brazilian Corporation of Agricultural Research) in partnership with the Institute for Society, Population and Nature.

Despite the dearth of information generally, one traditional group from the north and northeast of Brazil who continue to manage Brazil's wild plant resources has been widely studied: the 'catadoras de mangaba' (mangaba pickers; Figure 7.1). Information has been gathered on the resources that they use to survive, how they access and manage them, their habitats, related national

Figure 7.1 Mangaba pickers.
Credit: J.F. Silva Júnior.

legislation, labour, gender, sociability and exploitation of raw materials (Mota *et al.*, 2011).

A less studied but also long-lived and neglected useful species is *Butia odorata* (Barb. Rodr.) Noblick (butiá or jelly palm), whose genetic diversity and associated knowledge are under severe erosion and only a few natural areas remain. One strategy for sustainable conservation is the adoption of best management practices to ensure the survival of butiá for future generations and the sustainability of the ecosystem, with consideration of environmental, social and economic perspectives (Rivas, 2013).

In the following sections we describe successful experiences of good management practices for the *in situ* conservation of mangaba and butiá, two species of major importance for farming communities.

In situ conservation strategies and good practices for the mangaba tree

Mangaba (*Hancornia speciosa*) occurs naturally in open vegetation areas such as the coastal tablelands, dunes and savannas of Brazil. It is also found in Paraguay, Peru and Bolivia. The trunk of the tree is used for the extraction of latex for medicinal purposes. The fruit is a source of protein and iron. It has a delicious taste and colour and is used in agribusiness for making juices, sweets and ice creams.

In 2003, EMBRAPA started work on the conservation, characterization and use of mangaba in the Northeast region of Brazil, noting that the mangaba pickers, who were mainly women, were *de facto* responsible for conservation of mangaba in the areas identified. These women had accumulated significant knowledge concerning the management, reproduction and post-harvest activities of this fruit (Mota and Silva Júnior, 2003; Silva Júnior *et al.*, 2006). Despite the market demand for mangaba, it is overwhelmingly threatened by destruction of its natural habitats because of the expanding cultivation of sugarcane, coconut, eucalyptus, grassland, corn, cotton and soya beans and real estate development in tourist areas and coastal cities. The mangaba pickers' livelihoods are further threatened by the increased activity of shrimp farming, which destroys the mangrove ecosystem and drastically reduces mollusc harvesting, another activity carried out by many of the communities concerned.

To explore how best to sustainably conserve and use mangaba, a team of social and natural scientists was assembled and a large amount of information was generated about the history, mapping, profile and typology of pickers; threats to natural habitats; access to and management of sites; knowledge and endangered traditions; organization of communities; and sale and consumption habits for mangaba (Mota *et al.*, 2011).

Changes in the mangaba tree habitats and exploitation (economic interests of landowners) and in land use (tourism, agriculture, shrimp production) require efforts to be made for sustainable management and conservation of mangaba. Good practices already exist. Some practices contributing to community harmony are agreed among pickers, such as that each family can collect mangaba, but from different trees; farmers from other locations collecting in areas used for generations by specific groups are not well regarded. Fences on private land are commonly respected, though there are also transgressions, such as collecting fruit without permission from the landowners.

A set of good practices for conserving mangaba in natural ecosystems was jointly defined through expeditions to different sites, meetings, coaching, training and knowledge sharing among farmers and researchers. According to Mota *et al.* (2011), good practices are associated with the type of access that women pickers have to private areas or to common lands. The good practices are listed below as sets of recommendations for natural ecosystems, private lands and post-harvest practices.

Traditional management practices in natural ecosystems

- Maintain native vegetation because it is a source of livelihood for the communities and houses mangaba pollinators
- Do not cut or burn mangaba trees, although they have resistance to soft burning
- Do not break branches
- Use 'hooks' to harvest fruits as branches break if adults climb up the tree
- Do not collect unripe fruit, pick only semi-mature fruit or that which has fallen to the ground
- Extract latex (medicinal use) sparingly by performing superficial cuts on the trunk and branches to prevent death of plant
- Take special care of seedlings growing around the adult plant. The presence of animals can be beneficial as it ensures species dispersal
- Remove dry branches and weeds, such as mistletoe (*Psittacanthus* sp.), which can kill the mangaba tree.

Good management practices in private areas

- Produce mangaba seedlings for the enrichment of native vegetation
- Fertilize mangaba with crop residues and organic material
- Diversify crops with other fruit trees and plants of interest to the community
- Control the fungus *Lasiodiplodia theobromae*, which causes 'seca-da-mangabeira' disease or dry branches.

Post-harvest practices

Improved post-harvest practices have resulted in better quality fruit for sale, which has driven an increase in the demand for mangaba from agribusiness. Public policies have benefited women mangaba pickers as government programmes guarantee the purchase of fruit with good appearance and organoleptic qualities and also purchase through school food programmes, which buy fruit and frozen pulp. These improved practices are:

- Remove unripe, overripe, rotten or malformed fruit
- Wash fruit to remove sand, dirt and any remaining latex
- Dry fruit on cloths or mats, in a cool and shaded area
- Package harvested fruit in containers (wicker baskets or paper boxes lined with plastic) to promote ripening.

Most of these activities are part of the daily life of mangaba pickers; however, some improvements, such as the proper selection of fruits for agribusiness and the correct way to package it, were practices introduced by the researchers and disseminated among women pickers through joint courses.

In situ conservation and good practices of butiá

Popularly known as butiá, or jelly palm, the genus *Butia* includes 20 species distributed over Brazil, Paraguay, Uruguay and Argentina (Lorenzi *et al.*, 2010). *Butia odorata* occurs in southern Brazil. Fruiting starts between 6 and 15 years after plant emergence and continues every year, with variable production depending on the environmental conditions. Even centennial plants produce fruit. The fruit pulp has a high content of potassium, manganese and iron and is rich in vitamin C and carotenoids (Fonseca, 2012). The colour of the mature fruit varies from pale yellow to dark red and has a sweetish-acidic taste. The fruit is used to make juices, liqueurs, jams, ice creams and cakes and to fill chocolates. The leaves are used in crafts, in the production of baskets, bags, hats and other products. The plants are used in landscaping (Büttow *et al.*, 2009) and the seeds contain high-quality oil that could be used to develop cosmetic, pharmaceutical and food products (Rossato, 2007). Depending on the region, butiá is collected for local use or for commercial purposes (Figure 7.2). Because of the strong relationship with the local culture, it is very common to find butiá plants grown in gardens and urban backyards.

Until the 1970s, large natural populations of butiá were found in Rio Grande do Sul. Since then, due to the cultivation of extensive monoculture crops (rice, soya bean, eucalyptus, pine and acacia), grazing (or livestock) and the expansion of urban areas, they have undergone a rapid decline and only a few natural areas remain. A strategy for sustainable conservation is the adoption of best management practices to ensure the survival of butiá for future generations and the sustainability of the ecosystem, considering the environmental, social and economic perspectives (Rivas, 2013).

In areas where the predominant economic activity is livestock production, one option for sustainable development is conservation management, a methodology for promoting native grassland and butiá conservation by excluding grazing during the winter. Conservative management can allow the development of new plantlets of butiá and also the improvement of native grassland biomass.

Good practices using butiá for sustainable development are centred on methods for collecting fruit that protect human health, the environment and the butiá plants:

- Avoid picking fruit near roads (there may be contamination from toxic products derived from fuel combustion) or in areas where agrochemicals were applied to crops like rice and soybean (herbicides, insecticides and fungicides).
- Do not harvest unripe or damaged fruit, as it may serve as food for wildlife and can contribute to the conservation of the ecosystem.
- Define a collecting rotation plan, avoiding picking the fruit always from the same plants. This is important to ensure availability of seeds for regeneration of the population, production of new plantlets and food for wildlife.

Figure 7.2 Collecting butiá in natural populations.
Credit: P.S. Rocha.

- Do not collect all the fruit in a given area; leave some of them on each plant to encourage new seedling production.

Conclusions

In situ conservation and sustainable harvesting of mangaba and butiá are closely related to the work of mangaba pickers and farmers because of their economic, social and cultural dependency on them. Farmers' contributions to safeguarding biodiversity and traditional knowledge are critical to valorise, cultivate and use these resources, which in turn will enhance their livelihoods and will sustainably ensure resource availability for future generations. The strategy of linking good management techniques, developed over generations by traditional communities, with scientific knowledge has contributed to the conservation of natural ecosystems of these species and to the valorisation of farmers – especially women – as beneficiaries of public policies for social inclusion.

References

Büttow, M.V., Barbieri, R.L., Neitzke, R.S., and Heiden, G. (2009) 'Conhecimento tradicional associado ao uso de butiás (*Butia* spp., Arecaceae) no sul do Brasil', *Revista Brasileira de Fruticultura*, vol 31, no 4, pp. 1069–1075

Fonseca, L.X. (2012) *Caracterização de frutos de butiazeiro (Butia odorata Barb.Rodr.) Noblick & Lorenzi e estabilidade de seus compostos bioativos na elaboração e armazenamento de geleias*, Master's thesis, Universidade Federal de Pelotas, Pelotas, RS, Brazil, p. 68

Giacometti, D.C. (1993) 'Recursos genéticos de fruteiras nativas do Brasil', in *Anais do I Simpósio Nacional de Recursos Genéticos de Fruteiras Nativas*, 1992, EMBRAPA-CNPMF, Cruz das Almas, BA, Brazil, pp. 13–27

Lima, I.L.P. and Scariot, A. (2010) *Boas práticas de manejo para o extrativismo sustentável da mangaba*, EMBRAPA Recursos Genéticos e Biotecnologia, Brasília, DF, Brazil, p. 68

Lorenzi, H., Noblick, L., Kahn, F., and Ferreira, E. (2010) *Flora Brasileira*: Arecaceae (Palmeiras), Instituto Plantarum, Nova Odessa, SP, Brazil, p. 384

Mota, D.M. and Silva Júnior, J.F. da (2003) 'Populações tradicionais e formas coletivas de gestão das áreas de ocorrência de mangabeira', *Raízes*, vol 22, no 2, pp. 225–233

Mota, D.M. da, Silva Júnior, J.F. da, Schmitz, H., and Rodrigues, R.F. de A. (eds) (2011) *A mangabeira, as catadoras, o extrativismo*, EMBRAPA Amazônia Oriental, Belém, PA, Brazil and Embrapa Tabuleiros Costeiros, Aracaju, SE, Brazil, p. 297

Oliveira, W.L. de and Scariot, A. (2010) *Boas práticas de manejo para o extrativismo sustentável do pequi*, EMBRAPA Recursos Genéticos e Biotecnologia, Brasília, DF, Brazil, p. 84

Rivas, M.M. (2013) *Conservação e uso sustentável de palmares de Butia odorata (Barb. Rodr.) Noblick*, PhD thesis, Universidade Federal de Pelotas, Pelotas, RS, Brazil, p. 102

Rossato, M. (2007) *Recursos genéticos de palmeiras nativas do gênero Butia do Rio Grande do Sul*, PhD thesis, Universidade Federal de Pelotas, Pelotas, RS, Brazil, p. 136

Silva Júnior, J.F. da, Araujo, I.A., Barreiro Neto, M., Espíndola, A.M.C.M., Carvalho, N.S.G., and Mota, D.M. da (2006) 'Recursos genéticos nos tabuleiros costeiros e baixada litorânea do Nordeste', in J.F. da Silva Júnior and A. da Silva Lédo (eds), *A cultura da mangaba*, EMBRAPA Tabuleiros Costeiros, Aracaju, SE, Brazil, pp. 57–74

8 Good practice

Using intraspecific crop diversity to manage pests and pathogens in China

Keyu Bai, Huaxian Peng, Jie Wu, Yayun Yang, Enlai Zhang, Luyuan Dai, Chunmin Lu, Yunyue Wang, Jie Yuan, Chunlin Long, Paola De Santis and Devra I. Jarvis

Introduction

The widespread use of one or a few genetically homogeneous cultivars creates the conditions for the spread of arthropod pests and virulent strains of pathogens in both perennial and annual species (Marshall, 1977). The susceptibility of five major commercial banana varieties to the fungal disease black sigatoka led to a loss of nearly 47 per cent of banana yield in Central America (FAO, 1998). On the other hand, cultivar mixtures of perennial fruit tree crops have been shown to reduce pathogen spread for apple scab caused by *Venturia inaequalis* in apples in Europe (Didelot *et al.*, 2007; Parisi *et al.*, 2013).

The crop loss from arthropod pests and diseases is regarded as one of the major limitations for increasing crop productivity and safeguarding food security in China (Wang *et al.*, 2000). The resulting economic and food resource costs are, to a significant extent, a consequence of cultivars with uniform resistance structures, and the continuing evolution of new races of pests and pathogens that are able to overcome resistance genes introduced by modern breeding, creating the phenomenon of boom and bust cycles (Yu *et al.*, 2007; Bourke, 1993; Singh *et al.*, 2006). Pesticides have been used on a large scale since the 1960s to protect crops from damage inflicted by insects and diseases, but pesticide consumption has increased over the years: after the mid-1990s annual pesticide use in China passed 500,000 tonnes (Huang *et al.*, 2000). Crop breeding programmes exist in China to develop new varieties and to replace varieties that have lost their pest resistance, but maintenance costs of the current system can be high, particularly for developing countries (Strange and Scott, 2005). Most, if not all, known resistance to arthropod pests and pathogens in crops used in breeding programmes are derived from varieties collected from farmers who traditionally grew them in genetically diverse systems. Even so, the development of new cultivars grown as monocultures continues to be central to modern agriculture in China and worldwide (Leung *et al.*, 2003;

Finckh and Wolfe, 1997; Brown, 1999). The inherent instability of large scale monocultures and thus risk for farmers has led to a reliance on various generations of pesticides, a decrease in food safety, and the expending of funds to purchase new varieties usually every three to five years.

Integrated Pest Management (IPM), an ecosystem approach to crop production and protection that combines different management strategies and practices to grow healthy crops and minimize the use of pesticides with considerable success, is extensively used in China, particularly in the new and growing area of organic agriculture (You *et al.*, 2004). However, until recently these methods have concentrated on using agronomic techniques to modify the environment around predominantly modern cultivars and interspecific or crop diversity to reduce the need for pesticides. Limited use of managing the intraspecific diversity of local crop varieties themselves has been used within conventional IPM.

Farmers in the mountainous areas and more remote areas of southwestern China continue to grow food crops with a wide genetic base in the form of traditional crop varieties (Guo *et al.*, 2012). New practices are emerging to use such a diverse genetic base of crop resistance through the deployment of crop varietal diversity on farm to provide low-cost, stable management of pest and disease pressure. This method has the potential to both reduce crop loss for the current season and reduce the probability of future loss to new pathogen strains and pest biotypes.

In this chapter we describe the good practices and results gathered over the past eight years that use intraspecific diversity (in the form of crop varietal diversity) in the fields of smallholder farmers to reduce current pest and disease loss in nine agro-ecological sites in southwestern China for three main food subsistence crops: rice, maize and barley (Figure 8.1; Table 8.1; data are also included for faba bean). We also discuss: practices to measure vulnerability (which we define as the probability of future crop loss to pest and diseases),

Table 8.1 List of sites, crops and pests and diseases in China

Site	*Crop*	*Pests*
Yuanyang	Rice	Blast, brown leaf hopper
Chongmin	Barley	Net blotch, powdery mildew
Shilin (Kunming)	Rice	Blast, brown leaf hopper
Menghai (Xishuangbanna)	Maize	Southern leaf blight, stem borer
Zhongdian (Shangri-La)	Barley	Net blotch, powdery mildew
Menghai (Xishuangbanna)	Rice	Blast, brown leaf hopper
Chuxiong	Faba bean	Aphids, botrytis, brochus, rust
Zhaojue	Maize	Southern leaf blight, stem borer
Shehong	Rice	Blast, brown leaf hopper
Meitan	Rice	Blast, brown leaf hopper

Note: Two targeted crops – rice and maize – were included at the Menghai project site.
Source: as modified from Jarvis and Campilan (2006).

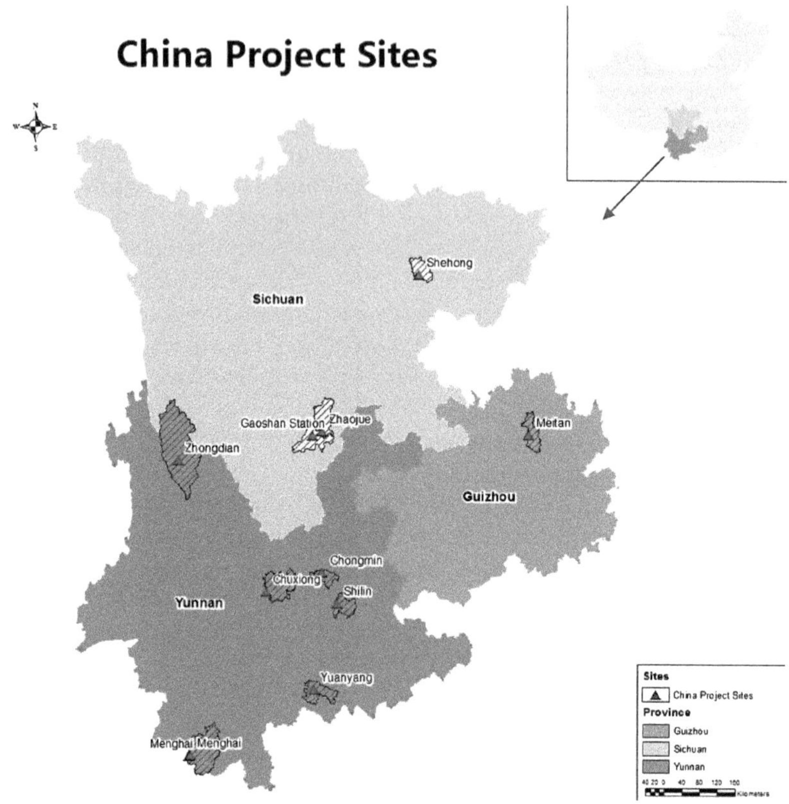

Figure 8.1 Project sites in China.

the role of seed access and the development of public awareness products and policy recommendations to support and benefit farmers' use of crop varietal diversity in China.

A common participatory diagnostic protocol

Information from farmers, in-field observations, and in-field and laboratory trials were collected using a common protocol (Jarvis and Campilan, 2006). This included a decision framework developed through a global project[1] with national partners from China, Uganda, Ecuador and Morocco to guide research questions for determining when and where crop genetic diversity on farm would be an appropriate option to manage pest and diseases (Box 8.1). Each step of the diagnostic process included assessment of farmers' beliefs and practices and measured data, including data from farmers' fields, on on-farm, on-station, greenhouse and laboratory trials.

Vital to the process was first testing whether variation for resistance exists among and within traditional crop cultivars. This required collecting a sample of the varieties grown to assess variation in resistance to target pests. Second was to determine whether this variation in resistance actually worked to reduce pest damage and vulnerability in farmers' fields. The cultivars were tested on farm, on station and in the laboratory to understand whether the resistance found was reducing damage or had the potential to reduce future damage from new mutations and pest biotypes within the crop population, or from new migrations from outside the community. The latter required the collection of isolates to test how the variation in the population structures of pests and pathogens varied over time and space in order to determine whether we were dealing with a variable pathogen or pest species, and the nature of the variation, and glass house and laboratory work was needed in order to measure richness and evenness of host–pathogen reactions.

Box 8.1 Six-step decision guide for determining when and where crop genetic diversity on farm would be an appropriate option within an IPM Strategy ***(from Jarvis and Campilan, 2006)***

Step 1 Are pest and diseases viewed by both farmers and scientists as a significant factor limiting production? If so –

Step 2 Does intraspecific diversity with respect to pests and diseases exist within project sites and, if not, do other sources of intraspecific diversity with respect to pests and diseases exist from earlier collections or from similar agroecosystems within the country? And/or –

Step 3 Does diversity with respect to pests and diseases exist but it is not accessed or optimally used by the farming communities? If so –

Step 4 Is there diversity in virulence and aggressiveness of pathogens and/or diversity in biotypes in the case of pests?

Step 5 Are pests and diseases moving in and out of the project sites, including the role of the local seed/propagation material systems? And if so, how?

Step 6 What 'genetic choices' do farmers make, including using or discarding new and old genotypes, selecting criteria for hosts that are resistant, and managing mixtures to minimize crop loss due to pests and diseases?

Characterizing host–pest/pathogen relationships for rice, barley and maize varietal diversity in farmers' fields using the participatory diagnostic approach

Participatory diagnostics were used to identify and characterize with the farming communities the diverse local and modern varieties of rice, barley and maize grown in their fields. Focus group discussions (FGDs) were first carried out, five per site, including groups of leaders, old men, young men, old women and young women. A standardized set of questions grouped under seven themes (Jarvis and Campilan, 2006) was used to ensure that all participants in the FGDs were asked the same set of questions. Questions were based on materials that farmers brought, including traditional and modern plant materials for the target crops. These materials were used as a physical basis for discussion among farmers and researchers to understand farmers' knowledge of the crop varietal diversity, pest and disease symptoms for each crop, the host–pest/pathogen differences in plant health among varieties within each crop, and the main sources of seeds for the community. The steps for FGDs include: (1) asking farmers the questions for each crop on the farmers' knowledge of varietal diversity, the traits that the farmers use to distinguish their varieties, and their value, be it agronomic, adaptive, or quality or use traits for the different target crop varieties; (2) then starting the process with varieties brought by the farmers being grouped by the farmers into varieties determined to be the same, and ensuring consistency of variety names among farmers in each FDG; (3) an individual farmer per variety to describe the specific variety, with inputs from the other farmers. The name or names given by the group to the variety, and whether the variety was traditional or modern, were written down by the researcher/rapporteur together with the morphological, agronomic, adaptive and quality traits used by the group to describe the variety; (4) all this information was then organized by the researchers in a table of traits versus varieties in front of the farmers; (5) the final step was to check this table with the farmers to ensure agreement across the groups. Once each variety was named and described, farmers were asked to divide the plant materials they brought to the discussion into two groups: healthy or non-healthy plants. The farmers were then asked to divide the group of non-healthy plants into what they perceived to be damage from different pests and diseases by the symptoms they recognized on the plants. Farmers were also asked to clearly define what they perceived as different growth stages of the plants. The farmers' descriptions of the symptoms for the diseases and pests were noted, including a list of the symptoms on the different plants and at different growth stages. Finally, the researcher showed pictures of other diseases and plants not brought to the meeting, and farmers were asked to identify and give any names they had for these diseases. After this step, farmers were asked which of the different pests and diseases identified caused the most severe damage, and then to rank varieties on their level of resistance to the complex of pests and diseases in their systems.

Table 8.2 Site level number of traditional and modern varieties, farmers' perceptions of resistance and source of seed

Project site	*Crop*	*Number of traditional varieties*	*Number of modern varieties*	*Percent area planted with traditional varieties%*	*Key resistant varieties to arthropod pests and diseases*		*Seed source of traditional variety*	*Seed source of modern variety*
					Rice blast	*Rice hopper*		
Menghai, Yunnan	Rice	17	3	93	Amoqie, Husikao-1, Duolong-1, Modelong-1	Molao, Qiena, Duolong-1, Zhangmeleng-1, Husikao-1	self	local agricultural extension
Shehong, Sichuan	Rice	5	34	10	ChuanXiangYou, GangYou,IIYou	ChuaXiangYou, GangYou	self, exchange and neighbours	local agricultural extension station
Meitan, Guizhou	Rice	3	8	33	Qianyou, Dalixiang	Dalixiang, Nuodao, Qianyou	self, exchange and neighbours	local seed company
Yuanyang, Yunnan	Rice	19	9	88	Huagu, Yuelianggu, Xiaogu	Zaogu, Yuelianggu	self, exchange and neighbours	local seed company
Shilin, Yunnan	Rice	2	20	6	Chujin-27, Xiangnuo	Yunjinyou-15	self, exchange and neighbours	local seed company and extension station
					NLB/SLB	*Stem borer*		
Menghai, Yunnan	Maize	4	4	86	Aduxiu, Aduniu	Selileng	self	local agricultural extension
Zhaojue, Sichuan	Maize	10	21	25	ChuanDan15, LiangDan4, LiangDan3, DanJiao73	LiangDan4, DanJiao73	self, exchange, neighbours	local agricultural extension station
					Net Blotch	*Powdery Mildew*		
Shangri-la, Yunnan	Barley	4	1	99	Duanbaiqingke, Changbaiqingke	Duanbaiqingke, Heiqingke	self, exchange, neighbours	local agricultural extension station
Chongmin, Yunnan	Barley	3	4	89	Midamai, Pijiudamai	Pijiudamai, Kaihuadamai	self, exchange, neighbours	local agricultural extension station

Note: NLB = northern leaf blight; SLB = southern leaf blight.
Source: Summarized by the author based on the household survey in 2008.

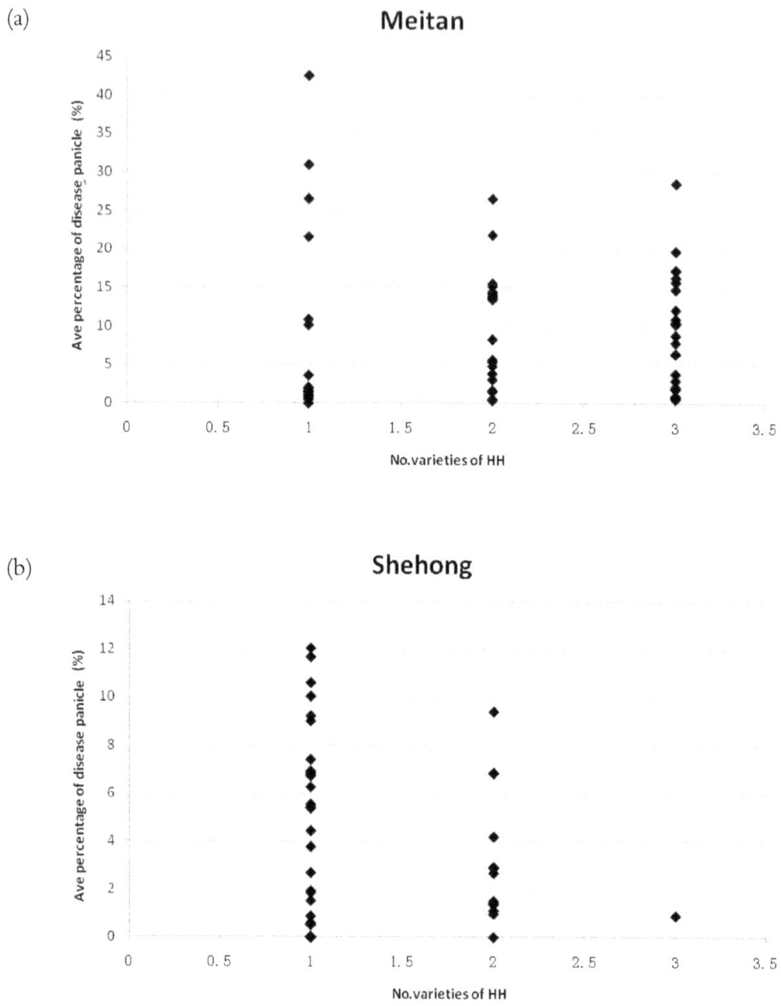

Figure 8.2 The relationship between number of varieties grown by farmers (horizontal axis) and the disease index (vertical axis) for rice blast. (a) Meitan; (b) Shehong; (c) Yuanyuang; (d) Menghai; (e) Shilin.

Reproduced with permission from: Qu *et al.* (2012).

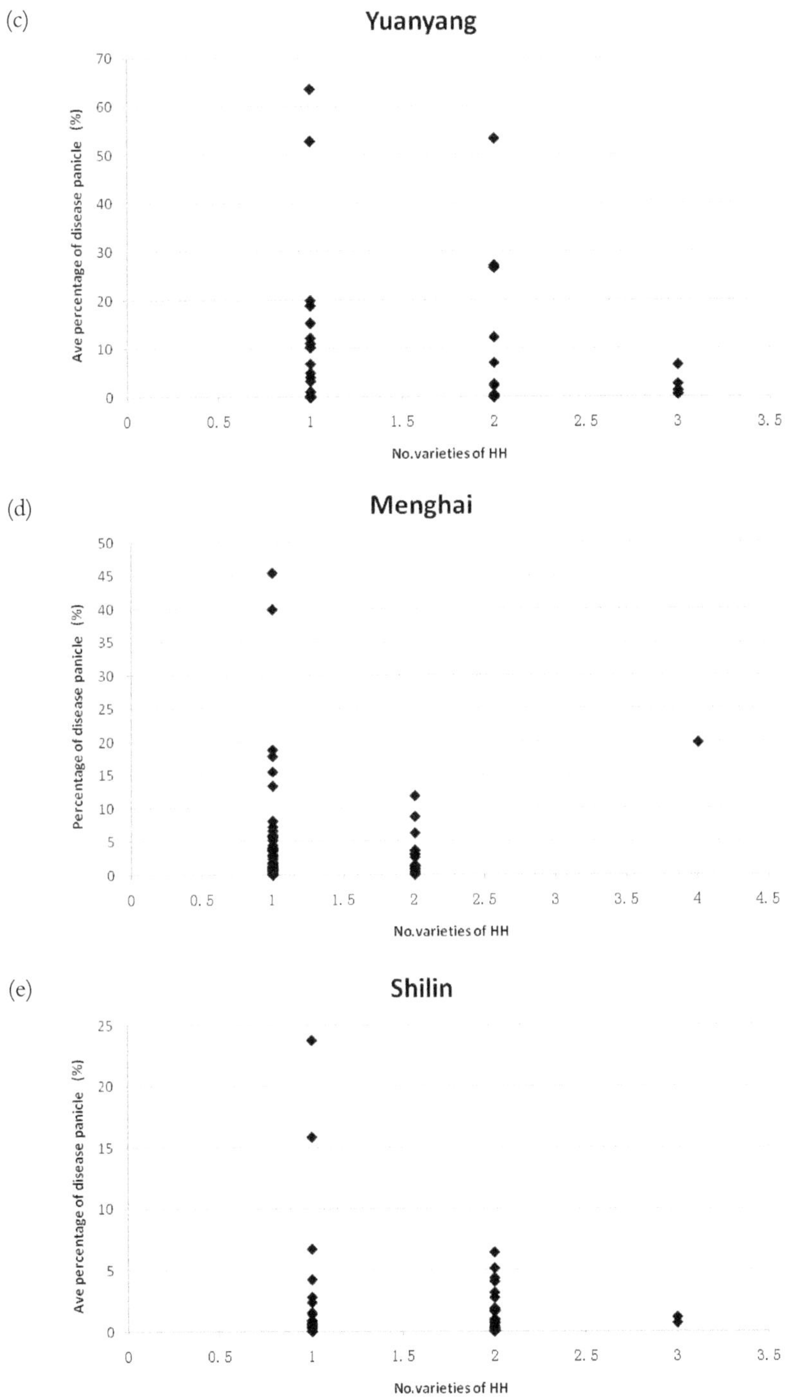

Figure 8.2 (continued)

FGDs not only provided information that was used as the basis of a concise household survey, but more importantly allowed farmers from different villages to see and exchange new crop planting materials and local information. Table 8.2 shows the number of variety names per crop, the key diseases and the number and names of local and modern varieties that farmers believed could be used to manage these diseases from the FGDs. These discussions led to identification of constraints farmers had in accessing or using diverse planting materials.

The FGDs were followed by household surveys of 60 households randomly selected at each site. The household survey was designed to complete the data collected from the FGDs and to link crop varietal diversity on farm to observations of damage by target pests and diseases in the farmers' fields. Five rice sites, two maize sites and two barley sites, making up a total of 540 households, were sampled and field observations carried out for damage by the targeted pests and diseases for each variety grown in each household. Information on the area planted of each variety by the farmers (based on the information collected through the FGDs) was collected for both modern and traditional varieties for each household. This was done through the interviewer asking the farmer to draw a map of his or her farmland and to mark, or have the interviewer mark, where each of the varieties of the target crop (rice, maize or barley) was planted, the area of each variety and the total area planted of each crop. Such a map enabled both farmer and researcher to visualize together the different varieties of the crop planted. In order to ensure a full picture of household knowledge, in half of the households per site women farmers were surveyed and in the other half men. The interviewer then asked the farmer to draw a farm map showing the boundaries and area of his/her land and mark this according to how he/she divides the farm into plots (writing the plot name or label if applicable).

On-farm observations for the targeted pests and diseases were then carried out for a minimum of 30 plants for each variety the farmer grew. Thus for each rice variety 30 observations of rice blast and rice plant hopper were taken; for each maize variety 30 observations of northern and southern leaf blight and maize stem borer were taken; and for each barley variety observations were made for powdery mildew and net blotch infection. This information enabled the researcher to compare the on-farm crop varietal diversity, in terms of the richness of varieties (number of varieties) and the frequency of evenness of each variety across the farmers' land, with the actual damage to the farmers' crop.

An example comparing rice diversity at household level with percentage of diseased plants to rice blast is shown in Figure 8.2.

Genetic diversity management choices by farmers to control pests and diseases

Genetic diversity management choices are defined practices that affect the evolution of crop populations with respect to pest and disease management. During the FGDs, farmers' management choices were classified into those choices that could affect the population structurc for the next generation (genetic management choices) and other management choices such as crop rotation, the use of pesticides, or traditional products for seed storage. Farmers and researchers together determined that genetic diversity management choices could be classified into (i) spatial/temporal arrangements and (ii) selection of planting materials, whether in the field, plot, single plant or post-harvest.

The farmers also follow some criteria for seed selection (Box 8.2). Seed selection after harvest was widely adopted by the farmers, especially in Menghai and Songming. More than 66 per cent of farmers used population selection and more than 50 per cent of farmers used plant selection (see Table 8.3).

Increasing access to diverse intraspecific crop planting materials to manage pests and diseases

A range of practices introduced by the project have been used to increase access to diverse planting materials and information on these materials for farmers. This has included agronomic trials and field demonstrations that compared the performance of local and commercially bred planting materials; diversity fairs; public awareness materials; and the setting up of three community seedbanks.

Table 8.3 Percentage of farmers who make selection choices that influence the population of seed to be planted the next year based on 60 households per site

Site	*Crop*	*Percentage of farmers based on 60 households per site*			
		No Selection	*Population selection: Selection of a particular population of plants to harvest seeds or planting materials for next season*	*Plant selection: selection of particular plants in the field or marking plants for seeds for next year season*	*Post-harvest seed selection*
Menghai, Yunnan	Rice	14	66	50	100
Meitan, Guizhou	Rice	31	71	80	78
Yuanyang, Yunnan	Rice	19	68	73	73
Zhaojue, Sichuan	Maize	20	67	91	95
Shangri-La, Yunnan	Barley	17	82	60	90
Chongmin, Yunnan	Barley	21	68	83	100

Note: Some data are not available on farm for 2008.
Source: Summarized by the author based on the household survey in 2008.

Box 8.2 Selection of planting materials for the following season

1. Seeds selected from particular plots of the population on the farm
    - 1.1 Select a population from fertile field
    - 1.2 Area of the plot/home garden set aside specifically for mother plants/plants for seeds
    - 1.3 Area planted specifically for harvesting plants
    - 1.4 Select healthy population with low occurrence of pest/disease by field
2. Selection of plants or parts of plants
    - 2.1 Spike selection
    - 2.2 Plant selection
    - 2.3 Mark the best plants by putting sticks around them
    - 2.4 Mark the best plants by putting scarecrow close to them
    - 2.5 Select seeds from healthy plants only
    - 2.6 Take out off types within crop
    - 2.7 Remove diseased plants
    - 2.8 Avoid selecting water suckers
3. Post-harvest methods for selecting seeds
    - 3.1 Treat seeds with pesticides
    - 3.2 Specific practices for seed preparation for storage
    - 3.3 Specific seed containers such as bags and wooden box
    - 3.4 Type of preparation of storage facilities
    - 3.5 Specific storage location
4. Criteria for choosing high-quality seed
    - 4.1 Select seed with normal colour (includes: uniform colour, standard colour for the crop, no discolouration)
    - 4.2 Select big seeds (includes: large panicle, big ears, large grain, clean corm)
    - 4.3 Choice of uniform and true to type seed
    - 4.4 Planting only healthy seed (includes: no tunnels for sword maiden (a pest), no holes, no spots, clean budding seed)

Field demonstration trials for rice and maize were held in three project sites in Yunnan and Sichuan in 2013. 44 local rice varieties and 12 local maize varieties were grown in the demonstration plots in these sites. Data on agronomic traits, yield and disease index were collected during the trial, and some varieties with good performance were regenerated to be planted the following year. Some promising materials were selected and recommended to the local farmers based on the performance in the trials. Participants were quite interested in comparing the characteristics of different varieties, and had a lively discussion in the maize field. After the visit, corn porridge of four landraces and one modern was cooked for people to taste. Onc tasty local landrace named Suoluoyiqu was identified by all participants.

Field demonstration trials for Menghai, Yunnan Province included four maize landraces and 35 rice landraces together with three modern maize varieties and one rice variety. The seeds for the trials were collected mainly from the local community, the provincial genebank in Yuanyang county (nine rice landraces), and from the local seed company (some popular modern varieties). The total area planted was 864 m^2 for rice and 168 m^2 for maize. In Zhaojue, Sichuan Province, five maize varieties (one modern and four traditional) were planted in a demonstration plot located in a farmer's field. These trials were used to assess the performance of different varieties for mixture trials in the future and to regenerate seeds for the community genebank. At the same time, local biodiversity fairs were conducted in 11 project sites to increase the awareness of local farmers about using intraspecific crop diversity to increase productivity and reduce crop loss. Related handbooks and year calendars with an introduction to diversified crop varieties have been distributed to local farmers, technicians, government officials, researchers and students as a source of information.

There are two phases for the preparation of a crop biodiversity fair. First is the preparatory phase. A pre-meeting was organized with village committees so that they could have a clear picture of the planning logistics of a biodiversity fair, give some good suggestions for organization of the fair, and introduce this activity to local farmers. The local media were invited to these activities. After the preparatory phase, crop biodiversity fairs were held in each of the project sites. The groups of villagers brought crops to display on tables provided, and the activity and scoring rules were introduced by the hosting organization. Then each group displayed the diversity of their crops and introduced their diversified varieties to the audience. An evaluation team scored each group. The evaluation team included representatives of farmers, scientists, local extension workers and local officials. Local farmers were encouraged to participate in asking questions to the groups as part of the input to the prize-evaluation phase. Traditional ethnic cultural dancing and singing events were performed by local farmers. Three top scoring farmers' groups were awarded with a cash prize (Plates 5 and 6).

The diversity fairs and demonstration fields attracted farmers' interest in traditional crop varieties, particularly seeing those rare varieties that had once been popular in this area. Participants were quite interested in comparing the characteristics of different varieties and had a lively discussion in the field. Some good performance varieties were multiplied by the provincial project partners as well as the local research station for planting the following year. Through such activities, the enthusiasm and participation of local people in managing local crop genetic diversity have greatly increased.

Based on the activities above, community seedbanks supported by the project were established in Menghai and Yuanyang, Yunnan, and in Zhaojue, Sichuan with the objective of collecting and safeguarding locally available varieties for conservation and use.

Until now, a total of 199 samples of crop varieties have been collected and conserved in the community seedbanks, including 171 samples of rice, 20 of maize and 8 of other crops (coix, broomcorn, sunflower, white gourd, etc.). Inventories of the collections were maintained. The community seedbanks will distribute the seed that farmers are interested in planting, and farmers who get the seed for free will commit to returning 1 to 2 kg of seeds back to the seedbank after the harvest. All local farmers were encouraged to take part in this activity to increase the number and quantity of varieties. Some traditional varieties have been described by farmers as being cold-tolerant, disease-resistant, or having high nutrient characteristics, which are good candidates for future research.

The lessons learnt from the efforts made for establishing the community seedbanks are: (1) most of farmers have no idea about the names of the local varieties they grow were passed down from their parents, and therefore more training is needed on the identification of local varieties; (2) the variety field demonstration is needed for farmers to better understand the characteristics of local landraces; (3) the main reason for continuing to grow a variety is the good taste of the variety rather than the need to protect the diversity of the crop germplasm resources; and (4) many of the crop varieties were grown within a limited area instead of a large-scale field plantation, which made it difficult to estimate the yield of those varieties. It is strongly recommended that the maintenance of community seedbanks should be included within local policies to encourage the farmers to protect the varieties on farm.

The role of public awareness in changing norms in China

During the past eight years since the study, several materials for raising public awareness have been developed and have played an important role in changing farmers' attitudes towards safeguarding and using the diversified varieties for pest and disease control. These include: (1) a technical booklet on Rice Germplasm Resources from Xiding township in Yunnan developed by Yunnan Academy of Agricultural Sciences; (2) an agrobiodiversity calendar developed by the Sichuan Academy of Agricultural Sciences; (3) several agriculture and biodiversity posters developed for raising public awareness of local farmers by the Sichuan Academy of Agricultural Sciences (SAAS) and the Yunnan Academy of Agricultural Sciences (YAAS); (4) a detailed map of planting plots in Zhaojue, Sichuan developed by the SAAS to give farmers a clear picture of the distribution of their planting plot and area; and (5) several training sessions held in our project sites with a focus on pest and disease diagnosis, farming management for reducing loss caused by pest and disease, and on increasing the yield of target crops and protecting the environment.

Concluding points

During the last eight years the participatory diagnostic approach used in the southwest of China has revealed that significant intraspecific diversity of major crops continues to exist and be managed by local farmers. Assessment, participatory field trials and demonstration plots have significantly increased interest in, access to and use of different local planting materials for the management of anthropod pests and diseases. Crop varietal mixtures, evaluation of pest and disease resistance across varieties, public awareness materials, community seedbanks, cross-site visits and demonstration fields have increased the awareness and knowledge of local farmers, agricultural extension workers and government officials about the value of their diversified varieties for managing these pests. These principles and methods for using intraspecific crop diversity to manage pests and diseases for annual plants have high potential to be applied also in the context of managing fruit tree diversity to combat pests and diseases.

References

Bourke, A. (1993) *The Visitation of God? The Potato and the Great Irish Famine*, Lilliput Press, Dublin, p. 230

Brown, A.H.D. (1999) 'The genetic structure of crop landraces and the challenge to conserve them in situ on farm', in S. Brush (ed.), *Genes in the Field: On Farm Conservation of Crop Diversity*, IDRC Books, Ottawa, Canada, pp. 29–48

Didelot, F., Brun, L., and Parisi, L. (2007) 'Effects of cultivar mixtures on scab control in apple orchards', *Plant Pathology*, vol 56, pp. 1014–1022

FAO (1998) *The State of the World's Plant Genetic Resources for Food and Agriculture*, FAO (Food and Agriculture Organization of the United Nations), Rome, Italy

Finckh, M. and Wolfe, M.S. (1997) 'The use of biodiversity to restrict plant disease and some consequences for farmers and society', in L.E. Jackson (ed.), *Ecology in Agriculture*, Academic Press, San Diego, CA, pp. 203–237

Guo, Y., Li, Y., Huang, Y., Jarvis, D.I., Sato, K., Kato, K., Suyuzaki, H., Chen, L.J., and Long, C.L. (2012) 'Genetic diversity analysis of hulless barley from Shangri-la region revealed by SSR and AFLP markers', *Genetic Resources and Crop Evolution*, vol 59, pp. 1543–1552

Huang, J., Qiao, F., Zhang, L., and Rozelle, S. (2000) 'Farm pesticide, rice production, and human health', *EEPSEA Research Report 2001-RR3*, IDRC, Singapore

Jarvis, D.I. and Campilan, D. (2006) *Crop Genetic Diversity to Reduce Pests and Diseases On-Farm: Participatory Diagnosis Guidelines. Version I*, Bioversity International technical bulletin no. 12, Bioversity International, Rome, Italy

Leung., H., Zhu, Y., Revilla-Molina, I., Fan, J.X., Chen, H., Pangga, I., Vera Cruz, C., and Mew, T.W. (2003) 'Using genetic diversity to achieve sustainable rice disease management', *Plant Disease*, vol 87, pp. 1156–1169

Marshall, D.R. (1977) 'The advantages and hazards of genetic homogeneity', *Annals of the New York Academy of Sciences,* Special Issue: The genetic basis of epidemics in agriculture (ed. P. Day), vol 287, pp. 1–20

Parisi, L., Gros, C., Combe, F., Parveaud, C.E., Gomez, C., and Brun, L. (2013) 'Impact of a cultivar mixture on scab, powdery mildew and rosy aphid in an organic apple orchard', *Crop Protection*, vol 43, pp. 207–212

Qu, W., Cao, F., Lu, C., and Wang, Y. (2012) 'Relating crop damage levels on-farm to crop varietal diversity measured by richness, evenness and diversity for rice in China', in D.I. Jarvis, C. Fadda, P. De Santis, and J. Thomspon (eds), *Damage, Diversity and Genetic Vulnerability: The Role of Crop Genetic Diversity in the Agricultural Production System to Reduce Pest and Disease Damage*, Proceedings of an International Symposium 15–17 February 2011, Rabat, Morocco; Bioversity International, Rome, Italy, pp. 160–162

Singh, R.P., Huerta-Espino, J., and William, M. (2006) 'Breeding for resistance to biotic stresses', in K.R. Lamkey and M. Lee (eds), *Plant Breeding: The Arnel R. Hallauer International Symposium*, Blackwell Publishing, Ames, IA, pp. 310–322

Strange, R.N. and Scott, P.R. (2005) 'Plant disease: A threat to global food security', *Phytopathology*, vol 43, pp. 83–116

Wang, X., Jin, D., Dai, F., and Liang, K. (2000) 'The diversity of resistance to disease and pest in crop germplasm and the sustainable developing on agriculture', *Journal of Agricultural Science and Technology (China)*, no 5, pp. 67–70

You, M., Liu, Y., and Hou, Y. (2004) 'Biodiversity and integrated pest management in agro ecosystems', *Acta Ecologica Sinica*, vol 24, no 1, pp. 117–122

Yu, D., Cao, Y., Yang, X., Yang, L., and Wang, S. (2007) 'Sustainable management of rice pest and adjustment of agricultural structure', *Journal of Agricultural Science and Technology (China)*, vol 9, no 5, pp. 68–72

Notes

1 The global project on using biodiversity to control pest and disease in support of sustainable agriculture was funded first by the United Nations Environment Programme and the Global Environment Facility along with the Swiss Development Cooperation and the UN Food and Agriculture Organization, then later by the International Fund for Agricultural Development titled 'Improving productivity and resilience for the rural poor through enhanced use of crop varietal diversity in IPPM'. It ran in four countries (Uganda, Morocco, China and Ecuador) from 2007 to 2014.

9 Kandyan home gardens

A time-tested good practice from Sri Lanka for conserving tropical fruit tree diversity

D.K.N.G. Pushpakumara, H.M.S. Heenkenda, B. Marambe, R.H.G. Ranil, B.V.R. Punyawardena, J. Weerahewa, G.L.L.P Silva, Danny Hunter and J. Rizvi

Introduction

In Sri Lanka, home gardens (HGs) have been identified as an integral part of the landscape and culture for centuries and remain today one of the major and oldest forms of land use in the country (Mahawansa, undated; De Silva, 1981; Jacob and Alles, 1987; FSMP, 1995; MFE, 1999; Pushpakumara *et al.*, 2010). Although the term Kandyan home garden (KHG), as a subset of HGs in Sri Lanka, is commonly used in literature, the term has several definitions (see Jacob and Alles, 1987; Perera and Rajapakshe, 1991). In this study KHG is defined, based on the historical Kandyan Kingdom, to include HGs in Kandy and adjacent districts, such as Badulla, Kegalle, Kurunegala, Matale, Nuwara Eliya and Rathnapura. This area largely falls in the wet zone of Sri Lanka but occasionally in the intermediate zone, where the climate and edaphic environment support luxurious growth of perennial trees. The area consists of deep soil (i.e. reddish brown latasolic, immature brown loam and red yellow podzolic soils). The rainfall is year-round, sufficient to meet the evaporation demand of the atmosphere, with a distinct dry spell of one to two weeks that triggers the flowering of perennial species (personal communication, B.V.R. Punyawardena, Department of Agriculture, Sri Lanka).

KHGs are considered a result of farmers' conception, investment and long-term planning. Through generations, KHGs in Sri Lanka have evolved to satisfy households' food and other needs while countering the resource constraints resulting from population pressure and shortage of arable lands and capital. The composition and structure of the plant and animal species found in KHGs are a result of a combination of farmers' selection, natural evolution, environmental suitability and occasional recommendations by researchers and extension workers and subsequent co-adaptation. They form a complex to suit context and environment. Thus, KHGs blend characteristics to suit the socio-economic, cultural and ecological needs of the area's diverse communities and landscapes. About 70 per cent of the households in Kandy and the adjacent districts have long-standing KHGs.

KHGs are managed through family labour as smallholdings with an average land area of 0.4 ha (range from 0.05 to 2.5 ha; Pushpakumara *et al.*, 2012). Despite their small average size, they are characterized by dense, multi-storeyed arrangements with a combination of mixed but compatible species. HGs and KHGs are tree-based systems where many of the trees (40%–50%) are tropical fruit trees (Ariyadasa, 2002; Heenkenda, 2014). As a result, HGs in Sri Lanka are a major contributor to fruit production in the country. Different canopy and root configurations and different requirements for light, nutrients, water and space maximize the resource use in the system. The layered structure and the composition of the KHGs are dynamic and change according to uses and cropping seasons, while largely maintaining their overall structure and functions. The dynamic nature of the system is illustrated by the age classes of tree species, which include seedlings, saplings and mature trees in production. Annuals are cultivated based on the season and land suitability throughout the year (McConnell and Dharmapala, 1973; Jacob and Alles, 1987; Wickramasinghe, 1995; Pushpakumara *et al.*, 2010; 2012).

A good practice for maintaining diversity (GPD) has been defined as a practice in a system, organization or process that over time and space maintains, enhances and creates crop genetic diversity and ensures its availability to and from farmers and other actors for improved livelihoods on a sustainable basis (Sthapit *et al.*, 2004). KHGs at a landscape level represent a land use system that over time and space maintains and, in some instances, enhances and creates crop genetic diversity (Wickramasinghe, 1995; Pushpakumara *et al.*, 2012) and so represents a GPD. As a result of the good practice, the system provides a wide range of products year-round (Figure 9.1). The combination of trees, crops and livestock with different production cycles and rhythms provides a relatively uninterrupted supply of food products, which helps to increase the self-reliance of households. In some instances, KHGs are used to develop new business ventures as a means of value addition to either the home gardens themselves or their products (Pushpakumara *et al.*, 2010). KHGs also provide many ecosystem services: provisioning (Jacob and Alles, 1987; Perera and Rajapakse, 1991; Mohri *et al.*, 2013), regulating (Krishnarajah and Sumanarathne, 1988; Raheem *et al.*, 2008; MENR, 2009; Dela, 2011; Kudavidanage *et al.*, 2012; Pushapakumara *et al.*, 2012; Mattsson *et al.*, 2013), cultural services and support services (Siddique *et al.*, 2007). They also reduce pressure on fragmented natural forests by connecting them with a biodiversity-friendly land use system. Hence, KHGs are crucially important in Sri Lanka, and in particular in the Kandy and adjacent districts, as they provide products and services and an attractive living environment for household members (Wickramasinghe, 1995; Pushpakumara *et al.*, 2010; 2012).

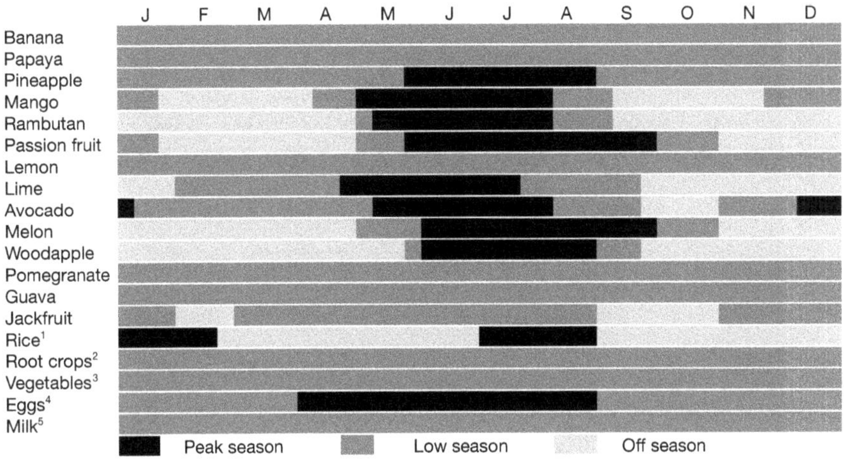

Figure 9.1 Availability and seasonality of fruit, rice, root and vegetable crop production and egg and milk production in Sri Lanka.

Sources: 1–3 derived from available information; 4 and 5 from Silva, G.L.L.P (an author of this chapter); all others extracted and prepared from SCS (2011).
Note: Fruit is widely grown in home gardens. Root crops and vegetables are also substantially grown in home gardens. Rice is mainly grown in other agricultural lands.

The historical introduction of a large number of well adapted, economically important exotic species and the country's agro-ecological diversity, biogeography, geographical location and cultural diversity, coupled with antiquity and dual agriculture (small *vs* large and subsistence *vs* commercial), and the long existence of a unique hydraulic civilization (one that has an agricultural system that is dependent on large-scale government waterworks or irrigation systems) have played a role in the evolution of today's agrobiodiversity in Sri Lanka (Pushpakumara and Silva, 2008). Despite the rich diversity including tropical fruit trees and thriving systems such as KHGs, concern has been growing over the last two decades about the loss of genetic diversity of field crops, fruit and vegetables, livestock and poultry, in the agricultural landscape (MFE, 1999; Pushpakumara and Silva, 2008). In Sri Lanka, HGs constitute the most significant production system for fruit (Heenkenda, 2014). Despite their thriving nature as a GPD, some KHGs have been subjected to fragmentation due to population pressure and recent replacement of low-yielding genetic resources of tropical fruit and vegetable crops by high-yielding varieties of such crops. This leads to loss of genetic resources (MFE, 1999; Dela, 2011; Pushpakumara *et al.*, 2012), although the extent has not yet been properly assessed (Pushpakumara and Silva, 2008).

Ex situ conservation of traditional varieties, landraces and underutilized fruit crops is limited in scope in Sri Lanka. So, it is generally agreed that *in situ* conservation of genetic resources is an indispensable complementary tool (UN,

1992). *In situ* conservation encompasses the maintenance of species in biotic environments that they belong to either as uncultivated plant communities or in farmers' fields (on-farm conservation). On-farm conservation seeks to maintain the process of evolution and adaptation of existing species to their environments and calls for active participation by farmers (Jarvis *et al.*, 1997). Although KHGs are important to maintain the unique agricultural biodiversity of Sri Lanka, few scientific studies on their role in the conservation of agricultural biodiversity have been carried out (Pushpakumara *et al.*, 2012). The study described in this chapter was conducted to investigate the role of KHGs in the conservation and use of fruit crop diversity in Sri Lanka as a GPD.

Materials and methods

The present chapter evaluated the existing scientific data regarding KHGs to identify their role in and contribution to the conservation and use of fruit crop genetic resources in Sri Lanka. The land extent of KHGs was obtained from HG areas of Kandy and adjacent districts of Badulla, Kegalle, Kurunegala, Matale, Nuwara Eliya and Rathnapura. Total land area, population density, forest cover and HG cover were obtained from the Forestry Sector Master Plan (FSMP) (FSMP, 1995) and IUCN and MENR (2007). The HG extent of all selected districts was compared with the national average. The tree cover was calculated by considering forest cover and HG cover of each district. The information on the species diversity of fruit crops in Sri Lanka was obtained from Dassanayake and Fosberg (1980–1991), Dassanayake *et al.* (1994), Dassanayake and Clayton (1995–2000), Ashton *et al.* (1997) and Pushpakumara and Silva (2008). The information on varietal diversity of fruit crop species was extracted from Mankotte (2011) and Heenkenda (2014). The occurrence or presence of fruit crop species and their varieties in KHGs was obtained from the literature related to KHGs, and also by the authors while conducting field surveys in each selected district. Custodian farmers – those farmers who maintain, adapt and disseminate unique fruit crop species and their varieties over time and space, with knowledge needed for their use and cultivation (Sthapit *et al.*, 2013) – were also recorded. Focus group discussions were also held with custodian farmers to identify reasons for use of local varieties of fruit crops.

Results and discussion

Land extent of Kandyan home gardens (KHGs)

The extent of HGs in Sri Lanka was reported as 858,100 ha in 1995, which represents 13.1 per cent of the total land area of the country (FSMP, 1995). The total land extent of the Kandy, Badulla, Kegalle, Kurunegala, Matale, Nuwara Eliya and Rathnapura districts represent 28 per cent total land area of the country; 17 per cent of this area is covered by KHGs. The average population density in the study areas is higher than that of the national average except in the

Kurunegala, Badulla and Matale districts (Table 9.1). Apart from the Matale and Nuwara Eliya districts, the forest cover in all the districts studied is lower than that of the national average, whereas the KHG cover is higher than that of the national average (Table 9.1). In the study area, the total tree canopy cover is similar to the national average tree cover. The data presented in this section indicate that KHGs are a dominant form of land use in these districts.

Diversity of fruit crop species and their production

The species diversity of fruit in Sri Lanka is represented by about 196 species belonging to 46 plant families. This species diversity is composed of 18 per cent endemic, 41 per cent indigenous and 41 per cent exotic fruit species. Of the 196 species, 56 species are considered wild relatives of fruit species (Tables 9.2 and 9.3).The main fruit species grown in Sri Lanka in terms of land extent are banana (*Musa* spp.), pineapple (*Ananas comosus*), papaya (*Carica papaya*), mango (*Mangifera indica*), avocado (*Persea americana*) and rambutan (*Nephelium lappaceum*) (Heenkenda, 2014). Besides the major fruit species, there are a large number of minor, underutilized fruit species (at least 50 species) grown in various parts of the country, which recently have begun to gain popularity at the national scale (Pushpakumara *et al.*, 2007; 2011). The majority of fruit plants in HGs have originated from seedlings (Weerakkody, 2004; Pushpakumara *et al.*, 2007; Heenkenda, 2014). Planting material dissemination of improved fruit plants is from recommended sources such as nurseries by the Department of Agriculture. Many farmers visit research stations and their demonstration sites to collect improved planting material. However, farmer-to-farmer exchange of planting material is also common in KHGs (Plate 7).

Table 9.1 Land area, population density, forest cover and home garden (HG) area of administrative districts where Kandyan home gardens (KHGs) exist more widely in Sri Lanka

District	*Area (km²)*	*Population density (per km²)*	*Forest cover (%)*	*Home garden cover (%)*★	*Tree canopy cover (%)*★★
Badulla	2,803	294	19.0	17.7	36.7
Kandy	1,906	704	17.0	30.4	47.4
Kegalle	1,693	468	9.5	23.2	32.7
Kurunegala	4,813	311	5.0	15.1	20.1
Matale	1,993	233	40.5	11.7	52.2
Nuwara Eliya	1,720	423	24.5	5.3	29.8
Rathnapura	3,255	325	20.0	15.8	35.8
All KHGs	18,183	394	19.0	17.0	36.0
Sri Lanka	65,610	314	23.5	13.1	36.6

Note: ★★ Tree canopy cover is defined to include forest and HG areas.
Sources: IUCN and MENR (2007) and ★ based on FSMP (1995).

Until the early 1990s, only seedlings were used as planting material of many fruit species. Thereafter vegetatively propagated improved material has been introduced and used for selected fruit species. Except for a few fruit species, such as mango and rambutan, usually both improved vegetatively propagated material and seedlings are used as planting material.

Conservation and use of fruit crop species diversity in KHGs

KHGs conserve about 50 per cent of the species diversity of fruit crops in Sri Lanka, of which 6 per cent are endemic, 34 per cent are indigenous and 60 per cent are of exotic origin. This suggests that the majority of fruit species conserved in KHGs are well-adapted species of exotic origin (Tables 9.2 and 9.3). Hence, KHGs, as a good practice for diversity (GPD), provide an important option for on-farm conservation of species diversity of naturalized exotic fruit crops. In addition, KHGs add value to the conservation of endemic and indigenous fruit species in the country by providing the option of a field genebank because they have natural populations in forests. Based on frequency of occurrence, the most common fruit tree species in Sri Lankan HGs are jackfruit (10.437 million trees), mango (5.607 million trees), cashew (3.001 million trees), citrus (2.484 million trees), guava (1.790 million trees), sweet orange (1.468 million trees), rambutan (1.178 million trees) and avocado (0.986 million trees), of which 57 per cent, 48 per cent, 47 per cent, 35 per cent, 43 per cent, 49 per cent, 46 per cent and 90 per cent of their populations, respectively, are recorded in the KHGs (Ariyadasa, 2002).

On average, one KHG conserves 10 to 20 fruit crop species. Fruit crops are perennials and usually require more space than herbaceous plants. Only one or two individual trees per species may be sufficient to provide fruit for family consumption. Therefore, their representation may be limited to one or two individuals of each species in each KHG. Although individual populations of species in KHGs may be small, at a landscape level they are a vital refuge for species that are neither grown in the wider agro-ecosystems nor found in the wild. The average density of trees in the KHGs of Sri Lanka is around 200 trees/ha (with a range from 20 to 475 trees/ha) of which 40–50 per cent are fruit trees (Ariyadasa, 2002).

Fruit species represent all layers of KHGs. In the understorey layer below 3 m, pineapple is the most common species, whereas in the lower stratum (3–10 m) banana, cacao, passion fruit, lime and lemon are common. In the middle stratum (10–15 m), papaya, avocado, mangosteen, bread fruit and some citrus species commonly occur. In the upper middle stratum, over 15 m, jackfruit, mango, durian, wild bread fruit and bread fruit are the dominant fruit crop species. Perera and Rajapakse (1991) reported that out of 39 fruit species recorded in KHGs, 32 species were lesser known. Hitinayake and Ekanayake (1999) also reported that out of 39 fruit species recorded in KHGs, 20 species are underutilized. However, in neither case were attempts made to intensify the garden's productivity by replacing these with improved or

Table 9.2 Species diversity of fruit crops in Sri Lanka and their endemic, indigenous and exotic status and occurrence in Kandyan home gardens (KHGs)

Category	*In Sri Lanka*		*In KHGs*		
	Number	*%* [a]	*Number*	*%* [b]	*%* [c]
Fruit species recorded	196		98	50	–
Plant families of fruit species	46		36	78	–
Wild relatives of fruits	54	29	11	20	–
Endemic fruit species	35	18	6	17	6
Indigenous fruit species	81	41	33	41	34
Exotic fruit species	80	41	59	74	60

Note: [a] = percentage out of total fruit species; [b] = percentage of species recorded in KHGs out of total fruit species in each category; [c] = percentage out of total fruit species recorded in KHGs.

vegetatively propagated more productive plants. In terms of *ex situ* conservation, out of 12,333 accessions over 125 plant species at the Plant Genetic Resources Centre of Sri Lanka, only 163 accessions are of fruit species. The fruit germplasm collection at the Horticultural Crop Research and Development Institute consists of 670 samples of 20 fruit species, but details of varieties are not available (Chithral, 2011). These examples clearly indicate the potential for KHGs to act as an already existing good practice for conservation and use of fruit crop species.

Conservation of fruit crop varietal diversity (intraspecific) in KHGs

A field survey with farmers revealed that the majority of known fruit crop varieties are recorded in KHGs (Table 9.4), indicating the conservation of a high level of genetic diversity of fruit crops in KHGs. However, few studies have been carried out on their genetic diversity. An assessment of morphological and genetic diversity of jackfruit using RAPD markers revealed that much of the genetic variation of the species is conserved in KHGs (Pushpakumara and Harris, 2007). This is mainly due to the seedling origin of most of the fruit crops in KHGs, which are highly heterogeneous as a result of the outcrossing nature of the species (Pushpakumara *et al.*, 1997). Confirming this, a wide range of variation has also been observed for the jackfruit population in the Kandy district in terms of fruiting season, fruit shape, number of fruit per tree, fruit weight, flesh thickness and hardness, flesh texture, aroma, colour and juiciness, and latex quantity. Although there has been no comprehensive analysis of genetic diversity of mango in Sri Lanka, the country has a large number of mango morphotypes. The distribution of mango morphotypes such as *gira amba* and *mee amba* in the KHGs of the Matale and Kandy districts, respectively (personal communication, Mr Leel Randeniya, Ministry of Environment and Renewable Energy, Sri Lanka), suggests that the bulk of the genetic diversity of many perennial fruit crop species is conserved through

Table 9.3 Species diversity of fruit in Sri Lanka and their occurrence in Kandyan home gardens

Family	*Botanical name*	*Common names*
Anacardiaceae (9, 5, 3/1, 2/1, 4/3)	***Anacardium occidentale* L.**	Cashew nut
	***Buchanania axillaris* (Desr.) Ramamoorthy+**	Kiri palu
	***Mangifera indica* L.**	Mango
	Mangifera pseudoindica Kosterm.*•	–
	***Mangifera zeylanica* (Blume) Hook.f.*•**	Atemba, Wal amba
	Semecarpus subpeltata Thw.*	Maha badulla
	***Spondias dulcis* Sol. ex Parkinson**	Amberella
	Spondias mombin L.	Hog plum tree
	Spondias pinnta (L.f.) Kurz+•	Hog plum, Ambarella
Annonaceae (8, 4, 1/0, 1/0, 6/4)	***Annona cherimola* Miller**	Cherimoyer
	Annona glabra L.	Wel artha
	***Annona muricata* L.**	Soursop
	***Annona reticulata* L.**	Bullock's heart, Weli artha
	***Annona squamosa* L.**	Custard apple, Seeni artha
	Enicosanthum acuminata (Thw.) Airy Shaw*	Mal laulu
	Polyalthia korinti (Dunal) Thw.+•	Ul kenda
	Uvaria grandiflora Roxb.	–
Apocynaceae (4, 3, 0/0, 2/2, 2/1)	***Carissa carandas* L.+•**	Maha karamba
	***Carissa grandiflora* A. DC.**	Damson
	Carissa inermis Vahl•	–
	***Carissa spinarum* L.+•**	Heen karamba
Arecaceae (8, 2, 1/1, 3/1, 4/0)	*Borassus flabellifer* L.	Palmyrah
	Calamus rotang L.+	Heen wewel
	Corypha umbraculifera L.	Talipot plant
	***Loxococcus rupicola* (Thw.) H. Wendl. & Drude*•**	Dothalu
	Nypa fruticans Wurmb+	Water coconut
	Phoenix dactylifera L.	Date palm
	***Phoenix pusilla* Gaertn.+•**	Indi
	Phoenix sylvestris (L.) Roxb.	Indi
Berberidaceae (3, 1, 1/1, 2/0, 0/0)	***Berberis ceylanica* Schneider***	Barberry
	Berberis tinctoria Leschen.+•	Barberry
	Berberis wightiana Schneider+•	Barberry
Bombacaceae (4, 2, 2/0, 0/0, 2/2)	***Adansonia digitata* L.**	Baobab
	Cullenia ceylanica (Gardner) K. Schum.*•	Kataboda
	Cullenia rosayroana Kosterm.*•	Kataboda
	***Durio zibethinus* Murr.**	Durian
Boraginaceae (2, 0, 0/0, 1/0, 1/0)	*Cordia dichotama* Forst. f.	Lolu
	Cordia myxa L.	Lolu
Bromeliaceae (1, 1, 0/0, 0/0, 1/1)	***Ananas comosus* (L.) Murr.**	Pineapple
Burseraceae (2, 0, 1/0, 1/0, 0/0)	*Canarium zeylanicum* (Retz.) Blume*	Kekuna
	Scutinanthe brunnea Thw.	Maha bulumora

Cactaceae (1, 1, 0/0, 0/0, 1/1)	***Hylocereus undatus* (Haw.) Britt & Ross.**	Dragon fruit
Caricaceae (1, 1, 0/0, 0/0, 1/1)	***Carica papaya* L.**	Papaya
Clusiaceae (7, 5, 4/2, 1/1, 2/2)	*Calophyllum calaba* L.*	Heen keena
	Garcinia quaesita* Pierre.	Goraka
	***Garcinia zeylanica* Roxb.*•**	Ela goraka
	***Garcinia mangostana* L.**	Mangosteen
	***Garcinia morella* (Gaertn.) Desr.+•**	Gamboge
	Garcinia terpnophylla (Thw.) Thw.*•	–
	***Garcinia xanthochymus* Hook. f.**	Rata goraka
Combretaceae (1, 1, 0/0, 0/0, 1/1)	*Terminalia catappa* **L.**	Kotamba
Cucurbitaceae (1, 1, 0/0, 0/0, 1/1)	***Citrullus lantanus* (Thunb.) Matsum & Nakai**	Melon
Elaeagnaceae (1, 0, 0/0, 1/0, 0/0)	*Elaegnus latifolia* L.+•	Katu embilla
Elaeocarpaceae (4, 1, 3/0, 1/1, 0/0)	*Elaeocarpus amoenus* Thw.*•	Titta weralu
	Elaeocarpus glandulifer (Hook.) Masters*•	Gal weralu
	***Elaeocarpus serratus* L.**	Indian olive, Weralu
	Elaeocarpus subvillosus Arn.*•	Gal weralu
Erythroxylaceae (1, 1, 0/0, 1/1, 0/0)	***Erythroxylum moonii* Hochur.**	Bata kirilla
Euphorbiaceae (13, 5, 3/0, 8/3, 2/2)	*Antidesma alexiteria* L.+•	Heen embilla
	***Antidesma bunius* (L.) Spreng.+**	Karawala kebella
	Antidesma ghaesembilla Gaertn.+•	Bu embilla
	Antidesma thwaitesianum Muell. Arg.+•	Karawala kebella
	Aporusa cardiosperma (Gaertn.) Merr.+	Kampottu
	Aporusa lanceolata (Tul.) Thw.*	Heen kebella
	Aporusa lindleyana (Wight) Baill.+	Kebella
	***Baccaurea motleyana* Mull.-Arg.**	Gaduguda
	Drypetes gardneri (Thw.) Pax & Hoffm.•	Eta wira
	***Drypetes sepiaria* (Wight & Arn.) Pax & Hoffm.**	Wira
	***Phyllanthus acidus* (L.) Skeels**	Rata nelli
	***Phyllanthus emblica* L.+**	Amla, Nelli
	Ptychopyxis thwaitesii (Baill.) Croizat*	Wal rambutan
Fabaceae (7, 3, 2/0, 2/1, 3/2)	*Adenanthera bicolor* Moon*	Mas mora
	Castanospermum australe Cunn. & Fraser ex W.J. Hook.	Australian chestnut
	***Cynometra cauliflora* L.**	Nam nam
	Dialium ovoideum Thw.*	Velvet tamarind
	***Pongamia pinnata* (L.) Pierre+**	Gal karanda
	Tadehagi triquetrum (L.) Ohashi+	Baloliya
	***Tamarindus indica* L.**	Tamarind

Table 9.3 continued

Family	*Botanical name*	*Common names*
Flacourtiaceae (6, 2, 3/0, 1/1, 2/1)	*Dovyalis hebecarpa* (Gardner) Warb.*	Ceylon gooseberry
	***Flacourtia indica* (Burm. f.) Merr.+**	Uguressa
	***Flacourtia inermis* Roxb.**	Lovi
	Flacourtia jangomas (Lour.) Rausch.	Rata uguressa
	Hydnocarpus octandra Thw.*•	Wal divul
	Trichadenia zeylanica Thw.*	Tolol
Hippocrateaceae (2, 0, 0/0, 2/0, 0/0)	*Salacia chinensis* L.+	Heen himbutuwel
	Salacia oblonga Wall. Ex Wight & Arn.+•	Chundan
Lauraceae (4, 1, 1/0, 2/0, 1/1)	*Crtptocarya membranaceae* Thw.*	Galmora
	Crtptocarya wightiana Thw.+	Galmora
	***Persea americana* Miller.**	Avacardo
	Persea macrantha (Nees) Kosterm.+	Ululu
Melastomataceae (1, 1, 0/0, 1/1, 0/0)	***Melastoma malabathricum* L.**	Bowitiya
Meliaceae (2, 1, 0/0, 1/0, 1/1)	***Sandoricum koetjape* (Brum. f.) Merr.**	Donga
	Walsura trifoliolata (A. Juss.) Harms+	Kirikon
Moraceae (6, 5, 1/1, 2/1, 3/3)	***Artocapus heterophyllus* Lam.**	Jackfruit
	***Artocarpus incisus* L.f.**	Breadfruit
	***Artocarpus nobilis* Thw.*•**	Waldel
	***Ficus racemosa* L.+**	Attikka
	***Morus alba* L.**	Mulberry
	Streblus asper Lour.+	Getanitul
Musaceae (3, 1, 0/0, 3/1, 0/0)	*Musa acuminata* Colla+•	Unel
	Musa bulbicianan Colla+•	Atikesel
	***Musa* spp. L.**	Kesel
Myrsinaceae (4, 1, 1/0, 3/1, 0/0)	***Ardisia elliptica* Thunb.+**	Baludan
	Ardisia solanaceae Roxb.+	Baludan
	Ardisia willisii Mez*	Baludan
	Embelia ribes Burm. f.+	Wel embilla
Myrtaceae (14, 8, 3/0, 6/5, 5/3)	*Cleistocalyx nervosum* (DC.) Kosterm.*	Batadamba
	Eugenia uniflora L.	Brazilian cherry
	***Psidium guajava* L.**	Guava
	***Psidium guineense* Sw.•**	Cheena pera
	Psidium montane Sw.	Embulpera
	***Rhodomyrtus tomentosa* (Ait.) Hassk.+**	Wild guava
	Syzygium aqueum (Burm. f.) Alston+•	Wal jambo
	***Syzygium caryophyllatum* (L.) Alston+**	Heen dan
	Syzygium cordifolium Walp.*•	Wal jambu
	***Syzygium cumini* Skeels+**	Madan, Jamun
	***Syzygium jambos* (L.) Alston+**	Rose apple
	***Syzygium malaccense* (L.) Merr. & Perry+**	Jambu
	***Syzygium samarangense* (Blume) Merr. & Perry**	Pini jambu
	Syzygium umbrosum Thw.*•	Heen damba

Nelumbonaceae (1, 0, 0/0, 1/0, 0/0)	*Nelumbo nucifera* Gaertn.+	Lotus
Oxalidaceae (2, 2, 0/0, 0/0, 2/2)	***Averrhoa carambola* L.**	Star fruit
	***Averrhoa bilimbi* L.**	Biling
Passifloraceae (4, 1, 0/0, 0/0, 4/1)	***Passiflora edulis* Sims**	Passion fruit
	Passiflora laurifolia L.	Water melon
	Passiflora mollissima (HBK) Bailey•	Banana passion fruit
	Passiflora quadrangularis L.	Desi puhul
Punicaceae (1, 1, 0/0, 0/0, 1/1)	***Punica granatum* L.**	Pomegranate
Rhamnaceae (5, 2, 2/0, 3/2, 0/0)	*Ziziphus lucida* Moon ex Thw.*•	Eraminya
	***Ziziphus mauritiana* Lam.+**	Masan
	Ziziphus napeca (L.) Willd.*•	Yak eraminiya
	Ziziphus oenoplia (L.) Miller+•	Heen eraminiya
	***Ziziphus rugosa* Lam.**	Maha eraminiya
Rosaceae (11, 4, 1/0, 3/0, 7/4)	*Duchesnea indica* (Andr.) Focke+	Indian strawberry
	***Eriobotrya japoica* (Thunb.) Lindley**	Japan batu
	***Fragaria vesca* L.**	Strawberry
	***Malus sylvestris* Miller**	Apple
	Prunus cerasoides D. Don+•	Indian cherry
	Prunus persica (L.) Batsch	Peach
	Prunus walkeri (Wight) Kalkman*•	Golumora
	***Pyrus communis* L.**	Pear
	Rubus ellipticus Smith+	False blackberry
	Rubus moluccanus L.	Blackberry
	Rubus rosifolius Smith	Wild raspberry
Rubiaceae (7, 1, 0/0, 6/0, 1/1)	*Anthocephalus chinensis* (L.) A. Rich. Ex Walp.+•	Ela bakmi
	Canthium coromandelicum (Burm. f.) Alston+•	Kara
	Ixora coccinea L.+	Ratambala
	***Ixora macrothyrsa* (Teys. & Binn.) Moore**	Ixora
	Ixora pavetta Andr.+,	Maha rathambala
	Morinda umbellata L.+	Kiri wel
	Nauclea orientails (L.) L.+•	Bakmi
Rutaceae (15, 10, 0/0, 6/1, 9/9)	***Aegle marmelos* (L.) Correa**	Baelfruit
	Atalantia ceylanica (Arn.) Oliver+•	Yakinaran
	Atalantia monophylla (Roxb.) DC.+•	Perukuruntu
	Atalantia rotundifolia (Thw.) Tanaka+•	–
	***Citrus aurantifolia* (Christm. & Panzer) Swingle**	Dehi, Lime
	***Citrus aurantium* L.**	Sour orange
	***Citrus grandis* (L.) Osbeck**	Pummelo
	***Citrus hystrix* DC.**	Gada dehi
	***Citrus limon* (L.) Burm. f.**	Lemon
	***Citrus medica* L.**	Citron
	***Citrus reticulata* Blanco**	Mandarin, Heen naran
	***Citrus sinensis* (L.) Osbeck**	Sweet orange
	Glycosmis pentaphylla (Retz.) A. DC.+•	Dodan pana
	***Limonia acidissima* L.+**	Woodapple
	Naringi crenulata (Roxb.) Nicolson+•	Wal beli

Table 9.3 continued

Family	*Botanical name*	*Common names*
Sabiaceae (1, 0, 0/0, 1/0, 0/0)	*Meliosma pinnata* (Roxb.) Maxim.+•	Wal bilin•
Sapindaceae (6, 4, 2/0, 3/3, 1/1)	*Dimocarpus gardneri* (Thw.) Leenh.★	Nurai
	***Dimocarpus longan* Lour.**+	Mora
	Glenniea unijuga (Thw.) Radlk★•	Wal mora
	***Nephelium lappaceum* L.**	Rambutan
	***Pometia pinnata* J.R. & G. Forst.+•**	Bulu mora
	***Schleichera oleosa* (Lour.) Oken**	Ceylon oak, Kon
Sapotaceae (8, 6, 0/0, 3/2, 5/4)	***Chrysophyllum cainito* L.**	Kos ata laulu
	Chrysophyllum oliviforme L.	Date plum
	***Chrysophyllum roxburghii* G. Don**	Laulu
	***Manilkara hexandra* (Roxb.) Dubard**+•	Palu
	***Manilkara zapota* (L.) P. van Royen**	Sapodilla
	***Mimusops elengi* L.+•**	Munamal
	***Pouteria campechiana* (Kunth) Baehni**	Canistel, Rata laulu
	Xantolis tomentosa (Roxb.) Raf.+	Mul makil
Solanaceae (3, 2, 0/0, 1/1, 2/1)	*Datura stramonium* L.	Thorn apple
	***Physalis micrantha* Link+**	Nalal batu
	***Physalis peruviana* L.**	Cape gooseberry
Sonneratiaceae (1, 0, 0/0, 1/0, 0/0)	*Sonneratia alba* J. Sm.	Kirala
Sterculiaceae (2, 1, 0/0, 1/0, 1/1)	*Sterculia foetida* L.+	Telabu
	***Theobroma cacao* L.**	Cocoa
Tiliaceae (4, 2, 0/0, 3/1, 1/1)	***Grewia damine* Gaertn.+**	Daminiya
	Grewia helicterifolia Wall. Ex G. Don+•	Bora daminiya
	Microcos paniculata L.+•	Keliya
	***Muntingia calabura* L.**	Jam tree
Ulmaceae (1, 0, 0/0, 1/0, 0/0)	*Holoptelea intergrifolia* (Roxb.) Planch.	Goda kirilla
Verbanaceae (3, 2, 0/0, 2/1, 1/1)	***Gmelina arborea* Roxb.+**	Athdemata
	Gmelina asiatica L.+	Asiatic beach berry
	***Lantana camara* L.**	Gandapana
Vitaceae (1, 1, 0/0, 0/0, 1/1)	***Vitis vinifera* L.**	Grape

Note: ★, + and • indicate endemic species, indigenous species and wild relatives of crops, respectively. Botanical names without any symbol indicate exotic species whereas botanical names in bold indicate species observed/recorded in KHGs. Values in parenthesis under each family name (1, 1, 0/0, 0/0, 1/1) represent fruit crop species recorded from the given family in Sri Lanka, number of species recorded in KHGs, endemic fruit species recorded in Sri Lanka/KHGs, indigenous fruit species recorded in Sri Lanka/KHGs, exotic species recorded in Sri Lanka/KHGs, respectively.

Source: Derived from Ashton *et al.* (1997); Dassanayake and Fosberg (1980–1991); Dassanayake *et al.* (1994); Dassanayake and Clayton (1995–2000).

Table 9.4 Varietal diversity of commonly grown fruit recorded in Sri Lanka and their presence in Kandyan home gardens (KHGs)

Species	*Varieties*
Banana	**Alukesel, Amban, Ambul kesel, Anamalu, Binkesel, Cavendish, Kolikuttu, Nethrappalan, Puwalu, Rathambala, Seenikesel, Suwadel, Wathabanga, Local types**
Papaya	**Rathna, Red lady, Local types**
Mango	**Alponso, Ambalavi,** Seedless, **Beti amba, Chembatan, Dampara, Gira amba, Karthakolomban,** Kalu kohu amba, **Kohu amba, Malwana, Mee amba, Petti amba, Piterprasand, Pol amba, Tom EJC, Velleikolomban, Villard,** Walu amba, **Local types**
Avocado	**Booth 7, Furete, Hass, Peradeniya purple, Pollock, Simonds, Tower 2, Local types**
Jackfruit	Arthur V Dies, Dahaata masya, **Fartherlong,** Ganegoda, Horana, Kalpitiya, **Kothmale, Kurukos, Kuruwita, Maharagama**, Mandoor, Pani waraka, **Rosakos/Hirosa**, Thellippalai, **Gannoruwa, Local types**
Guava	**Horana rosi, Horana sweet, Lanka giant, Red giant, Local types**
Rambutan	**Malayan red, Malwana special, Malayan yellow, Local types**
Durian	**Ambathenna, Gannoruwa, Kasun, Local types**

Note: Bold letters indicated varieties observed/occurred in KHGs.
Source: Survey data by authors (2014).

KHGs. Similar information has also been reported by Muthukuda and Wijerathne (2007) for several perennial fruit crops. Hence, as a GPD, KHGs constitute a valuable system for on-farm conservation of genetic diversity and facilitation of their gene flow.

Although the number of individuals of each species is limited, the presence of even a few trees in each KHG may preserve rare alleles related to elite characteristics allowing for present use and future selection and breeding. Uninterrupted maintenance of landraces and farmers' varieties in KHGs has prevented the erosion or extinction of most economically important varieties of fruit and other crop species in Sri Lanka. It was observed that custodian farmers safeguarded specific fruit crop varieties for several reasons (Table 9.5). Similar observations were made in focus group discussions with farmers, where it was revealed that some farmers appreciated certain quality characteristics for local food preparations, for example local jackfruit types (*pani waraka*, a sweet hard-fleshed type of which the immature fruit is used in the preparation of polos curry). Other qualities appreciated were medicinal properties (nelli, bael and pomegranate), premium marketability (hard-fleshed durian type), cultural reasons (use in New Year festivals) and lack of pest and disease problems (local jackfruit, mango and guava types).

Table 9.5 Custodian farmers with local varieties and landraces of fruit crop in Kandyan Home Gardens (KHGs) and reasons for maintaining such diversity

Custodian farmers of KHGs	*Fruit species and their diversity used in KHGs*	*Reason(s) for maintaining diversity*
D. Werake, Kundalagama, Kundasale	Local types of mango, passion fruit, guava and annona	1, 2, 3, 4, 5
K.M. Gunathilaka, Walawwaththa, Hondiyadeniya	Local types of mangosteen, guava, durian, passion fruit, mango, lime, mandarin, jackfruit (waraka), avocado	1, 2, 3, 4
A. Dharmakerthi, Mahaweli Uyana, Kundasale	Local types of mango, jackfruit (waraka), sweet orange, passion fruit, nelli (*Phyllanthus emblica*)	1, 2, 3, 4, 5
B.M. Perera, Teekawatta, Thannekumbura	Local types of rambutan, mango, sweet orange, mandarin, jackfruit (waraka), avocado, pummelo, jambu (*Syzygium malaccense*), uguressa (*Flacourtia indica*)	1, 2, 3, 4, 5
M.M. Kerthirathne, Malgammana, Nugawela	Local types of jackfruit (waraka), uguressa, annona, avocado, mango, banana	1, 2, 3, 4, 5
R.K. Wickramarathne, Yatihalagala, Katugastota	Local types of mango, jackfruit (waraka), avocado, mandarin, jambu, banana	1, 2, 3, 4
M. Heenkenda, Araliyawa, Arangala, Nattranpota	Local types of mango, jackfruit (waraka), avocado, mandarin, jambu, banana	1, 2, 3, 4
G. Wijewardena, Padiwatta, Kundasale	Local types of mango, jackfruit (waraka), avocado, mandarin, jambu, banana	1, 2, 3, 4, 5
G. Hennkenda, Karaththamada Road, Naththaranpotha	Local types of durian, avocado, breadfruit, mango, jackfruit (waraka), banana, annona, mangosteen, jambu, uguressa, cocoa	1, 2, 3, 4
G. Rathnayake, Gadaladeniya Road, Pilimathalawa	Local types of mango, jackfruit, guava, sweet orange, mandarin, banana, jambu, uguressa, avocado	1, 2, 3, 4, 5

Source: Survey data by authors (2014).

Note: 1 = believed that the quality (flavour, flesh character, medicinal value etc.) of local fruit varieties is superior to improved varieties; 2 = under dense canopy of HGs, improved varieties do not perform well but local varieties adapt to conditions; 3 = local varieties need low levels of technical and management input (pruning and training); 4 = in most local varieties fruit ripening is not synchronous and is thus supportive of a longer production period; 5 = less susceptible to pest and disease and to extreme conditions. All the custodian farmers listed are men because they are the owners of the KHGs. In Sri Lanka, while, in theory, women may own land, in practice, land ownership documents must be signed by the 'head of household' and this is usually deemed to be the man (Gomez and Tran, 2012).

Until the early 1960s, farmers were the custodians of the complete range of genetic diversity available in the landraces of the traditional cultivars of the majority of crops and fruits in Sri Lanka (Ganashan *et al.*, 1996; Jayasuriya and Rajapakse, 2004). By the late 1980s and early 1990s, especially with regard to landraces of some vegetatively propagated fruit crops, some landraces and local cultivars had been replaced with introduced or improved cultivars. At the same time, the survey revealed, some varieties of fruit crops had almost been wiped out from the KHGs as the result of: the use of improved varieties of grafted planting material (i.e. local cultivars of durian, rambutan and mango); lack of attention and care (bullock's heart or *weli anoda*, and *koholla laulu*, *laulu*); fragmented HGs due to population pressure (i.e. large trees of durian, mango and jackfruit) and damage by animals such as monkeys (many fruit species). However, still a large number of KHGs at a landscape level help to safeguard local varieties and landraces of many fruit species (Tables 9.3–9.5).

KHGs and sustainable landscape management

The KHG network in Sri Lanka provides a complementary resource base acting as a kind of field genebank for conservation of fruit crop genetic resources. The network also acts as a complementary option to mediate the *in situ–ex situ* gap through on-farm conservation and as a platform for continuation from natural vegetation to monoculture fields. Thus, KHGs can be used to conserve other crops, tree, livestock and poultry species as well as fruit crops in the landscape. The presence of KHGs provides environmental conservation, such as easing pressure on natural forests, increased multi-layered vegetation cover leading to a pleasant living environment, control of erosion and pollution, and fertility replenishment, in addition to food and nutritional security and biodiversity conservation with associated ecosystem services. The systems also bolster cultural identity, as traditional systems based on indigenous knowledge and species are part of the cultural patterns of the communities. Trees, which are the main structural feature of KHGs, have a positive effect on the global carbon balance (Mattsson *et al.*, 2013) and are climate resilient (Marambe *et al.*, 2012; Weerahewa *et al.*, 2012). Thus, it is a land use system that can produce goods and services while protecting and connecting environments in fragmented areas and is useful in the management of landscapes in Sri Lanka.

In order to protect the multi-storeyed and multispecies nature of KHGs under increasing intensification of land use practices, it will be increasingly important to find ways to increase the profitability of KHG systems while maintaining, as much as possible, their biodiversity benefits. This consideration is key to ensuring the sustainability of this system. It will also be important to increase the level of technology used in KHGs. Comparatively low levels of technology and crop management techniques are practised by KHG owners and their attention is also variable. As a result, the yield and quality of KHG fruit are comparatively low (Weerakkody, 2004; Heenkenda, 2014). Introduction of technology packages, for example induction of flowering and proper pruning of fruit trees to enhance productivity through rejuvenation, is essential for the conservation and sustainability of the system.

During the survey, many farmers indicated that they did not receive any incentives for on-farm conservation of genetic resources, although some farmers had received awards, certificates and cash prizes for management of KHGs with high diversity in the context of family food and nutrition security. The importance of HGs for genetic resource conservation is not yet widely recognized outside scientific circles, and little work has been done on custodian farmers in Sri Lanka with respect to reasons for maintaining unique genetic resources of fruit species.

KHG policy environment

Over the years, the number and total area of KHGs and HGs have increased annually, despite little policy support. During the last two decades the number of species and structure of HGs has remained constant because HGs and KHGs are now increasingly recognized as examples of traditionally developed agroforestry systems with excellent promise for facing present and future challenges. Having realized the importance of HGs, the national development policy framework of the government of Sri Lanka now includes strategies to expand and improve food and timber production in these landscapes (NCSD, 2009). In addition, the National Agriculture Policy of 2007 (MADAS, 2007) also highlights the need to promote HGs, especially focusing on the urban sector and the role played by women in HGs. Improvement of HGs in Sri Lanka has been the priority of many development programmes implemented over the past five to seven years, after development of 375,000 HGs was targeted under the "*Api Wawamu Rata Nagamu*" (Let us grow and uplift the nation) programme launched in 2007. The strengthening of 1.5 million HGs is the target of the "*Divi Neguma*" (Livelihood development) programme in order to achieve self-sufficiency in vegetable production leading to reduction in vegetable prices to make them affordable for all (Government of Sri Lanka, 2011).

The Sri Lankan policy framework includes establishing 'fruit villages' based on strengthening HGs in order to improve fruit production, and promoting fruit production activities at village level as a mean of ensuring village empowerment. To achieve this, planting material of identified species (jackfruit, durian, rambutan) and their varieties has been distributed to villages so that they can develop their own mechanisms to promote planting material production and marketing; and collection of fruit has been organized by villagers in the Kandy, Matale and Kurunegala districts. As yet, impacts of the fruit village concept and HG improvement programmes are yet to be identified and researched. Further production enhancement, genetic conservation, dissemination and exchange of old and new germplasm to farmers, fruit crop mating systems and selection, and effects of change of socio-economic status of householders on genetic diversity of fruits are also to be identified. These are priority research areas to understand the impacts on conservation of fruit crops in KHGs.

Conclusion and way forward

Sri Lankan HGs in Kandy and adjacent districts, such as Badulla, Kegalle, Kurunegala, Matale, Nuwara Eliya and Rathnapura, are defined and popularly known as Kandyan home gardens (KHGs) or Kandyan forest gardens. KHGs represent a scattered but important human-made land use system that increases the tree cover of this area of Sri Lanka. Out of 196 fruit species recorded in Sri Lanka, more than half of the species are recorded from 17 per cent of the area of the above districts. The KHGs, therefore, are an important land use system for Sri Lanka in terms of the percentage of land area occupied, conservation of fruit crop species and their genetic diversity, and provision of other environmental services, while helping to generate income and food and nutrition security of households. This study shows that KHGs represent a land use system that, over time and space, in most instances maintains and in some instances enhances and creates crop genetic diversity; hence they can be regarded as a good practice for maintaining diversity (GPD). Year-round production of a wide range of products required by householders, new business ventures through value addition, provision of many ecosystem services and easing pressure on natural forests have been identified as key elements of KHGs. Having said that, relatively little attention has been paid to assessing the ecosystem services and dynamics of KHGs under the influence of rural transformation to commercialization, land degradation and the impact of fragmentation of KHGs on social, cultural and ecological sustainability. Hence, a community-based long-term, multidisciplinary and participatory research programme is needed to understand the dynamics of conservation and use of species and genetic diversity of fruits. On-farm assessment of genetic diversity using temporal quantitative data against the changing social dimensions in society brings inferences at the level of individual KHGs as well as the level of the landscape on the ecosystem services of KHGs.

References

Ariyadasa, K.P. (2002) *Assessment of Tree Resources in the Home Gardens of Sri Lanka*, FAO Regional Office for Asia and the Pacific, Bangkok, Thailand

Ashton, M., Gunatilleke, S., de Soyza, N., Dassanayaka, M.D., Gunatilleke, N., and Wijesundara, S. (1997) *A Field Guide to the Common Trees and Shrubs of Sri Lanka*, WHT Publication (Pvt) Ltd., Colombo, Sri Lanka

Chithral, G.M.W. (2011) 'Quality seed in Sri Lanka: Production, processing, legal and quality control and marketing system', in N. Huda and I. Saiyed (eds), *Quality Seed in SAARC Countries: Production, Processing, Legal and Quality Control and Marketing System*, SAARC Agriculture Centre, Bangladesh, pp. 303–454

Dassanayake, M.D. and Clayton, W.D. (1995–2000) *A Revised Handbook to the Flora of Ceylon*, vols 10–14, Oxford & IBH Publishing Company Private Limited, New Delhi, India

Dassanayake, M.D. and Fosberg, F.R. (1980–1991) *A Revised Handbook to the Flora of Ceylon*, vols 1–7, Amerind Publishing Company Private Limited, New Delhi, India

Dassanayake, M.D., Fosberg, F.R., and Clayton, W.D. (1994) *A Revised Handbook to the Flora of Ceylon*, vols 8–9, Oxford & IBH Publishing Company Private Limited, New Delhi, India

Dela, J.D.S. (2011) 'Impact of monkey–human relationships and habitat change on *Semnopithecus vetulus nestor* in human modified habitats', *Journal of the National Science Foundation of Sri Lanka*, vol 39, no 4, pp. 365–382

De Silva, K.M. (1981) *A History of Sri Lanka*, C. Hurst & Company, London

FSMP (1995) *Sri Lanka Forestry Sector Master Plan*, Forestry Planning Unit, Forest Department, Ministry of Agriculture, Lands and Forestry, Colombo, Sri Lanka

Ganashan, P., Balendira, S., and Dassanayake, M.D. (1996) *Sri Lanka: Country Report to the FAO International Technical Conference on Plant Genetic Resources (Leipzig)*, Department of Agriculture, Peradeniya, Sri Lanka

Government of Sri Lanka (2011) 'Improving family nutrition, a main objective of *Divi Neguma* Programme'; http://www.priu.gov.lk/news_update/Current_Affairs/ca201107/20110713improving_family_nutrition.htm, accessed on 10 May 2012

Heenkenda, H.M.S. (2014) *Better Future Through Fruit Crop Research*, Fruit Research and Development Institute, Horana, Sri Lanka

Hitinayake, H.M.G.S.B. and Ekanayake, U. (1999) 'Utilization of underutilized fruit tree species grown in Kandyan home gardens', in H.P.M. Gunasena (ed.), *Proceedings of the Tenth National Workshop on Multipurpose Trees: Fruit for the Future, Kandy, Sri Lanka*, 5 November 1999, pp. 186–212

IUCN and MENR (2007) *The 2007 Red List of Threatened Fauna and Flora of Sri Lanka*, IUCN and Ministry of Environment and Natural Resources, Colombo, Sri Lanka

Jacob, V.J. and Alles, W.S. (1987) 'The Kandyan gardens of Sri Lanka', *Agroforestry Systems*, vol 5, pp. 123–137

Jarvis, D., Hodgkin, T., Eyzaguirre, P., Ayad, G., Sthapit, B.R., and Guarino, L. (1997) 'Farmer selection, natural selection and crop genetic diversity: The need for a basic dataset', in D. Jarvis and T. Hodgkin (eds), *Strengthening the Scientific Basis of In Situ Conservation of Agricultural Biodiversity On-farm: Options for Data Collecting and Analysis*, Proceedings of a Workshop to Develop Tools and Procedures for *In Situ* Conservation On-farm, International Plant Genetic Resources Institute, Rome, Italy

Jayasuriya, A.H.M. and Rajapakse, R.M.T. (2004) 'Plant genetic resources activities in Sri Lanka', in Bhag Mal, P.N. Mathur, V. Ramanatha Rao, and A.H.M. Jayasuriya (eds), *Proceedings of the Sixth Meeting of South Asia Network on Plant Genetic Resources (SANPGR)*, held at PGRC, Peradeniya, Sri Lanka, 9–11 December 2002; IPGRI South Asia Office, New Delhi, India, pp. 22–38

Krishnarajah, P.S. and Sumanarathne, H.D. (1988) 'Magnitude and extent of soil erosion under different landuse in the mid country region of Sri Lanka', *Krushi*, vol 11, no 1, pp. 18–21

Kudavidanage, E.P., Wanger, T.C., de Alwis, C., Sanjeewa, S., and Kotagama, S.W. (2012) 'Amphibian and butterfly diversity across a tropical land-use gradient in Sri Lanka: implications for conservation decision making', *Animal Conservation*, vol 15, no 3, pp. 253–265

McConnell, D.J. and Dharmapala, K.A.E. (1973) *The Economic Structure of Kandyan Forest Garden Farms. Small Forest Garden Farms in the Kandy District of Sri Lanka*, Farm Management Diversification Report No. 7, UNDP/SF/FAO Diversification Project, Peradeniya, Sri Lanka

MADAS (2007) *National Agriculture Policy*, Ministry of Agriculture Development and Agrarian Services, Sri Lanka

Mahawamsa (undated) *The Great Chronicle of the History of Sri Lanka*; www.mahawamsa.org, accessed on 6 September 2012

Mankotte, K. (2011) *Department of Agriculture in Retrospect (1912–2011)*, Department of Agriculture, Peradeniya, Sri Lanka

Marambe, B., Weerahewa, J., Pushpakumara, G., Silva, P., Punyawardena, P., Premalal, S., Miah, G., and Roy, J. (2012) *Vulnerability of H Garden Eco-systems to Climate Change and its Impacts on Food Security in South Asia*; http://www.apn-gcr.org/resources/archive/files/3ddf57b875774091a38f95bdfee0b6c9.pdf, accessed on 25 July 2012

Mattsson, E., Ostwald, M., Nissanka, S.P., and Marambe (2013) 'Exploring the potential of home gardens as a multifunctional land-use strategy in Sri Lanka with focus on carbon sequestration', *Ambio*, vol 42, pp. 892–902

MENR (2009) *Fourth Country Report to the Convention on Biological Diversity*, in J.D.S. Dela (compiled), Ministry of Environment and Natural Resources, Sri Lanka; https://www.cbd.int/doc/world/lk/lk-nr-04-en.pdf, accessed on 7 November 2014

MFE (1999) *Biodiversity Conservation in Sri Lanka: A Framework for Action*, Ministry of Forestry and Environment, Battaramulla, Sri Lanka

Mohri, H., Lahoti, S., Saito, O., Mahalingam, A., Gunatilleke, N., Irham, Hoang, V.T., Hitinayake, G., Takeuchi, K., and Herath, S. (2013) 'Assessment of ecosystem services in home garden systems in Indonesia, Sri Lanka and Vietnam', *Ecosystem Services*, vol 5, pp. 124–136

Muthukuda Arachchi, D.H. and Wijerathne, P.M. (2007) *Country Report on the State of Plant Genetic Resources for Food and Agriculture: Sri Lanka*, FAO Government Cooperative Programme: The Status of the PGRFA in Sri Lanka, Department of Agriculture, Sri Lanka

NCSD (2009) *National Action Plan for Haritha Lanka Program*, National Council for Sustainable Development, Presidential Secretariat, Colombo 1, Sri Lanka

Perera, A.H. and Rajapakshe, R.M.N. (1991) 'A baseline study of Kandyan forest gardens of Sri Lanka: structure, composition and utilization', *Forest Ecology and Management*, vol 45, pp. 269–280

Pushpakumara, D.K.N.G., Boshier, D.H., and Harris, S.A. (1997) 'Mating system in *Artocarpus heterophyllus* Lam', *Tropical Agricultural Research*, vol 9, pp. 1–14

Pushpakumara, D.K.N.G. and Harris, S.A. (2007) 'Potential of RAPD markers for identification of fruit types of *Artocarpus heterophyllus* Lam. (jackfruit)', *Journal of National Science Foundation*, vol 35, part 3, pp. 175–179

Pushpakumara, D.K.N.G., Gunasena, H.P.M., and Singh, V.P. (eds) (2007) *Underutilized Fruit Trees in Sri Lanka,* World Agroforestry Centre, South Asia Office, New Delhi, India, vol 1

Pushpakumara, G. and Silva, P. (2008) *Agrobiodiversity in Sri Lanka*, Biodiversity Secretariat, Ministry of Environment and Natural Resources, Battaramulla, Sri Lanka

Pushpakumara, D.K.N.G., Wijesekara, A., and Hunter, D.G. (2010) 'Kandyan home gardens: A promising land management system in Sri Lanka', in C. Belair, K. Ichikawa, B.Y.L. Wong, and K.J. Mulongoy (eds), *Sustainable Use of Biological Diversity in Socio-ecological Production Landscapes: Background to the 'Satoyama Initiative for the Benefit of Biodiversity and Human Well-being*, Technical Series no 52, Secretariat of the Convention on Biological Diversity, Montreal, Canada

Pushpakumara, D.K.N.G., Gunasena, H.P.M., and Singh, V.P. (eds) (2011) *Underutilized Fruit Trees in Sri Lanka*, World Agroforestry Centre, South Asia Office, New Delhi, India, vol 2

Pushpakumara, D.K.N.G., Marambe, B., Silva, G.L.L.P., Weerahewa, J., and Punyawardena, B.V.R. (2012) 'A review of research on home gardens in Sri Lanka: The status, importance and future perspective', *Tropical Agriculturist*, vol 160, pp. 55–118

Raheem, D.C., Naggs, F., Preece, R.C., Mapatuna, Y., Kariyawasam, L., and Eggleton, P. (2008) 'Structure and conservation of Sri Lankan land-snail assemblages in fragmented lowland rainforest and village home gardens', *Journal of Applied Ecology*, vol 45, pp. 1019–1028

SCS (2011) *List of Information on Department of Agriculture Registered Fruit Nurseries*, Seed Certification Services, Department of Agriculture, Ministry of Agriculture Development and Govijanaseva, Colombo, Sri Lanka

Siddique, I., Gutjahr, C., Seneviratne, G., Breckling, B., Ranwala, S.W., and Alexander, I.J. (2007) 'Changes in soil chemistry associated with the establishment of forest gardens on eroded, acidified grassland soils in Sri Lanka', *Biology and Fertility of Soils*, vol 44, no 1, pp. 163–170

Sthapit, B.R., Sajise, P., Ramanatha Rao, V., Quek, P., de Cruz, F., and Bellon, M. (2004) 'Selection of good practices of *in situ* conservation of tropical fruit tree diversity: Methodology and key practices, Appendix K', GEF/UNEP Project document (unpublished)

Sthapit, B.R., Lamers, H., and Ramanatha Rao, V. (2013) *Custodian Farmers of Agricultural Biodiversity: Selected Profiles from South and South East Asia*, Proceedings of the Workshop on Custodian Farmers of Agricultural Biodiversity, 11–12 February 2013, New Delhi, India; Bioversity International, New Delhi, India

UN (1992) *Convention on Biological Diversity*, United Nations

Weerahewa, J., Pushpakumara, G., Silva, P., Daulagala, C., Punyawardena, R., Premalal, S., Miah, G., Roy, J., Jana, S., and Marambe, B. (2012) 'Are home garden ecosystems resilient to climate change? An analysis of the adaptation strategies of home gardeners in Sri Lanka', *APN Science Bulletin*, vol 2, pp. 22–27

Weerakkody, W.A.P. (2004) 'Horticulture in Sri Lanka', *Chronica Horticulturae*, vol 44, pp. 23–27

Wickramasinghe, A. (1995) 'The evolution of Kandyan home gardens: An indigenous strategy for conservation of biodiversity in Sri Lanka', in P. Halladay and D.A. Gilmour (eds), *Conserving Biodiversity Outside Protected Areas: The Role of Agroecosystems*, IUCN, Gland, Switzerland and Cambridge, UK, pp. 164–182

10 Amazonian fruits

How farmers nurture nutritional diversity on farm and in the forest

Patricia Shanley, Charles R. Clement, José Edmar Urano de Carvalho, Alfredo Kingo Oyama Homma and Antonio José Elias Amorim de Menezes

Introduction

Within the Amazon region an estimated 14,000 species of vascular plants occur, of which approximately 3,500 have reported uses (Lleras, 2012). Native Amazonians have been responsible for domesticating to differing degrees 83 native plant species, 71 of which are woody perennials (Clement, 1999). These species underwent the lengthy process of domestication by indigenous populations in pre-Columbian times and were part of agro-ecological management systems that capitalized on species and genetic diversity, while enriching soils (Glaser and Birk, 2012), enhancing biodiversity and transforming landscapes (Balée, 2013). They continue to nourish rural and urban Amazonians today.

The number of species domesticated in the Amazon is comparable with that of other major regions of crop domestication. However, a greater variety of fruits, fibres and spices have entered the global market from Asia than from Amazonia. In Amazonia, the evolution from extraction to domestication is still underway for numerous species, thus offering an opportunity to witness the domestication process (Homma, 2012).

In contrast to the historical context of apparently biodiversity-enhancing farming systems, current Brazilian agribusiness employs imported genetic materials, such as soybean, oil palm, sugar cane, oranges and cattle, as well as agrochemicals that degrade rather than enrich soil. Agricultural expansion was the single largest driver of deforestation in the tropics from 2000 to 2010, accounting for 73 per cent of tropical deforestation (Hosonuma *et al.*, 2012). Clement (1999) estimates that changed land use practices, such as commercial logging and industrial agriculture in Amazonia, have contributed to 90 per cent of the genetic diversity loss that has occurred since the European conquest.

Faced with escalating deforestation and genetic erosion, the federal government has instituted laws and programmes to support sustainable agriculture and conserve wild crop relatives (EMBRAPA, 2009). In practice, however, investments in these programmes are negligible. Most national research institutes concentrate on large-scale export commodities over local and regional products. For this reason, scientists at the Brazilian Agricultural Research Corporation (EMBRAPA) and the National Institute for Amazonian Research (INPA) are working with farmers on locally preferred native fruit species to improve diets with nutrient rich foods and to restore degraded areas.

Urban consumer demand for indigenous, 'lesser known' Amazonian forest fruits has increased over the last two decades due to rural migration to the cities and this in turn has spurred innovation among smallholders to increase productivity (Menezes *et al.*, 2012). Such innovations in management are most prominent in peri-urban areas, where the species occur naturally, where there are markets and a history of management and use. Nutritious forest fruits, once considered 'fruits of the poor', are now purchased at high prices by urbanites who once denigrated them (Shanley and Gaia, 2004).

In this chapter, we focus on six native, nutritious fruit species that demonstrate a range of management practices from historic, indigenous initiatives to recent, farmer-led endeavours. Pupunha (*Bactris gasipaes*) is an example of a crop that was fully domesticated before the European conquest, while the others are incipient domesticates with growing markets now benefiting from attention by EMBRAPA (Plate 8). Management techniques for the six species are most often developed independently by farmers, with long-term, farmer-generated knowledge of species ecology and use forming a foundation for management systems.

Agro-ecological and nutritional context

A high degree of agrobiodiversity in farming systems and the presence of nutrient rich dark earths are inheritances from agro-ecological traditions practised by pre-Columbian Amazonians (Fraser *et al.*, 2011). Depending on numerous factors, including soil conditions, state of forest resources, distance to market and their cultural and experiential background, smallholders manage fruit trees along a gradient from low management to high-intensity management, generally with little to no input from agricultural extension agents. Within remote areas of Amazonia, long distances to market and lack of transportation are two principal obstacles for smallholders to invest time and energy in domestication.

The majority of the fruits that present-day farmers are focusing on, and those domesticated in pre-Columbian times, are not sweet and juicy, but composed of starches, oils and phytonutrients with relatively low levels of sugar (Clement, 1999). In contrast to contemporary urban diets comprised of high-energy carbohydrates, oil and sugar, each of the locally favoured trees and palms produce nutrient rich, culturally preferred foods that enhance family and societal well-being (Johns *et al.*, 2013).

An unforeseen consequence of industrialized agriculture and processed foodstuffs has been the burgeoning not only of diabetes, obesity and cardiovascular disease, but also of nutrient deficiencies. Small increases in dietary diversity can ameliorate essential-nutrient deficiencies. Many underutilized species are rich not only in nutrients, but also in compounds such as carotenoids that act as antioxidants and prevent damage to cells (Johns and Sthapit, 2014). Among school children in Brazil, for example, palm fruits of buriti (*Mauritia flexuosa*), bearing high amounts of beta-carotene (β-carotene) have been used to improve eyesight, especially night blindness, and poor eye health caused by lack of vitamin A (Lima, 1987; Santos, 2005). Doctors also recommend consuming pupunha (*Bactris gasipaes*) and tucuma (*Astrocaryum aculeatum*) to improve vision, as these too offer high levels of β-carotene. Tucuma has three times the β-carotene of carrots and one tucuma fruit satisfies the daily requirement.

For children, bacuri's (*Platonia insignis*) combination of calcium, phosphorous and iron is recommended as an excellent food to fortify the growth of bones and teeth. The protein content of Brazil nut (*Bertholletia excelsa*) is almost equivalent to that of cow's milk, containing high levels of the amino acid methionine, an element often lacking in Amazonian diets. Brazil nut also has an extremely high level of selenium that is indicated to elevate moods and boost the body's immune system to prevent disease. Assai (*Euterpe oleracea*) has garnered international attention due, in part, to its wide range of phytonutrients, such as anthocyanin. A regionally embraced fruit, uxi (*Endopleura uchi*), offers an impressive array of minerals for fortifying the body, such as iron, phosphorus and magnesium, and its phytosterols help to reduce cholesterol levels in the blood stream as well as having possible anticarcinogenic effects (Table 10.1).

The six tree and palm fruit species demonstrate a gradient of management practices from forest-gathered to monoculture and a range of uses from subsistence to export. For each species we list the densities and yields of managed and unmanaged populations (Table 10.2), as well as the management practices used by select smallholders (Table 10.3). Management systems vary from extractivism, in mature, relatively unmanaged forests (i.e. Brazil nut, uxi, bacuri), to managing one or numerous species within enriched forests or home gardens (assai, pupunha, tucuma, uxi), to intensive management of fallows and/or flooded forests to favour single species (bacuri, assai). Data describing the densities and fruit yield of many Amazonian fruit species are difficult to obtain and the numbers that are available are often inconsistent, reflecting both the physiology of tropical fruit trees and that the studies are conducted within distinct geographies, ecosystems and soil types.

Fruit trees: Bacuri, *Platonia insignis*; Brazil nut, *Bertholletia excelsa*; Uxi, *Endopleura uchi*

Bacuri, Platonia insignis *Mart.*

Over the last three decades, the soft aromatic flesh of bacuri has become prized in rural and urban areas and is eaten fresh and in ice creams, cakes, jams and

Table 10.1 Available nutritional information per 100 g of fruit of six Amazonian tree and palm fruits

	Vit C (mg)	*Ca (mg)*	*K (mg)*	*Mg (mg)*	*Fe (mg)*	*Protein (%)*	*Calories (Kcal)*	*Other*
Assai *Euterpe oleracea* [1,2]	9–84	50–480	740	149	329	1.2–4.3	80–247	P – 140 mg Zn – 1.01 mg Cu – 2.04 mg Flavonoids Anthocyanin – 0.50–1.67% Vit. B1, Vit. E
Bacuri *Platonia insignis* [2,3]	2.4–33	17–20	150	22	0.45–2.2	1.4–3.8	105	P – 10.7–36 mg Zn – 1.04 mg Cu – 0.38 mg
Brazil nut *Bertholletia excelsa* [4,5,6]	NA	180–205	512-715	310	2.8–9.3	12–20	690–751	P – 563 mg Zn – 4.2–6.9 mg Cu – 1.35 mg Mn – 5.9 mg Se – 0.03 – 512 ppm Vitamin B1 – 150 mg Fat – 61 g
Pupunha *Bactris gasipaes* [7]	Lost in cooking	10–25	200–290	17	470–740	1.8–2.7	160–300	Zn – 258–278 μg Mn – 82–115 μg Se – 3.3–11.4 μg
Tucuma *Astrocaryum aculeatum* [8]	NA	88	86	7.3	576	3.39	320	P – 5.63–46 mg Zn – 213 μg Cu – 204 μg Mn – 189 μg Cr – 388 μg
Uxi *Endopleura uchi* [9, 10]	22	64–96	53–460	53–70	460	0.9–2.2	252–284	P – 39–46 mg Na – 22 mg Vit. B1 – 0.13 mg Vit. B2 – 0.10 mg Vit. E – 1–6.8 mg

Sources: [1]Teixeira *et al.* (2012); [2]Rufino *et al.* (2010); [3]Rogez *et al.* (2004); [4]Villachica (1996); [5]Chang *et al.* (1995); [6]Silva *et al.* (2010); [7]Yuyama *et al.* (2003); [8]Leitão (2008); [9]IBGE (1981); [10]Marx *et al.* (2002).

Table 10.2 Reported tree densities and fruit yields with and without management

	Recorded density without management	*Recorded density with management*	*Average annual yield without management*	*Average annual yield with management*
Acai *Euterpe oleracea*	300–400 clumps/ha[1] (good soils) 100–200 clumps/ha[1] (poor soils)	1,200 clumps/ha – intensive management	4,200 kg/ha[1]	8,400 kg/ha[1]
Bacuri *Platonia insignis*	1 tree/ha[2]	100 trees/ha[7]	400 fruits/tree[2]	20,000 fruits/ha[6]
Brazil nut *Bertholletia excelsa*	0.1–4 trees/ha[2]	4–15 trees/ha[2]	up to 1,000 nuts/tree[5]	45 kg in shell/tree[5]
Pupunha *Bactris gasipaes*	100 palms/ha[8] (traditional swiddens)	400 palms/ha in monoculture, but agroforestry systems more common, always with fewer trees[8]	2–3 bunches/tree[8]	6–8 bunches/tree[8]
Tucuma *Astrocaryum aculeatum*	10 palms/ha[4] (forest)	20–43 palms/ha[3] (disturbed areas)	12 kg/palm[3]	720 fruits/palm
Uxi *Endopleura uchi*	0.03–1 tree/ha[2]	35 trees/ha[2]	400–1000 fruits/tree[2]	700–2,000 fruits/tree[2]

Sources: [1] Brondízio (2008); [2] Shanley *et al.* (2011a); [3] Schroth *et al.* (2004); [4] Costa *et al.* (2002); [5] Homma *et al.* (2014); [6] Homma *et al.* (2010); [7] Menezes *et al.* (2012); [8] Clement *et al.* (2004).

puddings, with the cost of fruit and pulp increasing five- to tenfold since the 1980s (Plate 9).

Bacuri is principally sourced through extraction from forests, having defied efforts at rapid modern domestication. In mature forests, densities are as low as 1 tree/ha. Average yield is 300–400 fruit, though in good years a vigorous tree can produce 800 to over 1,000 fruit (Shanley *et al.*, 2011a). Bacuri's expansive roots can grow abundantly in poor, sandy soils and are efficient in recycling nutrients (Carvalho and Müller, 2007; Menezes *et al.*, 2012). In degraded areas, 1,800 root sprouts can grow per hectare (Ferreira and Medina, 2004), so to become productive trees the density of saplings requires radical thinning. Recently, in degraded areas surrounding cities where natural regeneration occurs, some smallholders, supported by technical assistance from

EMBRAPA, are beginning to manage spontaneous seedlings in response to skyrocketing sales and growing prices. Prompted by consumer demand, on-farm trials with EMBRAPA and encouragement from neighbours, farmers are now selecting the best trees, thinning to 4–8 metres and cutting, burning, and intercropping the area (Menezes *et al.*, 2012; Homma *et al.*, 2008). After a decade of management in communities, production today is such that families can consume and market bacuri (Menezes *et al.*, 2012). In these areas, bacuri is becoming a key species to transform fallows into functioning agroforests (Homma *et al.*, 2010). Parallel to the farm trials, meticulous experiments in germplasm selection and hybridization are underway at EMBRAPA's experiment station in Belém (Carvalho and Müller, 2007).

Brazil nut, Bertholletia excelsa *Bonpl.*

Considered the meat of the plant kingdom, Brazil nut trees are widely valued for their nutritious seeds, which contain 12–20 per cent protein. Similar to uxi and bacuri, production is highly variable between trees and from year to year. Groups of Brazil nut trees occurring in primary forest are considered remnants of centuries-old management, with collectors traversing several kilometres to harvest seeds. Brazil nut can also be a viable part of agroforestry systems, particularly if planted in young fallows. At present, in spite of its nutritional, economic and subsistence value, little active planting of the species appears to take place, with today's harvests being a legacy of plantings undertaken decades ago (Salomão, 2014). Local and regional demand for Brazil nut currently exceeds supply. Prior to protection by law in 1965, Brazil nut trees were cut down in many regions, including Maraba, Pará, where an economic assessment of loss to producers over a 10-year period as a result of the decline in Brazil nut tree populations was estimated to be US$5 million a year (Homma, 2004).

Uxi, Endopleura uchi *(Huber) Cuatrec.*

The woody flavour and oily, grainy, textured pulp of uxi is relished by traditional Amazonians and increasingly popular among urban consumers. Traditionally consumed fresh from the tree, uxi is now being processed into ice creams, popsicles and juices. Earlier derided as 'poor man's fruit', uxi is an excellent source of calories, low in sugar and high in fibre, and containing vitamins C, E and B, and iron. Rising demand has driven up prices and catalyzed interest in its production in peri-urban areas (Carvalho *et al.*, 2007). Uxi is difficult to domesticate, because of its slow germination rate, decade-long juvenile period and the difficulty of transplanting spontaneous seedlings (Carvalho and Müller, 2005). However, smallholders living near the city of Belém, stimulated by its vibrant markets, are overcoming such obstacles using traditional management practices. These practices include: thinning competing vegetation; protecting seedlings (by specifically not cutting when clearing);

selecting promising germplasm; transplanting seedlings; clearing beneath trees to spot fallen fruits; and fertilizing with natural compost gathered during clearing, resulting in an increase in species density and production.

Without any outside intervention, these traditional management techniques result in rich agroforestry systems composed entirely of economically useful species including palms such as assai, pupunha and buriti *(Mauritia flexuosa)*, and fruit trees such as cupuassu (*Theobroma grandiflora*), cacao (*Theobroma cacao*) and bacuri. In unmanaged forests, densities of uxi are low (0.03–1 tree/ha), while in managed forests densities can reach as high as 35 trees/ha (Shanley *et al.*, 2011a) (Plate 10). Corresponding research by EMBRAPA is also deepening understanding of uxi's germination, growth rate and productivity (Carvalho *et al.*, 2007; Carvalho and Müller, 2005).

Palms: Assai (*Euterpe oleracea*), Pupunha (*Bactris gasipaes*), Tucuma *(Astrocaryum aculeatum)*

Assai, Euterpe oleracea *Mart.*

A multi-stemmed palm common to the Amazon estuary and floodplain forests, assai palms bear purple, pebble-sized fruit, which is processed into a non-alcoholic drink called 'assai wine'. The fruit drink has always been a mainstay among rural poor and urban citizens alike; eastern Amazonian meals are not complete without assai and farinha (manioc flour). Today, ice cream, smoothies and energy bars are also flavoured with assai. Traditionally, assai was managed by increasing densities of the palm through techniques such as maintaining wildlings, pruning most of the young stems and thinning competing vegetation. The palms can produce as much as 4,200 kg/ha/yr in upland dry forest and up to 8,400 kg/ha/yr in flooded forest (Brondízio, 2008). In the 1990s entrepreneurs discovered the potential to market assai as a health and energy drink, causing demand to skyrocket (Nogueira and Homma, 1998). The innovative management techniques of smallholders were appropriated and scaled up by industries, claiming that they had instructed smallholders in management practices. Currently, to supply rising demand, traditional agricultural systems have intensified, turning the Amazon estuary into a monoculture of assai, crowding out other native species and leaving smallholders with a decreasing share of the profits (Brondízio, 2008).

Pupunha, Bactris gasipaes *Kunth*

Pupunha is a multi-stemmed, spiny palm bearing nutritious, brightly coloured red, yellow or orange fruit, which require cooking before consumption. One of the first plants domesticated by Native Amazonians in pre-Columbian times, its fruit was originally oily but was selected for increasing levels of starch, and it was consumed as a major source of energy and fermented for festivities (Clement *et al.*, 2004). Pupunha grows well on Amazonian dark earths, but also produces on the poor clay soils typical in Amazonia. As a domesticate, pupunha is always planted, principally in home gardens and in swiddens; in swiddens it may be very abundant and remains during the fallow, where palms are managed for many years. By pruning young shoots, farmers take advantage of the palm heart for food while renewing the vigour of the palm cluster (Clement *et al.*, 2004). Although pupunha was a staple before the European conquest and is still important in rural areas, decades of research and development failed to transform it from an underexploited crop into a global market success like assai (Clement *et al.*, 2004). Nonetheless, this all-purpose, nutritious food for people and fodder for animals continues to nourish Amazonians today, principally in rural areas but also as a snack in major Amazonian cities.

Tucuma, Astrocaryum aculeatum *G. Mey*

A tall palm bearing long spines on its trunk, tucuma offers meaty, nutritious fruits appreciated by both wildlife and people. Over the past 20 years, the popularity of tucuma has supplanted that of pupunha (Lleras, 2012) and sales are soaring as cafés in Manaus have begun serving tucuma sandwiches. Loaded with calories, protein and β-carotene, tucuma sandwiches now represent 60–80 per cent of sandwiches sold in regional breakfast cafés (Shanley *et al.*, 2011a). Ability to grow in poor soils, tolerance to fire and abundant fruit production make it ideal for regeneration in fallows and secondary forests. When farmers burn a piece of land to prepare agricultural fields, the heat helps tucuma seeds to germinate; farmers then tend the seedlings that regenerate. Smallholders manage spontaneous regeneration of tucuma by clearing vegetation beneath palms, monitoring productivity, maintaining high-quality fruit producers, eliminating individuals bearing poor-quality fruit, leaving some fruits for fauna (agoutis, pacas – rodents) and favouring shorter trees whose fruit bunches can be more easily reached (Schroth *et al.*, 2004). The management of spontaneous populations of palms that grow freely in pastures and secondary forests (*in situ* domestication) does not require financial investment and helps to improve the native population (Schroth *et al.*, 2004).

Threats to indigenous fruit trees – the invisibility of locally valued species

Farmers value each of the six species not only for their use as fruit, but because they possess multiple functions. However, over the last 25 years, timber companies have harvested these long-lived, nutritious fruit and medicinal tree species, most of which occur in low densities (Shanley and Luz, 2003). During the course of ten years, logging and fire reduced the number of productive bacuri and uxi trees in three smallholder communities by 81 per cent and 83 per cent, respectively (Shanley *et al.*, 2011a). For the remaining individuals, lowered rates of regeneration as well as declining pollinator frequency and reliability can lead to loss of genetic diversity through low fruit set, loss of vigour and mortality (Dawson *et al.*, 2014). Thus, deforestation and a corresponding decline in species that support human well-being can be an important motivator to identify and conserve locally valued but potentially vulnerable species, and the smallholder management practices that sustain them.

One obstacle to conserving indigenous fruit trees is limited understanding of the role these species play in livelihoods (Johns and Sthapit, 2014). Invisibility is problematic as locally used species provide essential nutrition, subsistence, livelihood and ecosystem functions. For incipiently domesticated species, invisibility is particularly deleterious as deforestation and fire associated with expanding agricultural and logging frontiers can eradicate valuable germplasm of species with high nutritional value. Documentation of species' function and smallholders' management, harvesting and processing expertise is essential as a lens to understand which species are prioritized, how and why.

Managing native species for diversity, food security and health

In areas close to markets demand often exceeds supply, and former 'wild' species are being successfully managed by farmers. Management of incipiently domesticated and sometimes semi-domesticated species can transform the forest, increasing both the suite of economic species and their densities, contributing important ecosystem services and retaining high levels of biodiversity (Wiersum, 2004). In these forests, fruit, medicinal, fibre and latex tree species predominate, filling the forest with economically useful products year round. These anthropogenic forests serve as a socio-economic mainstay and ecological buffer in times of stress (Dawson *et al.*, 2014) (Table 10.3).

Subsistence and commercial management of Brazil nut, uxi, bacuri, tucuma, assai and pupunha grown in traditional agricultural systems are based on generations of observation and experimentation, resulting in complex bodies of silvicultural, ecological, physiological and socio-cultural knowledge. For these species, research and extension have built upon traditional knowledge to enhance productivity as well as to improve desired qualities. R&D extension efforts were mounted for pupunha in the 1970s and assai in the 1990s, increasing yield of both the palm heart and fruit (Clement *et al.*, 2004; Oliveira *et al.*, 2012), thus nourishing families throughout the region but falling short of reaching a global market. In the case of assai, expanding to an international market has undermined traditional systems that supplied local and regional markets by usurping land, raising local prices for assai, diminishing the biodiversity-rich estuary and appropriating the management systems of smallholders for corporate gain (Brondízio, 2008). Where indigenous fruit production has scaled up to meet local and regional demand, rather than international markets, greater socio-economic and environmental benefits have resulted. In the Zona Bragantina in the state of Pará, for example, farmers learned new methods to manage the remarkably vigorous sprouting of bacuri through experimentation

Table 10.3 Indicative stand level silvicultural practices in managed areas of six Amazonian tree and palm fruits

Species	*Protect regener-ation (by not cutting)*	*Thin competing vegetation &/or target species*	*Select germ-plasm*	*Plant seeds &/or seedlings*	*Clean beneath trees*	*Remove non-productive individuals*	*Spread mulch as fertilizer*
Assai *Euterpe oleracea*	✓	✓			✓	✓	✓
Bacuri *Platonia insignis*	✓	✓	✓		✓	✓	
Brazil nut *Bertholletia excelsa*	✓			✓	✓		
Pupunha *Bactris gasipaes*	✓	✓	✓	✓	✓	✓	✓
Tucuma *Astrocaryum aculeatum*	✓	✓	✓			✓	
Uxi *Endopleura uchi*	✓		✓	✓	✓	✓	✓

and workshops offered by EMBRAPA. Scientists predict that vast areas (50,000 ha) of degraded land and fallow in the Northeast of Pará and the island of Marajo could be transformed into productive landscapes by actively managing bacuri sprouts and preventing fire (Homma *et al.*, 2010). Transforming degraded areas with economic species can improve soil, local economies and food security (Plate 11).

Many smallholder farmers in Amazonia, women farmers in particular, consider forests, food, medicine and health as interrelated (Shanley *et al.*, 2011b). Throughout Brazil, social movements expound the belief that healthy forests sustain healthy, autonomous families. These convictions affirm what research illustrates: countries that maintain traditional food systems prevent the onset of chronic disease associated with societies in which agro-industrial food predominates (Popkin *et al.*, 2001).

Conclusion

In light of decades of deforestation and biodiversity loss, the recent trend of increasing consumer interest in indigenous Amazonian fruits as well as a political climate more receptive to a conservational policy agenda offers an opportunity to be realized. As our study indicates, management of fruit trees in which wild, semi-domesticated and cultivated plants coexist along a gradient of management intensity while retaining high diversity is ecologically and socially sustainable, while leading to significantly increased fruit density and yields. Further research must be devoted towards optimizing those practices that maximize output while sustaining local agricultural and ecosystem biodiversity. This will in turn help create an economically profitable system that provides farmers with incentives to continue cultivating local varieties of native species that play such an integral role in maintaining ecosystem and human health. Likewise, innovations developed by both farmers and researchers must be documented and shared so that best practices can be implemented on local, regional and national scales. Monoculture agribusiness interests can only be balanced with environmentally sound and socially just policies if sustainable techniques are seen as feasible, popular and profitable alternatives. Local farmers, the agricultural practices they employ and the genetically diverse portfolios of species they cultivate are an invaluable resource for current and future generations.

References

Balée, W. (2013) *Cultural Forests of the Amazon: A Historical Ecology of People and Their Landscapes*, University of Alabama Press, Tuscaloosa, AL

Brondízio, E.S. (2008) 'The Amazonian Caboclo and the Açai Palm – forest farmers in the global market', *Advances in Economic Botany*, vol 16, p. 403

Carvalho, J.E.U. de and Müller, C.H. (2005) 'Avaliação de metodos para aumentar, acelerar e uniformizer a germinação de sementes de uxi (*Endopleura uchi* Huber, Cuatr.)', *Ciencia e tecnología com inclusão social: anais*, UFRA, EMBRAPA, Belém

Carvalho, J.E.U. de and Müller, C.H. (2007) 'Propagacao do bacurizerio', in M.C. Lima (org), *Bacuri: agrobiodiversidade*, Instituto Interamericano de Cooperação para a Agricultura, São Luiz, Brazil, pp. 29–46

Carvalho, J.E.U de, Müller, C. H., and Benchimol, R. L. (2007) *Uxizeiro: botânica, cultivo e utilização*. 1. ed., EMBRAPA Amazônia Oriental, Belém, p. 107

Chang, J.C., Gutenmann, W.H., Reid, C.M., and Lisk, D.J. (1995) 'Selenium content of Brazil nuts from two geographic locations in Brazil', *Chemosphere*, vol 30, no 4, pp. 801–802

Clement, C.R. (1999) '1492 and the loss of Amazonian crop genetic resources', I. The relation between domestication and human population decline', *Economic Botany*, vol 53, no 2, pp. 188–202

Clement, C., Weber, J.C., van Leeuwen, J., Astorga Domian, C., Cole, D.M., Arevalo Lopez, L.A., and Arguello, H. (2004) 'Why extensive research and development did not promote use of peach palm fruit in Latin America', *Agroforestry Systems*, vol 61, pp. 195–206

Costa, J.A., Duarte, A.P., and the Indigenous Community of Apurina (2002) 'Metodologia para manejo comunitário da espécie "tucumã" (*A. aculeatum*) na terra indígina Apurinã do km 45 da BR 317/AM brasil: um modelo replicavel para conservação da biodiversidade e incremento da renda em areas extrativas'; www.amazonlink.org/apuina

Dawson, I.K., Leakey, R., Clement, C.R., Weber, J.C., Cornelius, J.P., Roshetko, J.M., Vinceti, B., Kalinganire, A., Tchoundjeu, Z., Masters, E., and Jamnadass, R. (2014) 'The management of tree genetic resources and the livelihoods of rural communities in the tropics: Non-timber forest products, smallholder agroforestry practices and tree commodity crops', *Forest Ecology and Management*, vol 333, pp. 9–21

EMBRAPA (2009) *National Report on the State of Brazil's Plant Genetic Resources for Food and Agriculture*, in A. da Silva Mariante, A, M.J. Amstalden Sampaio, and M.C. Valadares Inglis (eds), EMBRAPA, Ministério de Agricultura, Pecuária e Abastecimento, Brasília

Ferreira, M.S.G. and Medina, G. (2004) 'Bacuri: Gold in the market, delicious in the mouth', in M. Alexiades and P. Shanley (eds), *Productos Forestales, Medios de Subsistencia y Conservación: Estudios de caso sobre sistemas de manejo de productos forestales no maderables*, CIFOR, Bogor, Indonesia, pp. 5–8

Fraser, J.A., Junqueira, A.B., Kawa, N.C., Moraes, C.P., and Clement, C.R. (2011) 'Crop diversity on anthropogenic Dark Earths in Central Amazonia', *Human Ecology*, vol 39, no 4, pp. 395–406

Glaser, B. and Birk, J.J. (2012) 'State of the scientific knowledge on properties and genesis of anthropogenic Dark Earths in Central Amazonia (terra preta de Índio)', *Geochimica et Cosmochimica Acta*, vol 82, pp. 39–51

Homma, A.K.O., Carvalho, J.E.U., Menezes, A.J.E.A., Rebello, F.K., Matos, G.B. de, Perotes, K.F., Santos, W.N.M. dos, and Pereira, P.R.S. (2008) *Viabilidade Técnica e Econômica da Formação de Bacurizal Mediante Manejo de Rebrotamento*, EMBRAPA, Belém, Brazil

Homma, A.K.O. (2004) 'Cemitério das Castanheiras', *Ciencia Hoje*, vol 34, no 202, pp. 60–63

Homma, A.K.O. (2012) 'Plant extractivism or plantations: What is the best option for the Amazon?', *Estudos Avançados*, vol 74, no 26, pp. 167–186

Homma, A.K.O., Menezes, A.J.E.A., Carvalho, J.E.U., Souto, G.C., and Gibson, C. da P. (2010) *Manual de Manejo de Bacurizeiros*, 2nd edition, EMBRAPA, Belém, Brazil

Homma, A.K.O., Menezes, A.J.E.A., and Maues, M.M. (2014) 'Castanheira-do-pará: Os desafios do extrativismo para a domesticação', *Bol. Mus. Para. Emílio Goeldi. Cience. Nat.*, vol 9, no 2, pp. 233–246

Hosonuma, N., Herold, M., De Sy, V., De Fries, R. S., Brockhaus, M., Verchot, L., Anglesen, A., and Romijn, E. (2012) 'An assessment of deforestation and forest degradation drivers in developing countries', *Env. Res. Letters*, vol 7, no 4. DOI:10.1088/1748-9326/7/4/044009

IBGE (1981) *Tabelas de composição de alimentos*. IBGE, Estudos Nacional Despesa Familiar, Publicações Especiais, Rio de Janeiro, vol 3, p. 213

Johns, T. and Sthapit, B.R. (2014) 'Biocultural diversity in the sustainability of developing country food systems', *Food and Nutrition Bulletin*, vol 25, no 2, pp. 143–155

Johns, T., Powell, B., Maundu, P., and Eyzaguirre, P.B. (2013) 'Agricultural biodiversity as a link between traditional food systems and contemporary development, social integrity and ecological health', *Journal of the Science of Food and Agriculture*; www.soci.org

Leitão, A.M. (2008) *Caracterização morfológica e fisico-química de frutos e sementes de Astrocaryum aculeatum Meyer (Arecaceae) de uma floresta secundária*, Tese (Doutorado em Botânica), INPA/UFAM, Manaus, p. 91

Lima, M.C.C. (1987) *Atividade de Vitamina A do doce de buruti (Mauritia vinífera Mart.) e seu efeito no tratamento e prevenção da Hipovitaminose A em crianças*, Universidade da Paraiba, Brazil, p. 125

Lleras, E. (2012) 'Plant diversity in Amazonia and world genetic heritage', in A. Borém, M.T.G. Lopes, C.R. Clement, and H. Noda (eds), *Domestication and Breeding: Amazonian Species*, Editora da Univ. Fed. Viçosa, Viçosa, Brazil, pp. 39–52

Marx, F., Andrade, E.H.A., Zoghbi, M. das G.B., and Maia, J.G.S. (2002) 'Studies of edible Amazonian plants. Part. 5: Chemical characterization of Amazonian *Endopleura uchi* fruits', *European Foods Research Technology*, vol 214, pp. 331–334

Menezes, A.J.E.A., Homma, A.K.O., and Schoffel, E.R. (2012) *Do Extrativismo á Domesticaçao: o Caso do Bacurizeio no Nordeste Parense e na Ilha de Marajó*, Documentos 379, EMBRAPA, Brazil, p. 66

Nogueira, O.L. and Homma, A.K.O. (1998) 'Importância do manejo de recursos extrativos em aumentar o carrying capacity: o caso de açaizeiros (*Euterpe oleracea* Mart.) no estuário amazônico', *Poematropic*, vol 2, pp. 31–35

Oliveira, M.S.P., Mochiutti, S., and Farias Neto, J.T. (2012) 'Domestication and breeding of assai palm', in A. Borém, M.T.G. Lopes, C.R. Clement, and H. Noda (eds), *Domestication and Breeding: Amazonian Species*, Editora da Univ. Fed. Viçosa, Viçosa, Brazil, pp. 209–236

Popkin, B.M., Horton, S., and Kim, S. (2001) 'The nutrition transition and prevention of diet-related diseases in Asia and the Pacific', *Food Nutrition Bulletin*, vol 22, pp. 51–58

Rogez, H., Buxant, R.L., Mignolet, E., Souza, J.N.S., Silva, E.M., and Larondelle, Y. (2004) 'Chemical composition of the pulp of three typical Amazonian fruits: araça-boi (*Eugenia stipitata*), bacuri (*Platonia insignis*) and cupuaçu (*Theobroma grandiflorum*)', *European Food Research and Technology*, vol 218, pp. 380–384

Rufino, M., Do, S.M., Alves, R.E, Brito, E.S. De, Jiménez, J.P., Calixto, F.S., and Mancini-Filho, J. (2010) 'Bioactive compounds and antioxidant capacities of 18 non-traditional tropical fruits from Brazil', *Food Chemistry*, vol 121, pp. 996–1002

Salomão, R. D. P. (2014) 'A castanheira: História natural e importância socioeconomic', *Bol. Mus. Para. Emílio Goeldi. Cience. Nat.*, vol 9, no 2, pp. 259–266

Santos, L.M.P. (2005) 'Nutritional and ecological aspects of buriti or aguaje (*Mauritia flexuosa* Linnaeus filius): A carotene rich palm fruit from Latin America', *Ecology of Food and Nutrition*, vol 44, no 5, pp. 345–358

Schroth, G., Mota, M. S. S., Lopes, R., and Freitas, A.F. (2004) 'Extractive use, management and in situ domestication of a weedy palm, *Astrocaryum tucuma*, in the central Amazon', *Forest Ecology and Management*, vol 202, no 1, pp. 161–179

Shanley, P. and Luz, L. (2003) 'Eastern Amazonian medicinals: Marketing, use and implications of forest loss', *BioScience*, vol 53, no 6, pp. 573–584

Shanley, P., Cymerys, M., Medina, G., and Serra, M. (2011a) *Fruit Trees and Useful Plants in Amazonian Life*, FAO/CIFOR/PPI, Rome, Italy

Shanley, P., Silva, F.C. da, and MacDonald, T. (2011b) 'Brazil's social movement, women and forests: A case study of the National Council of Rubber Tappers', *International Forestry Review*, vol 13, no 2, pp. 233–244

Shanley, P. and Gaia, G. (2004) 'A "fruta do pobre" se torna lucrative: A *Endopleura uchi* Cuatrec. Em areas manejadas próximo a Belém, Brasil', in M. Alexiades and P. Shanley (eds), *Productos Forestales, Medios de Subsistencia y Conservacion*, vol 3, America Latina, CIFOR, Bogor, Indonesia, pp. 219–240

Silva, R.F. da, Ascheri, J.L.R., Souza, J.M.L. de (2010) Influência do processo de beneficiamento na qualidade de amêndoas de castanha-do-brasil. *Ciênca e agrotecnologia*, 34: 445–450

Teixeira, G.H.A., Pessoa, J.D.C., Lopes, V.G., and Lima, K.M.G. (2012) 'Tecnologia NIRS para análise dos frutos de açaizeiros e juçara,' in J.D.C. Pessoa and G.H.A. Teixeira (eds), *Tecnologia para inovação nas cadeias Euterpe*, EMBRAPA, Brasília, pp. 81–115

Villachica, H. (1996) *Frutales y Hortalizas promisorios de la Amazonia*, Tratado de Cooperacion Amazonica, Secretaria Pro-Tempore, Lima, p. 367

Wiersum, K.F. (2004) 'Forest gardens as an intermediate land-use system in the nature–culture continuum: Characteristics and future potential', *Agroforestry* Systems, vol 61, no 1, pp. 123–134

Yuyama, L.K.O., Aguiar, J.P.L., Yuyama, K., Clement, C.R., Macedo, S.H.M., Fávaro, D.I.T., Afonso, C., Vasconcelos, M.B.A., Pimentel, S.A., Badolato, E.S.G., and Vannucchi, H. (2003) 'Chemical composition of the fruit mesocarp of three peach palm (*Bactris gasipaes*) populations grown in Central Amazonia, Brazil', *International Journal of Foods Sciences and Nutrition*, vol 54, no 1, pp. 49–56

Part 3

Case studies of the project 'Conservation and Sustainable Use of Tropical Fruit Tree Diversity' from South and Southeast Asia

Case studies

Propagation and planting materials

11 A suite of propagation and management techniques for *Garcinia* in the central Western Ghats region of Karnataka, India

Vasudeva R, B.M.C. Reddy and Bhuwon Sthapit

Context and introduction

GPD 'passport'

GPD code:	2
Focus area:	Propagation and planting materials
Character:	Techniques
Species and varieties involved:	*G. gummi-gutta* and *G. indica* (kokum) and *G. morella*
Name of location:	Salkani, Kadakeri villages in Uttara Kannada district, Karnataka, India
GIS reference of location(s):	N 14°42′51″; E 74°40′44″ Elevation: 546 masl
Name of farmers (data resource):	Mr Dattatreya Hegde, Bhairimane; Ms Bhrathi D. Hegde, Bhairimane; Mr Ramesh Hegde, Onikere; Mr Eshanna, Amchimani; Ms Lalita V. Hegde, Amchimani; Mr Krishna Hegde, Onikere; Mr Shantaram, Onikere; Mr M.B. Nayak, Siddapur; Mr Manju Pujari, Kanchikai

Uttara Kannada district, situated in one of the world's hotspots of biological diversity – the Western Ghats of India – is one of the largest districts in Karnataka state. The district has varied geographical features, with thick forests, perennial rivers, abundant flora and fauna, and a long coastline of about 140 km. In its 1,025,000 ha of geographical area, a large part (802,800 ha) is under forest and only about 120,000 ha (roughly amounting to 12 per cent) is under cultivation. The tropical climate of this region is strongly influenced by monsoons and moderated by proximity to the sea. During the monsoons, the region receives one of the heaviest rainfalls in the world. Average rainfall in the district is 2,835 mm. However, the western coastal and crest-line regions of the district receive heavy rainfall exceeding 4,000 mm annually. Because of the heavy rainfall, the lateritic soils are rather poor with respect to nutrition. The average temperature at sea level is 33°C during the summer and 20°C during the winter. The vegetation in the region is mainly moist deciduous forests, wherein valuable timberwood is found. Evergreen and semi-evergreen formations are fragmented in the crest-line of the Ghats. Deforestation and poaching have been the cause of conservation concern in recent years. Arecanut (betelnut) and rice are the main crops of the irrigated region, along with cardamom, vanilla and black pepper.

The study site falls under the high-rainfall region surrounded by the forested area. The lifestyle and the culture of the communities are closely associated with the resources of the forest. The communities are involved in the collection of non-timber forest products (NTFPs) such as wild pickle-mango and *Garcinia* fruit, gum, resin and leaves. More than 40 per cent of the indigenous communities who traditionally engage in the collection of NTFPs are engaged in the harvesting of *Garcinia gummi-gutta*; each such household earns an average of INR35,000 (around US$540) a year solely from this collection (Vasudeva *et al.*, 2010). Other species of *Garcinia* such as *G. indica*, *G. morella*, *G. pictorius* and *G. talbotii* are equally important to the NTFP harvesters and contribute significantly to their income.

This good practice for diversity (GPD) refers to a suite of propagation techniques that help conserve and maintain genetic diversity of *Garcinia* through increased and efficient use:

1. Selecting White *Garcinia indica* types
2. Top working of mother trees of *G. indica* to produce non-plagiotropic shoot growth that can be used while grafting
3. Maintaining an optimum ratio of male and female trees
4. Development of interspecific grafts of *G. indica* and *G. gummi-gutta*.

This GPD contributes to the improvement in total productivity as the innovative propagation techniques can lead to wide production of these rare types. Further, this suite of techniques could contribute to the conservation of economically important *Garcinia* species such as *G. gummi-gutta* and *G. morella*. The suite of techniques essentially helps domesticate and deploy a whole range of variability of *Garcinia* species that have high economic value and hence may have good impact on the livelihoods of communities. Further, the usage of *Garcinia* is strongly associated with local food and health traditions. Because of its excellent anti-gastric reflux properties, it is used by the communities as a home remedy. Adopting the suite of techniques could lead to the establishment of grassroots-level processing units and multi-species gardens, which can support important ecosystem services.

Methodology

The study was conducted in Sirsi, in the Uttara Kannada district of Karnataka state, southern India. A group of scientists interacted with progressive farmers and nursery experts in the focal communities to document the good practices they follow to sustainably use mango and *Garcinia* species. The progressive farmers were identified following baseline data gathered from more than 500 households in the focal communities. All four aspects of good practices – propagation and nursery management, production management, linking of farmers with markets and consolidation of the community's role in management of tropical fruit tree genetic resources (TFTGR) – were included in an unstructured questionnaire that was adopted to elucidate the responses.

In order to identify and document good practices, seven criteria were assessed: (1) diversity of target species, (2) sustainability, (3) improved livelihoods, (4) impacts, (5) potential for scaling up, (6) addressing at least one aspect of good practice and (7) applicability to more than one site (Sthapit *et al.*, 2008). Wherever possible, the practice, method and other details were documented photographically. Local youth clubs and women's self-help groups (SHGs) were also included in the process of documentation. Women harvesters were interviewed separately to get additional inputs. Each good practice was described using descriptors developed by Paul Quek (in Sthapit *et al.*, 2004) and its relevance to the enhancement of livelihood assets was noted for further evaluation.

Description of GPD

Selection of White **Garcinia indica** *types from natural population*

While *G. indica* normally produces red-coloured fruit, there are a few natural mutants that produce pale yellow fruit known as 'White *Garcinia*' (Plates 12 and 13). White *Garcinia* is reputed to have medicinal properties. Ayurveda is one form of traditional medicine that relies on White *Garcinia* to treat severe gastric reflux and Ayurvedic practitioners all over Karnataka state fight over the small White *Garcinia* supply. Due to the high demand, the market price of White *Garcinia* is nearly three times that of Red *Garcinia*. White *Garcinia* is so rare that many people have never heard of it. Even when one man walked more than 500 km through the Sirsi forest range looking for White *Garcinia*, he found only about 20 trees. Because of its rarity, White *Garcinia* is vulnerable to local extinction and current supply cannot meet demand. A new strain of disease or pest problem could be enough to kill all the White *Garcinia* trees in the area or destroy all the fruit in a season.

G. indica is normally seed propagated through sexual reproduction. However, in order to fix the desirable characteristics of the White *Garcinia* mother plant, vegetative propagation is practised by farmers by adopting grafting techniques (Vasudeva, 2013). This also helps maintain the optimal ratio of male to female trees that *G. indica* needs as a dioecious tree. Generally one male tree is necessary for every five female trees in an orchard. White *Garcinia* is extremely difficult to graft because its scions are very thin. The grafts must be perfect or the plant will die. However, farmers from Sirsi have developed a way to successfully grow and graft White *Garcinia*. Dattatreya Hegde from Salkani village (Bhairimane) explains the reason why:

> There is a culture of experimentation here. We are innovators. We are always looking for new ways to improve our farming methods. I saw that I could have a lot of business selling White *Garcinia* so I began experimenting.

The path to growing White *Garcinia* trees was, however, long and difficult. Initially, there was a short supply of White *Garcinia* scions for farmers to experiment with: 'There were only 12 White *Garcinia* trees in Sirsi area. Some of these trees were too young and small for us to take scions from'.

Top working of mother trees of G. indica *to obtain perfect scions*

After they had perfected their grafting technique, the farmers ran into another problem common to many *G. indica* grafts: their scions were growing at an awkward horizontal angle (plagiotropic) instead of growing vertically straight. The trees grew in a bush-like fashion, severely reducing growth and productivity (Plates 14 and 15). Two farmers were able to successfully standardize the technique of top working that produces straight-growing scions. Through experimenting, they discovered a particular way of cutting the mother tree at a height of around 1.5 m. Their scions now grow straight. These few farmers have been able to successfully grow and produce White *Garcinia* fruit and earn a considerable income compared with other farmers (Vasudeva *et al.*, 2013).

Maintaining an optimal ratio of male and female trees

Innovative management techniques have also been developed by the farmers of Sirsi for other *Garcinia* types besides White *Garcinia*. For example, all *Garcinias* are dioecious in nature, so it is essential that male trees occur in a plantation or in natural populations in some critical frequency. Removal of male trees would hamper the fruit set. Farmers have found that maintaining about 20 per cent male trees in the orchard improves the yield.

Development of interspecific grafts of G. indica *and* G. gummi-gutta

G. indica normally grows in drier areas whereas *G. gummi-gutta* prefers moist areas, and so each has a limited geographic spread. By making a graft of *G. gummi-gutta* onto *G. indica* rootstock, farmers have shown that *G. gummi-gutta* can be grown under relatively lower soil-moisture conditions. This interspecific grafting helps to expand the growing areas and to conserve diversity.

Impact on diversity

This suite of propagation techniques helps maintain genetic diversity of *Garcinia* through increased and efficient use as well as on-farm conservation. The suite of techniques essentially helps to domesticate and deploy a range of variability from the wild to the farm. With these techniques, farmers are easily able to cultivate White *Garcinia*, thus preserving this valuable rare species on farmlands. As a testimony to this, today several farming communities have started cultivating the white type of *Garcinia* on their farms. The GPD also contributes to the improvement in total productivity as the innovative propagation techniques can lead to wider production of these rare types. Standardization and adoption of the propagation techniques would greatly help to maintain these, and other, rare types. Furthermore, this suite of techniques could contribute also to the conservation of other economically important *Garcinia* species such as *G. gummi-gutta* or *G. morella* and it could provide the technological innovation required for the increase in area under *Garcinia* species, thus contributing tremendously to rapid domestication.

Impact on livelihoods

This GPD provides an opportunity for building technical capacity among communities, contributing to enhanced human capital through awareness of techniques, natural capital by providing multispecies crops and financial capital through production of grafts of rare varieties or sale of fruit. The GPD has the potential to enhance human capital by capacity building and mobilizing awareness of the market potential of *Garcinia* grafts and fruit, thus contributing to livelihood asset creation.

The techniques provide a technological and innovative basis to independently set up private nurseries of White *Garcinia* and other traditionally and economically important *Garcinia* species. As there is a good demand in the local markets for the grafted plants of these types, farmers can also sell them to earn a small cash income.

Farmers are able to make a substantial profit from producing White *Garcinia* and Ayurvedic practitioners can get their much valued White *Garcinia* fruit. Furthermore, the usage of *Garcinia* is strongly associated with local food and health traditions. Because of its excellent anti-gastric reflux properties, it is used by the communities as a home remedy.

The GPD has strength in increasing on-farm crop diversity, which may reduce the risk of income loss. Furthermore, validated techniques of the GPD could potentially help farming communities to commit to cultivating *Garcinia*. It creates income on a more regular basis during the whole year; hence this practice reduces economic vulnerability.

Sustainability and other benefits

Adopting this suite of techniques has several advantages: (a) the increased availability of rare and valuable *Garcinia* types can lead to the establishment of village-level fruit processing units to produce diverse products such as juice concentrates or jams that can be sold locally; (b) the farmers obtain higher yield and income, hence contributing to their resource capital; and (c) with the adoption of interspecific grafting, the area under *G. gummi-gutta* could be increased through the establishment of plantations even under areas with limited moisture; and (d) the species-rich home gardens support various ecosystem services such as enhancing the local pollinator insects and sequestering soil carbon, providing provisional services to the local people. Such multispecies and diverse home gardens are more resilient for climate change.

Factors favouring or hindering successful functioning of GPD

A major driving force for standardizing a suite of propagation and management techniques for *Garcinia* species has been the innovations by small groups of farmers in Uttara Kannada. Because this GPD has been standardized by innovative farmers themselves, it is likely that these techniques could easily be adopted by other farmers. For the successful functioning of the practice, it is essential that piloting on a larger scale be set up. One factor to consider that might hinder the adoption of the GPD is the fluctuating prices of the *Garcinia* products such as dried rinds and edible butter. However, the flair for experimentation among the farming communities would help in successful functioning.

Recommendations for a way forward

To support the success of this GPD, vigorous grassroots-level training, pilot demonstrations of the techniques and their popularization through media and video documentation are important actions. Demonstration of the tangible benefits derived from the adoption of the GPD is crucial for scaling up. Scaling up can also be effectively carried out with the help of the Department of Horticulture. Indications have already come to suggest that, due to the local cultural importance of White *Garcinia*, the GPD could be successful.

References

Sthapit, B., Sajise, P., Ramanatha Rao, V., Quek, P., Cruz, F.D., and Bellon, M. (2004) 'Selection of good practices of *in situ* conservation of tropical fruit tree species diversity: Methodology and key practices', project document (unpublished), Bioversity International, India

Sthapit, B.R., Shrestha, P.K., Shrestha, P., Upadhyay, M.P., and Eyzaguirre, P. (2008) 'Mobilizing and empowering community in biodiversity management', in M.H. Thijssen, Z. Bishaw, A. Beshir, and W.S. de Boef (eds), *Farmers, Seeds and Varieties: Supporting Informal Seed Supply in Ethiopia*, Wageningen International, Wageningen, The Netherlands, pp. 160–166

Vasudeva R, Hegde, N., and Sthapit, B. (2013) 'Promoting community management of *Garcinia* resources in the Central Western Ghats, India: Initiatives of Tropical Fruits Diversity Project,' in W.S. de Boef, A. Subedi, N. Peroni, and M.H. Thijssen (eds), *Farmers, Biodiversity and Resilience: Empowering Communities to Manage Their Livelihoods*, Centre for Development Innovation, Wageningen, The Netherlands, pp. 163–169

Vasudeva R (2013) 'The master grafting expert and bare-foot breeder of local mango varieties,' in B. Sthapit, H. Lamers, and V. Ramanatha Rao (eds), *Custodian Farmers of Agricultural Biodiversity: Selected Profiles from South and South East Asia*, Proceedings of the Workshop on Custodian Farmers of Agricultural Biodiversity, 11–12 February, New Delhi, India; Bioversity International, New Delhi, India

Vasudeva R, Janagoudar, B.S., Reddy, B.M.C., Sthapit, B., and Singh, H.P. (eds) (2010) '*Garcinia* genetic resources: Linking diversity, livelihood and management', in *Proceedings of National Symposium on* Garcinia *Genetic Resources*, College of Forestry, Sirsi, India, 8–9 May, pp. 1–188

12 A set of interconnected practices which enhance and conserve mango diversity in Malihabad, India

Shailendra Rajan, Hugo A. H. Lamers and Barsati Lal

Historical, cultural and agro-ecological context

In the great Hindu epic the *Ramayana*, the poet Valmiki writes of forests of mango trees spread across the land of Rama and his forebears about 4000 years ago.[1] This land is Uttar Pradesh, still one of the centres of mango production and diversity in India. There are three main centres of varietal diversity in India, where wide variability in cultivated types is still available: the Lucknow–Saharanpur belt of Uttar Pradesh, the Murshidabad area of West Bengal and the Hyderabad area of Andhra Pradesh (Yadav and Rajan, 1993; Ram and Rajan, 2003). Uttar Pradesh produces nearly 24 per cent of the mangoes in India, which is 3.6 million tonnes, more than any other state in the country (Yadav and Rajan, 1993; Ram and Rajan, 2003). Lucknow is the current capital of Uttar Pradesh and the former capital of the Nawabi rulers,[2] who were part of the Moghul Empire in the eighteenth and nineteenth centuries. The Nawabs played a major role in making Lucknow a centre of varietal diversity by establishing mango orchards with varieties collected from all over India (Mukherjee, 1953). Similarly, the mango plantations in the subdistrict of Malihabad were developed by Pathans,[3] influential trading families under the patronage of the Nawabs of Lucknow (Rajan *et al.*, 2013b).

Mango plays a significant role in Indian culture. Several Urdu philosophers and poets of the nineteenth century, such as Nazeer Akbar Abadi, Ghalib and Iqbal, have written about mangoes. Ghalib, known as a great mango connoisseur, is said to have loved eating mangoes more than composing his couplets.[4] Mango blossom is used even today for the worship of the goddess Saraswati and mango leaves are strung over doorways on auspicious occasions and as protection against evil spirits.

Malihabad subdistrict, located 20 km northwest of Lucknow, has a population of more than 16,000 people, of which the majority claim Pashtun descent. It is the area of origin of the mango variety Dashehari, which has dominated mango production in Uttar Pradesh over the last four or five decades. Malihabad

GPD 'passport'

GPD code:	03
Focus area:	*All four categories* • Propagation and planting materials • Production and crop management • Commercialization and home use • Collective action and social networking
Character:	System of three interacting practices
Species and varieties involved:	*Mangifera indica* including: 51 named farmer varieties all gathered in one orchard with trees more than 80 years old. Ramkela, Gola, Surkhi, Sundari, PaudaGaj, Deshi Bombaiya, Machhli, Pan, Matka Gola, Chandni, Bhura, Nauraj, Surkha Matiyara, Nazir Pasand, Baudi, Kamal Pasand and others. About 2,971 seedling trees that can all be considered as varieties as they are all genetically distinct. Amin, August, Bhagwanta, Bhola, Bhuzada Anees, Egrohiya, Jabjanat, Khala Khan, Lambauri, Seedling, Surkha, Suwaswala, Tuhiya Pahad, Tuhur, Tukuroo, Zardalu. More than 600 distinct named varieties or seedling types that have been displayed during several diversity fairs by farmers from Malihabad.
Name of location:	Malihabad, Lucknow
GIS reference of location(s):	N 26°91″; E 80°71″ Elevation: 128 masl
Name of farmers (data resource):	Heritage orchard: Mr Nawab Hassan Farmers planting seedling trees: Mr Chhote Lal Kashyap, Mr Raja Ram, Mr Amir, Mr Affak, Mr Anish Ahamad, Mr Shadab Ahmad, Mr Maiku Lal Nursery owners and mango experts of Malihabad: Abdullah Nursery, Mr Monish Ahmed, Mr Shadab Ahmad

is richly covered by mango trees and harbours many private and public sector nurseries. About 25,000–30,000 hectares are under mango production alone, comprising a major portion of annual mango production in Uttar Pradesh and India (Plate 16). Nowadays about 80 per cent of the tree population is of Dashehari trees only, 13 per cent of the trees are Lucknow Safeda and 5 per cent are seedling varieties. The remaining 2 per cent comprise all other farmer varieties (Gajanana *et al.*, 2014). Trees that do not have specific names are called just 'biju' (i.e. seedling) or 'deshi' (i.e. local).

The subdistrict is located in the Central Gangetic Plains, which have a subtropical climate featuring three distinct seasons: summer, monsoon and winter. Winter commences usually in the month of November and extends until March, followed by summer (April to mid-June) and then monsoon starts and lasts up to September or October. The maximum temperature is during the month of May (40–42°C) and the minimum temperature during January (5–7°C). Yearly average rainfall is about 1,014 mm, of which 90 per cent falls during the monsoon. The subtropical climate enables synchronous flowering and fruiting (Ram and Rajan, 2003) and suits varieties that require chilling before flowering.

Soils in Malihabad are deep and have developed from the alluvium deposits of rivers. The soils are neutral to moderately alkaline and calcareous especially at lower depths. Some of the land types around the basins of the Gomti River are sandier and less fertile. The specific climate and soil conditions of Malihabad, with extremely hot, rainless summers, help in developing premium quality Dashehari fruit, which are of better quality compared with other districts (Rajan, 2009). In 2009, the Dashehari mango from Malihabad subdistrict received Geographic Indication (GI) recognition as proof of its distinct quality.

Mango is one of the major income sources in the region, being exported to many neighbouring states. About 70 per cent of the mangoes in Malihabad are sold through pre-harvest contractors for a fixed price per quintal (100 kg) negotiated during fruit setting in April. Besides mango, people grow wheat, rice, pulses, chickpea, sugarcane or vegetables and often keep cows and buffalo for milk and manure. The average orchard size in Malihabad is 1.05 ha, with a few large farmers holding 4–8 ha. Farmers normally plant around 90 trees per hectare. The average household income of mango growers is INR102,131 (US$1,621) per annum and average income per hectare of mango is estimated as INR54,488 (US$865).

Methodology used for data collection

Focus group discussions were conducted with community representatives, mostly male farmers, to gather information on good practices in Malihabad. The participatory rural appraisal technique, comprising semi-structured interviews with key informants, was used for further information collection. Informants were identified based on the project baseline data (those farmers with high diversity or many seedlings types) or based on recommendations arising from the focus groups. Ten nursery experts, 42 farmers with mango orchards and 51 farmers who grew seedling types on farm boundaries were interviewed to collect information on the practices that contributed to the rich diversity in Malihabad. These discussions and the semi-structured interviews were often combined with other project activities, such as meetings or training courses. 15 custodian farmers in four communities were identified (see Chapter 4 for information about custodian farmers) and information obtained from them was added to further improve and validate the good practice descriptions.

Description of GPDs

The historical background of mango production in Malihabad has accumulated in several socio-culturally embedded good practices in the management and use of mango diversity that have contributed to the informal seed system in Malihabad. This chapter describes three practices that have evolved in Malihabad: (1) tradition of maintaining heritage orchards, (2) establishment of seedling types in marginal environments and along land borders and (3) the practice of organizing mango festivals to facilitate taste and trait evaluation and the exchange of knowledge and grafts of preferred seedlings for further multiplication by farmers and nurseries.

Despite the prevalence of the Dashehari variety, some farmers retain an interest in old farmer varieties or seedling types as the market for Dashehari seems saturated and prices have declined in the last few years. Reinvigorating old planting practices could help ensure conservation of these varieties and facilitate access to them, and could form the basis for the exploration of niche markets for mango diversity in Malihabad. The system of exchange of planting materials (grafts) has benefited the spread of different varieties, as evidenced by the display of farmer varieties and best seedling selections during diversity fairs or gifted to influential noble or business people. Influential nursery families in Malihabad screen the varieties presented at diversity fairs for potential planting material that could be taken up in their mother blocks. Traditional systems such as thick planting of seedling types along the boundary of commercial orchards are considered a good practice that supports on-farm management of mango diversity in the region.

Heritage orchards

In the sixteenth century the Moghul emperor Akbar ordered one lakh (100,000) of mango trees to be planted in one estate near Darbhanga in Bihar, which was known as Lakhibagh (Mukherjee, 1953; Singh, 1960). Similarly, the Nawabs of Lucknow, during their reign in the eighteenth and nineteenth centuries, planted a large number of mango trees of different elite varieties in several orchards surrounding Lucknow, including Malihabad subdistrict. These orchards were established as a sign of status and pride, but also for their economic value. The fruit was widely praised and relished in the Indian subcontinent and popular among rich and poor across all religions. The Nawabs were very fond of this special fruit and took pride in collecting as many varieties as possible in their orchards, ordering and collecting varieties from all over the Indian continent.

Mr Nawab Hassan from Kasmandi Kalan village in Malihabad continues this long family tradition in the cultivation of mango started by his ancestors. He is the proud owner of an old orchard where 40 different varieties are planted in one block (Rajan *et al.*, 2013b), with other varieties planted in other plots, all in all totalling 51 named varieties maintained. The general tendency in the last two decades to plant only commercial varieties such as Dashehari and Lucknow Safeda has not convinced him to cut down the old trees he inherited from his ancestors. Several 100-year-old trees, planted at the time when mango varieties were owned for family pride, feasts and exchange by the Nawabs of Lucknow, are still thriving in his orchard (Rajan *et al.*, 2013b).

Mr Hasan Ahmed, the father of Mr Nawab Hasan, purchased a piece of land about 60 years ago in Kasmandi Kalan. The plot contained a collection of lesser-known mango varieties, including traditional farmer varieties collected not only from Malihabad but also from other mango-growing areas such as Shahbad and Sandila, about 160 km northwest of the state capital, Lucknow. These varieties were collected and planted by the then owner of the land, a wealthy citizen of Lucknow, who had a flair for collecting and maintaining the diversity in his orchard. The collection contains a wide range of grafted non-commercial varieties. Most people were not interested in this plot of land; only people with knowledge of mango varieties admired the richness of the collection. The collection includes some of the unique varieties of Malihabad. Several years ago Mr Nawab Hassan considered cutting down the trees and replacing them with commercial types, but due to his curiosity, family pride and an emerging market interest for rare varieties, he has now been convinced to maintain this old heritage orchard.

Impact on diversity

Mr Nawab Hasan maintains 51 different traditional farmer varieties in his orchard, most of which are selections collected from different mango-growing areas. Varieties such as Surkhi, Sundari, Pauda Gaj, Deshi Bombaiya, Machhli, Pan, Matka Gola, Chandni, Bhura, Nauraj, Surkha Matiyara, Nazir Pasand, Baudi and Kamal Pasand may not be available in other orchards in Malihabad (Dinesh *et al.*, 2014). A few of them are the only living trees of these varieties on earth. The age of the trees ranges from 50 to more than 100 years, thus showing a continuous process of augmentation of varieties over decades. The trees are all grafted plants, providing an unbiased collection of trait-specific germplasm, which can be considered like a farmer's field genebank.

More generally, nursery experts and orchardists have been the curators of mango varieties in Malihabad for several generations. Maintaining diversity-rich mother blocks adds to the status of the nurseries and their owners (Sthapit, 2010a; 2010b). Traditional nursery families have collected and maintained mother plants of a wide range of traditional and lesser-known farmer varieties for several generations in their nurseries. These superior trees are maintained to take scions for grafting and the multiplication of saplings. Nursery experts have collected these elite materials from farmers all over India and even abroad. For example, Abdullah Nursery in Malihabad has grafted scions of more than 300 distinct varieties on a single tree both to save space and out of curiosity (Sthapit, 2010a; 2010b). Established nurseries such as Abdullah Nursery earned their reputation by maintaining a large number of varieties, while new nurseries tend to be limited to producing commercial varieties only.

Impact on livelihoods

The unique varieties maintained in the heritage orchards linked to the Nawabs were mostly used as showpieces during celebrations or used as special gifts in the form of the fruit or a sapling – a custom that has been eroded but still exists. However, many of these varieties were not known to the general consumer, received a low sale price and gradually the owners' enthusiasm declined. This led to the replacement of many old trees with commercial varieties or other crops. Mr Nawab Hasan, too, was reluctant to keep up the old orchard because of low returns. However, for the last four or five years he has been able to get a better price for the lesser known varieties. His efforts to sell the fruit to selected traders with an interest in old varieties, rather than the general mandi (government-controlled wholesale market), have been successful and at present he feels satisfied with his earnings. The higher price for the fruit has changed Mr Hasan's views about conserving the varieties. Initially the varieties were conserved because of affection and attachment to the trees planted by his ancestors, whereas now their conservation is also supported by a fair price for the unique varieties available in his orchard.

Sustainability and other benefits

So far the practice of maintaining heritage orchards has not been sustainable, as many orchards have been lost or are now limited to a few commercial varieties (e.g. Dashehari, Lucknow Safeda). The loss of the heritage orchards is due mostly to conversion to commercial varieties, but is also a result of the encroachment of Lucknow, where land prices have increased substantially which, in turn, has led to the conversion of orchards into residential areas. If these generations-old mango farming families, such as Nawab Hasan's, received more recognition for their conservation role and could find niche markets for some of their heritage farmer varieties, they would be more inclined to maintain and conserve them.

Establishment of seedling types in marginal environments and along land borders

Mr Chhote Lal Kashyap is a farmer in Gopramau village and belongs to the middle-income group. There are 19 people in his family, including 12 grandchildren. About five decades ago, Mr Chhote Lal, owner of approximately 2 ha, became interested in mango cultivation after seeing the orchards of the mango farmers in other villages in Malihabad subdistrict. His land is located in the Gomti River basin, where it was not common to grow mango. Nevertheless, he made efforts to plant commercial varieties such as Dashehari, but failed to grow the grafted saplings as a result of the poor sandy soils, undulating landscape and lack of irrigation facilities in his village. He met some farmers that planted seedling mango in a similar environment in the Malihabad-Kakori-Mal area and tried adopting the practices they recommended in his land. Seedling types are trees grown from seed, because of their multipurpose character, for a wide range of uses and their ability to thrive on land where grafts are difficult to establish. Eventually his efforts in planting seedling types were successful, resulting in an orchard of more than 100 seedlings. Initially he planted 150 Dashehari grafts that he purchased from the Malihabad commercial nursery, of which about 35 survived. Dead grafts were replaced with seedling types. Out of these seedling trees he selected the better ones and removed the rest, increasing the number of varieties to 135. Experimenting in this way, he gradually became the owner of an orchard that is very rich in mango genetic diversity (Rajan *et al.*, 2013a).

Mr Maiku Lal, another farmer, started mango cultivation about 35 years ago in Sarsanda village (Rajan *et al.*, 2013c). He developed a flair for the cultivation of seedling types. In addition to planting seedling varieties in sandy soils, he planted grafted saplings of commercial varieties in loamy soils, which retain more water and nutrients. He is convinced that his way of organizing and managing his orchard, including a good mix of common grafted varieties and lesser-known seedling varieties, provides him with the best income possibilities and benefits for his family's livelihood (Rajan *et al.*, 2013c).

Several other farmers (Mr Amir, Mr Affak, Mr Anish Ahamad and Mr Shadab Ahmad) plant a row of seedling types around their orchard of commercial varieties as a fence or to mark their land borders, as orchards are planted close to each other in Malihabad (Plate 17). Normally farmers plant 100 grafted Dashehari trees per hectare, but when including a hedgerow of seedlings they increase the number of trees to approximately 150 trees per hectare. Mr Amir observed that the seedlings are much more vigorous compared with grafted plants; they can grow twice as fast. Seedling trees grow taller and can survive much better without irrigation or application of compost or fertilizers. They also often require less spraying for pest control; grafted Dashehari is often sprayed two or three times a season, whereas seedlings are sprayed only once for pest management.

Impact on diversity

Mango, being a cross-pollinated and highly heterozygous plant, has high intraspecific diversity (Mukherjee, 1953; Ram and Rajan, 2003). The traditional practices of multivarietal orchards and the planting of seedling types in orchards or as hedgerows allows for cross-pollination across genetically distinct types when multiple varieties or seedlings are planted close to each other (Degani *et al.*, 1997). The combination of such planting practices allows evolutionary breeding and generates rich variability from which the farmer can select. Seedling trees in home gardens, as boundaries or in orchards provide an opportunity for the selection of promising types (Plate 18). Farmers and nursery owners have over generations evaluated seedling trees regarding their performance and fruit quality, and subsequently selected the best plants to take some stones and plant them again as a seedling or, when convinced about its unique quality, taken a scion from the tree for grafting. Almost all the commercial cultivars of mango in India have arisen as a result of farmers' and nursery experts' selection from seedlings.

Mr Chhote Lal now maintains about 135 different seedling types and three grafted varieties in his orchard, including Biju Deshi Dashehari, Deshi Chausa, Tukmi Heera, Sunehra, Badamba, Gola and Dil Pasand (Rajan *et al.*, 2013a). He has named some of the seedling types on the basis of their resemblance to known parent cultivars or because of the similarity of particular shapes or colours of the fruit. Seedling trees whose seeds were taken from a popular variety are often named after them. For example, Deshi Dashehari means a seedling of Dashehari. He selects the most attractive and high-quality fruits when selecting seeds or stones for planting seedlings. Traits used for this selection are: colour (more yellow than green at the ripe stage), shape (more uniform and less asymmetrical), pulp colour (orange), sweetness and pleasing aroma, as these traits are preferred by local consumers. In this way, this practice contributes not only to the maintenance but also to the enhancement and increase of diversity on farm. This practice allows a kind of evolutionary breeding for the tropical mango crop species.

About 40 different mango varieties developed from seedlings are maintained in the orchard of Mr Maiku Lal. Next to commercial varieties such as Dashehari, he maintains lesser-known varieties such as Tukmi Chausa, Gulab Jamun, Kism Safeda, Deshi Taimuriya, Gola Seb, Lambauri Chausa, Deshi Langra, Tukmi Surkhi, Lambauri, Safeda, Chonha Gola and several unnamed seedlings in about 2 ha of land.

Impact on livelihoods

Mr Maiku Lal considers the non-commercial varieties a better option for household consumption because of their high digestibility, juiciness and variation in taste and aroma. He also thinks that seedling types have a higher nutritional value because of their digestible fibres and suitability for making juice. Some of the seedling types are used for making pickle or are made into a powdered food ingredient (aamchoor) to impart a sour taste. According to Mr Maiku Lal, the commercial variety Dashehari provides fruit for only about one month. Other farmer varieties are available even after the end of the Dashehari season and a continuous supply of fruit from the orchard is possible for a longer period. In recent years he has noticed that fruit of some seedlings even get a better market price than commercial varieties. He considers an orchard based on seedlings or some lesser-known varieties as the best option for sandy soils with limited irrigation facilities.

Mr Chhote Lal maintains seedlings as the strong tap root system gives them higher survival rates in sandy soils compared with grafted saplings. Over the years he has also observed that harvest of seedling types is possible for a longer period than Dashehari, thus providing a prolonged supply of fruit for home consumption and sale. His income starts with the early sale of fruit from seedling trees suitable for pickle in May and June and continues even after July, which is the end of harvesting of commercial varieties like Dashehari. Seedling varieties mature at different times, thus allowing farmers to avoid the dip in market price due to the Dashehari glut. Recognizing these advantages, he is no longer eager to replace his seedling trees with commercial varieties. The productivity of seedling types is sometimes higher than commercial types; he has noticed that his orchard is much more productive under challenging pest conditions. Under conditions of water scarcity, Dashehari and other varieties produce small fruit, whereas several seedling types ripen late and develop a good fruit size because of rains before their harvest. During the mango season, Mr Chhote Lal devotes most of his time to the orchard. He also has a second occupation, as a tailor, when there is not much work in the orchard.

Sustainability and other benefits

It was a difficult task for Mr Chhote Lal to establish the orchard on sandy undulating land where irrigation facilities were not available. Cattle and wild animals damaged the saplings and he often used thorny bushes and shrubs to fence off the plants from grazing animals during the initial years of orchard establishment. However, nowadays Mr Chhote Lal gets half of his income from the orchard with its unique seedling types and prefers to plant new trees as seedlings instead of using grafted saplings. Mr Maiku Lal, together with his son Mr Raja Ram, wishes to continue the cultivation of a mixed and diverse group of varieties in his orchard for income and home use but also to conserve diversity richness. Mr Maiku Lal says it is important to find specific markets where they can obtain good prices for these traditional farmer varieties, as in the general mandi (market) prices and interest are generally low.

Mango festivals to exchange grafts and fruit

In the eighteenth and nineteenth centuries, during the harvest season, mango feasts were organized[5] by the then Nawabi rules and the best selections of

Plate 1 Photographs of custodian farmers

Plate 2 Typical commercial vegetable garden in Niamey, Niger

Plate 3 A gardener harvesting vegetables for sale

Plate 4 Selected mango accessions collected from farms in Kenya

Plate 5 Award certificate and bonus

Plate 6 Another award certificate and bonus

Plate 7 A typical Kandyan home garden in Sri Lanka

Plate 8 Roadside fruit stall near Manaus, Brazil

Plate 9 Sweet, white flesh of bacuri

Plate 10 Senhor Roxinho holding uxi fruit

Plate 11 Managing bacuri trees to restore degraded pasture

Plate 12 Two forms of Garcinia found in India

Plate 13 Tree with White Garcinia fruits

Plate 14 Normal tree

Plate 15 Plagiotropic bush with wrong scion selection

Plate 16 Mango orchard in Malihabad

Plate 17 Mango seedlings grown as hedgerow

Plate 18 Mango fruit variability

Plate 19 Rangpur lime (*C. limonia*) rootstock

Plate 20 Rough lemon (*C. jambhiri*) rootstock

Plate 21 Galgal (*C. pseudolimon*) rootstock

Plate 22 Nagpur mandarin (*C. reticulata*)

Plate 23 Maintenance of Rangpur lime in farmer's orchard

Plate 24 Maintenance of Rangpur lime in farmer's orchard

Plate 25 Nursery of Vasant Wankhade

Plate 26 Production systems of mandarin (*C. reticulata*) in Indonesia

Plate 27 Production systems of mandarin (*C. reticulata*) in Indonesia

Plate 28 Production systems of mandarin (*C. reticulata*) in Indonesia

Plate 29 Production systems of mandarin (*C. reticulata*) in Indonesia

Plate 30 Citrus market in Indonesia

Plate 31 Citrus market in Indonesia

Plate 32 Citrus propagation in South Kalimantan, Indonesia

Plate 33 Suradet Tapuan, innovative grafter

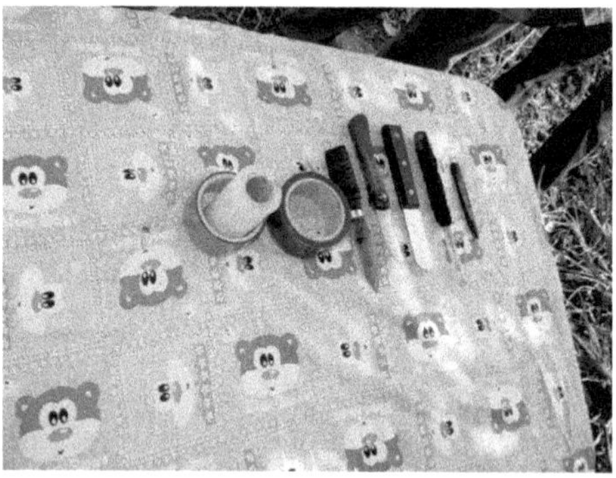

Plate 34 Equipment needed for side-grafting

Plate 35 New mango orchard in sloping land

Plate 36 Fruiting in side-grafted branch

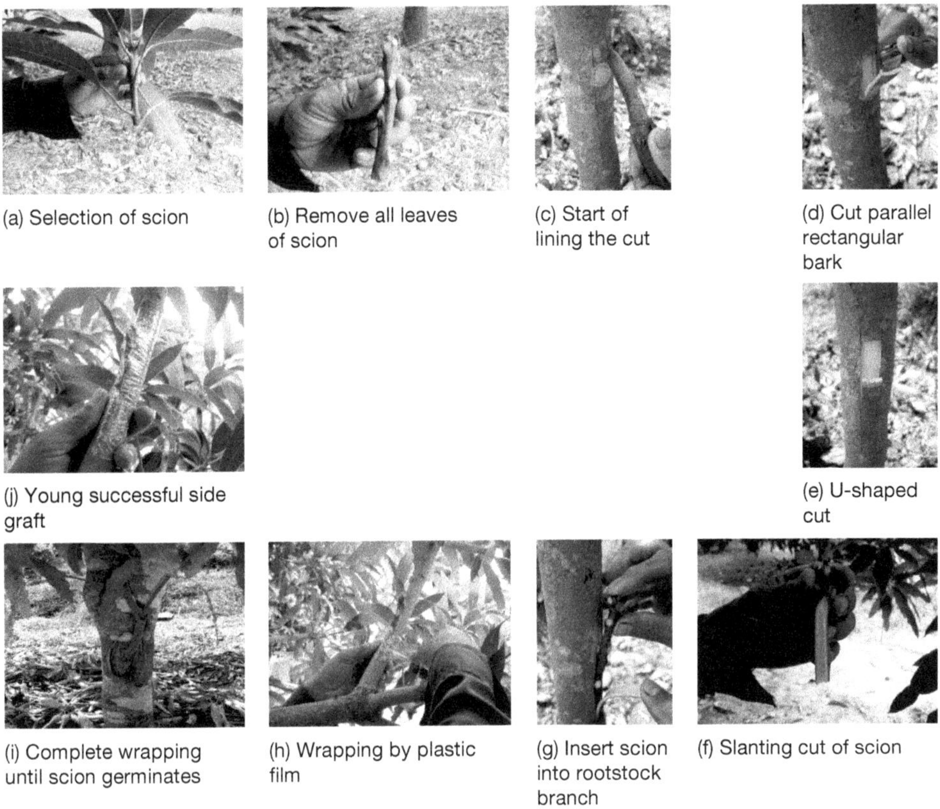

(a) Selection of scion

(b) Remove all leaves of scion

(c) Start of lining the cut

(d) Cut parallel rectangular bark

(e) U-shaped cut

(f) Slanting cut of scion

(g) Insert scion into rootstock branch

(h) Wrapping by plastic film

(i) Complete wrapping until scion germinates

(j) Young successful side graft

Plate 37 Step by step side-grafting technique

(a) Typical erect tree of Asam gelugor (orthotopic growth)

(b) Asam gelugor tree with (plagiotropic growth) branching

(c) Asam gelugor fruits

(d) Sun drying of sliced Asam gelugor for dried gelugor

(e) Asam gelugor curry

(f) Chutney from Asam gelugor

Plate 38 Morphotypes of asam gelugor trees, fruits and their uses

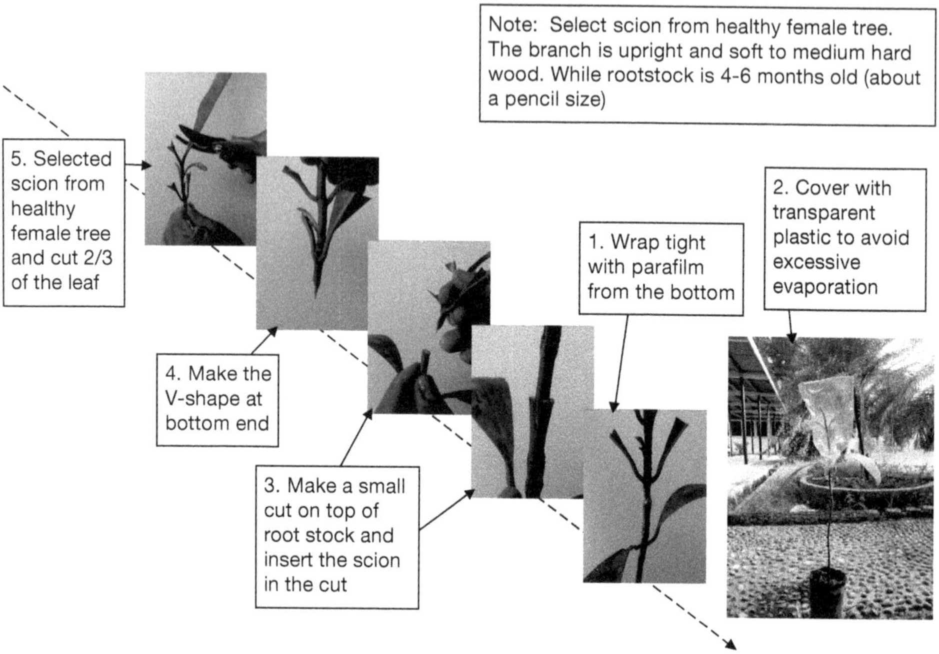

Plate 39 Farmers' practice of cleft grafting of *G. atroviridis*

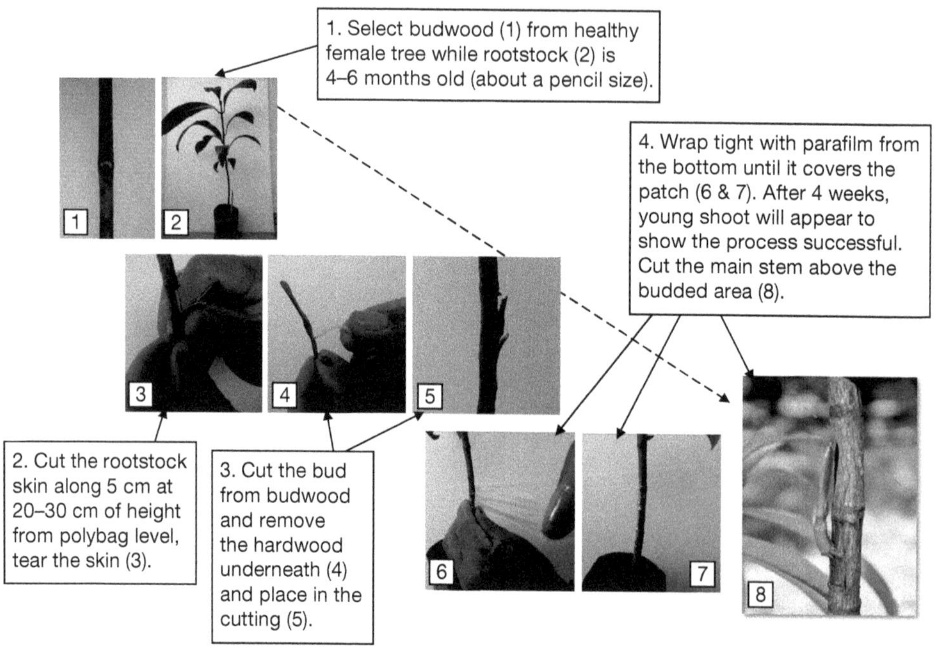

Plate 40 Farmers' practice of patch grafting of *G. atroviridis*

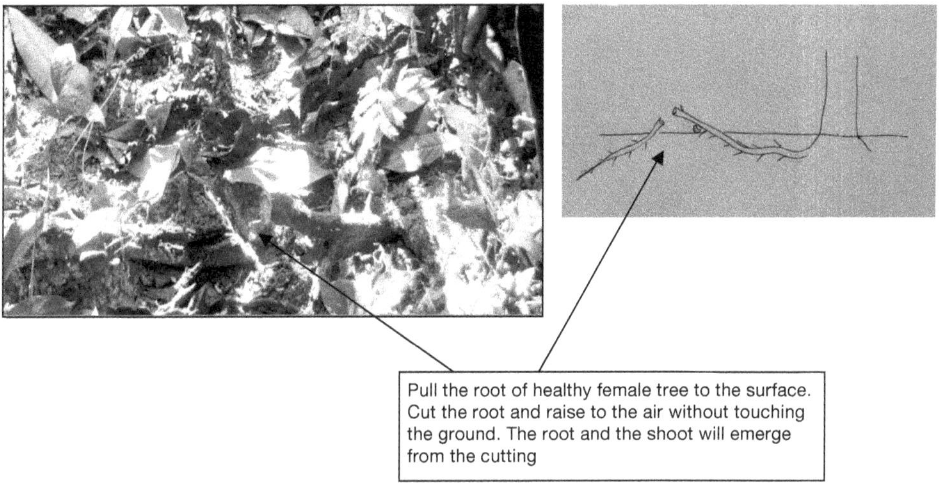

Plate 41 Traditional marcotting technique through root cuttings

(a) Significance of pomelo amongst landpoor homestead

(b) Local fruit market of pomelo

(c) Offering of pomelo and other fruits during Chhath puja

(d) Ladies holding offerings to God during Chhath puja

(e) Artificial waterbody prepared for Chhath puja

(f) Ladies performing puja during Chhath puja

Plate 42 Celebrating Chhatha Puja in India

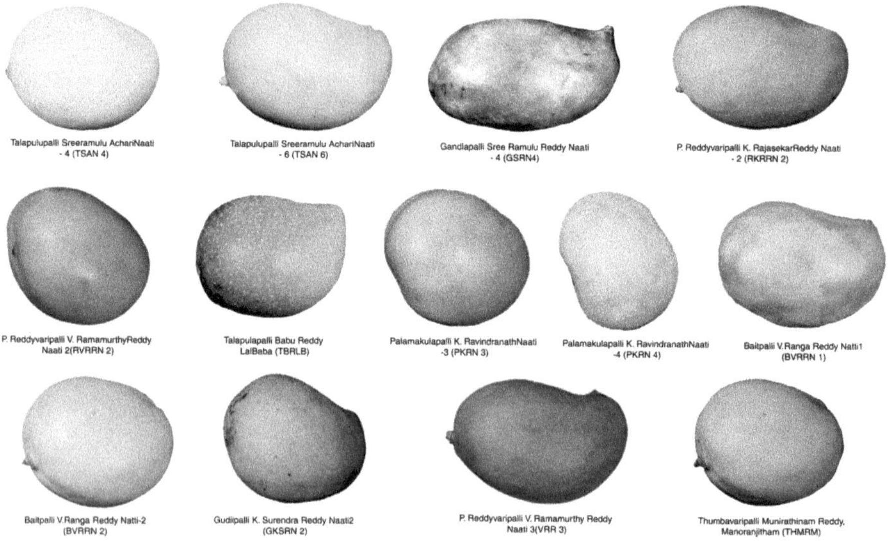

Plate 43 Mango fruit morphological diversity

Plate 44 A typical integrated home garden in East Java, Indonesia

(a) Madan shrubs in river bank habitat

(b) Natural habitat along the Thap Than River, Sisaket

(c) Planted along the boundary of home garden to reduce pressure on natural habitat

(h) Community participated in wild Madan planting at the degraded land along the river

(d) Harvested Madan stick before peeling

(g) Madan nurseries ready for community plantation

(f) Chicken grilled into Madan skewer

(e) Peeled stick of Madan

Plate 45a–h Production and management of Madan

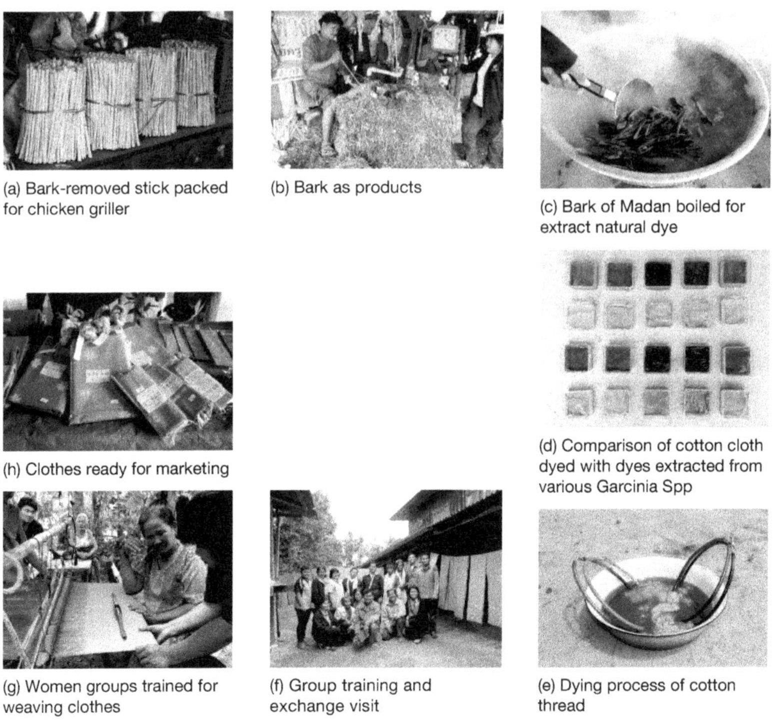

(a) Bark-removed stick packed for chicken griller

(b) Bark as products

(c) Bark of Madan boiled for extract natural dye

(h) Clothes ready for marketing

(d) Comparison of cotton cloth dyed with dyes extracted from various Garcinia Spp

(g) Women groups trained for weaving clothes

(f) Group training and exchange visit

(e) Dying process of cotton thread

Plate 46 Farmer practices of value addition of Madan

(h) Multi-species rich home garden.

(a) Aroi Tulen fruit on tree.

(b) Varieties/ species.

(g) Upscaling of the solar-drying cabinet to a solar-drying house. Credit: L. Jousim

(f) Dried rinds.

(e) Product label for packaging dried Aroi-Aroi rind designed by Agriculture Department to enhance value.

(d) Farmer using portable solar drying cabinet.

(c) Traditional sun drying rind.

Plate 47 Aroi Tulen fruit on tree

Plate 48 Awareness programme at the fruit diversity garden

Plate 49 Awareness programme at the fruit diversity garden

Plate 50 Kampung Kakeng jungle trek after upgrading

Plate 51 Kampung Kakeng jungle trek after upgrading

Plate 52 Tourist guide training for the community of Kampung Kakeng

Plate 53 Tourist guide training for the community of Kampung Kakeng

Plate 54 Tourist guide training for the community of Kampung Kakeng

Plate 55 Tourist guide training for the community of Kampung Kakeng

Plate 56
A typical Kiriwong village landscape

Plate 57
A typical Kiriwong village landscape

Plate 58 Plant materials required

Plate 59 Capacity building

Plate 60 Market outlet of diverse fabrics

Plate 61 Production of mangosteen soap

Plate 62 Diverse products

Plate 63 Market outlet in Kiriwong

Plate 64 Homestay facilities and surroundings

Plate 65 Homestay facilities and surroundings

Plate 66 Mixed fruit cropping orchard

Plate 67 Mixed fruit cropping orchard

Plate 68 Mixed fruit cropping orchard

Plate 69 Mixed fruit cropping orchard

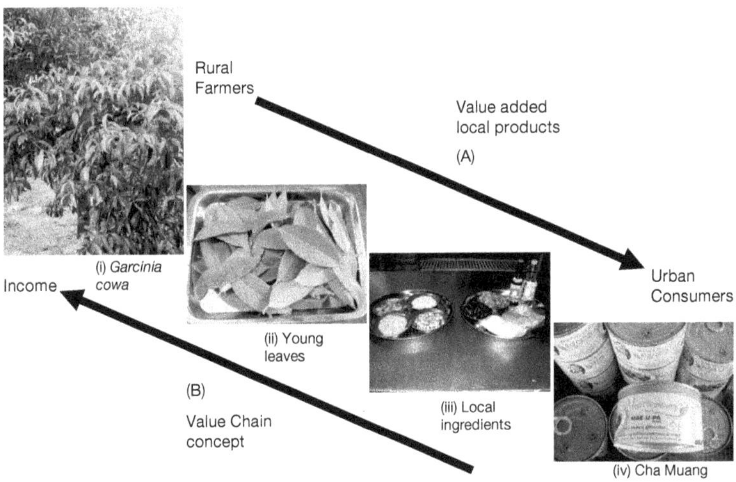

Plate 70 Processing of young leaves of *G. cowa* in Thailand

Plate 71 Traditional fruit rind dryer

Plate 72 Improved energy-efficient fruit rind dryer

Plate 73 Fruit morphotype

Plate 74 Mr Eshanna, custodian farmer with grafting expertise

Plate 75 Appemidi mango pickle stored in brine

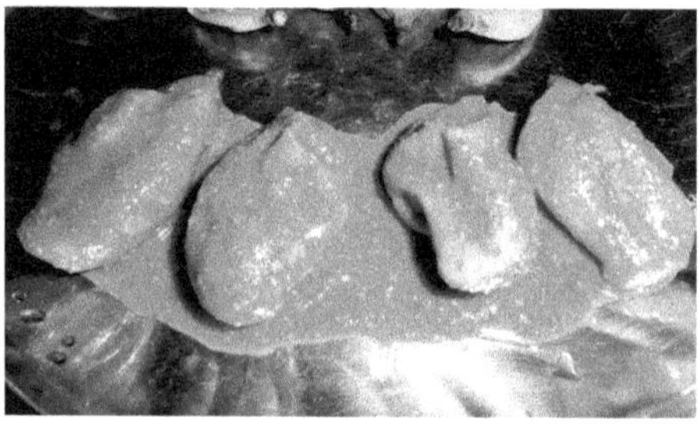

Plate 76 Ready to use pickle

Plate 77 Participatory methods to build social capital

Plate 78 Awareness through Padyatra

Plate 79 Women SHS group meeting

Plate 80 Men SHS group meeting

mangoes were often shared as a gift of pride with influential families, friends and other noble families. For the last two or three decades, similar mango festivals (mango mela) have been organized, where a wide range of varieties are displayed and visitors can buy boxes of their favourite varieties. As costs of organization and publicity are substantial, nowadays mango melas are organized mostly by the government, trade associations or tourism boards in major cities such as Lucknow, Bangalore or Delhi. Examples include the yearly Mango Mela, which was organized for the twenty-sixth time by the tourism board in New Delhi during July 2014. Similar melas are organized yearly in Hyderabad, Bangalore and Pinjore (Haryana) to attract consumers and buyers.

Lucknow has a similar tradition of yearly mango festivals organized in the city. In addition, several mango diversity fairs were organized by the Society for the Conservation of Mango Diversity (SCMD) in Malihabad in collaboration with the Central Institute for Subtropical Horticulture (CISH) in 2011 and 2012, in which the farmers played the central role.[6] In 2014 the Mango Mela of Lucknow was organized for the first time more traditionally inside the Habibullah Estate Orchards in Saidanpur village of Barabanki district, about 54 km outside of Lucknow.[7] The key organizers were an NGO called Agribusiness Systems International (ASI) in collaboration with private sector sponsors (Hindustan Times, a newspaper company; Maaza, a soft drink manufacturer; the Taj Hotel). It was organized as part of a horticultural market development programme targeting women (Sunhara, India).[8]

Impact on livelihoods and diversity

These local festivals contribute to the promotion and maintenance of several lesser-known farmer varieties or superior seedling selections. During such events, consumers, farmers and nursery experts taste, evaluate and 'discover' new varieties or seedling selections with market potential. Visitors are interested in discovering new tastes and varieties, while nursery experts are interested in finding potential new plant material to include in their nurseries' mother blocks. These village festivals or mango melas in the cities provide farmers with the opportunity to sell directly to consumers or retailers their fruit, their processed products or both, especially lesser-known varieties. This increases the price and the margins farmers can obtain, although turnover is often relatively small, especially for lesser-known varieties. Such melas are an ideal place to test and try new varieties and products with customers and to make improvements based on their direct feedback.

In the last few years, the market for saplings of some traditional varieties has been increasing slowly in Malihabad. A few farmers sell, at a good price directly in urban markets, unique farmer varieties such as Ramkela (preferred for pickling), Gola, Katchameetha or Husnara. Reinvigorating a market for seedlings and lesser-known farmer varieties would help to conserve the wide range of diversity found in Malihabad.

Conclusion

The above cases illustrate that the combination of several traditional farmer and nursery practices have generated rich intraspecific diversity of mango in Malihabad despite the prevalence of Dashehari and other commercial varieties. Over generations, a large number of commercial mango varieties have been developed from this diversity, including the now most popular variety Dashehari, which supports the livelihoods of thousands of farmers in Malihabad and beyond. The traditional seed system for mango in Malihabad is built upon the combination of several practices including: (1) tradition of maintaining heritage orchards, (2) establishment of seedling types in marginal environments and along land borders and (3) the practice of organizing mango festivals. These practices promote cross pollination and gene flow that allows the process of evolutionary selection to happen in the informal seed system. Interestingly, these practices were observed separately, as practised by the individuals interviewed, but when put together they sustain a local seed system for a perennial species that is strongly embedded within the socio-cultural traditions of the region.

References

Degani, C., Yutko, O., El-Batsri, R., and Gazit, S. (1997) 'Outcrossing rate in adjacent "Maya" and "Tommy Atkins" mango blocks', *Scientia Horticulturae*, vol 70, no 1. DOI: 10.1016/S0304-4238(97)00038-1

Dinesh, M.R., Rajan, S., Singh, I.P., Singh, A., Singh, S.K., Vasudeva R, Vinoth, S., Patil, P., Parthasarathy, V.A., Reddy, B.M.C., and Sthapit, B.R. (2014) *National Fruit Catalogue of Tropical Fruit Tree Diversity (Mangifera, Citrus and Garcinia)*, ICAR, Bioversity, India

Gajanana, T.M., Lamers, H.A.H., Rajan, S., Singh, A., Singh, S.K., Singh, I.P., Vasudeva R, Hegde, N., Dinesh, M.R., and Dakshinamoorthy, V. (2014) *Baseline Report India for the UNEP-GEF Project Entitled: Conservation and Sustainable Use of Cultivated and Wild Tropical Fruit Diversity: Promoting Sustainable Livelihoods, Food Security and Ecosystem Services*, GEF, UNEP, Bioversity and ICAR, India

Mukherjee, S.K. (1953) 'The mango – its botany, cultivation used and future improvement, especially as observed in India', *Economic Botany*, vol 7, pp. 130–162

Rajan, S. (2009) 'Impact assessment of climate change in mango and guava research', in P.M. Govindakrishnan, J.P. Singh, S.S. Lal, V.K. Dua, S. Rawat, and S.K. Pandey, (eds), *Information Technology Applications in Horticultural Crops*, Central Potato Research Institute, Shimla (HP), pp. 36–42

Rajan, S., Kishore, R., Ahmad, S., and Vijay (2013a) 'Chhote Lai Kashyap: Limited resources led farmers to create and conserve mango varieties, Malihabad', in B.R. Sthapit, H. Lamers, and V. Ramanatha Rao (eds), *Custodian Farmers of Agricultural Biodiversity: Selected Profiles from South and South East Asia*, Proceedings of the Workshop on Custodian Farmers of Agricultural Biodiversity, 11–12 February, New Delhi, India; Bioversity International, New Delhi, India

Rajan, S., Kishore, R., Ahmad, S., and Vijay (2013b) 'Nawab Hasan: Conserving century-old mango varieties in Kasmandi Kalan: The story of the Nawabs', in B.R.

Sthapit, H. Lamers, and V. Ramanatha Rao (eds), *Custodian Farmers of Agricultural Biodiversity: Selected Profiles from South and South East Asia*, Proceedings of the Workshop on Custodian Farmers of Agricultural Biodiversity, 11–12 February, New Delhi, India; Bioversity International, New Delhi, India

Rajan, S., Kishore, R., Ahmad, S., and Vijay (2013c) 'Maiku Lai: Finding the balance between commercial and seedling trees in Sarsanda', in B.R. Sthapit, H. Lamers, and V. Ramanatha Rao (eds), *Custodian Farmers of Agricultural Biodiversity: Selected Profiles from South and South East Asia*, Proceedings of the Workshop on Custodian Farmers of Agricultural Biodiversity, 11–12 February, New Delhi, India; Bioversity International, New Delhi, India

Ram, S. and Rajan, S. (2003) *Status Report on Genetic Resources of Mango in Asia-Pacific Region*, IPGRI Office South Asia, New Delhi, India

Singh, L.B. (1960) *The Mango: Botany, Cultivation and Utilization,* Leonard Hill, London, p. 438

Sthapit, B.R. (2010a) 'The rarest of rare mango in Malihabad: Three hundred genotypes of mango scion grafted in a single tree', flyer, Bioversity International, India

Sthapit, B.R. (2010b) 'The birthplace of the Dashehari mango', flyer, Bioversity International, India

Yadav, I.S. and Rajan, S. (1993) 'Genetic resources of Mangifera', in K.L. Chadha (ed.), *Advances in Horticulture*, vol 1, Malhotra Publishing House, New Delhi, pp. 77–94

Notes

1 http://www.yearofthedurian.com/2013/07/indias-mango-belt.html?m=1
2 Nawab was an honorific title ratified and bestowed by the reigning Mughal Emperor to semi-autonomous Muslim rulers in princely states before British colonization (Wikipedia).
3 Pathan is Hindi for Pashtun, the largest Muslim community in north India that migrated to this area from the tenth century onwards from Afghanistan and the Central Asia region.
4 www.boloji.com/index.cfm?md=Content&sd=Articles&ArticleID=6143
5 See en.wikipedia.org/wiki/Dasheri and www.indovacations.net/english/lucknow.htm for more about mango festivals.
6 www.scmdmalihabad.org/index.php?page=diversity_fair
7 www.facebook.com/LucknowMangoFestival
8 www.asintl.org/ourexperience-sunhara-india.html and scmdmalihabad.org/index.php?page=diversity_fair

13 Maintenance of mother blocks of *Citrus* rootstocks by farmers and nurseries for production of high-quality planting materials

Indra Pal Singh

GPD 'passport'

GPD code:	04
Focus area:	Propagation and planting materials
Character:	System with techniques
Species and varieties involved:	Rough lemon (*C. jambhiri*) and Rangpur lime (*C. limonia*) rootstock used for raising plants of Nagpur mandarin (*C. reticulata*) (Plates 19–22)
Name of location:	Amravati, Maharashtra, India
GIS reference of location(s):	N 21°27′35″; E 78°13′11″ Elevation: 407 masl
Name of farmer (source of information):	Mr Udhav Futane and nursery growers (Village Tiwasaghat, Warud, Amravati) and farmers of Jarud community of Maharashtra, India

Introduction

The average productivity of *Citrus* in India is substantially low in spite of the fact that *Citrus* ranks third amongst fruit crops grown in the country. For *Citrus*, the availability of high-quality planting material makes all the difference between achieving, or not, a productivity level equal to any of the frontline citrus-growing countries. To get high-yielding, standardized planting material with desired size and traits, it is necessary to raise a nursery of budded, grafted or layered plants, depending upon their suitability to specific site conditions. Hence, establishment of rootstock foundation seed blocks is important. Extensive surveys undertaken in different parts of the country revealed that most *Citrus* nurseries do not possess their own foundation blocks either for scion cultivars or for rootstocks.

Nagpur mandarin is a unique mandarin variety grown in Central India in a tropical climate, where the temperature reaches 47°C. No other mandarin is grown at such high temperatures. Its flavour is also unique. Nagpur mandarin is grown mainly in Maharashtra state (Amravati, Nagpur, Wardha and Yavatmal districts) but also in Madhya Pradesh and Rajasthan. As far back as 1977, Nagpur mandarin was reported as one of the best mandarin varieties grown in Central India under the Ponkan mandarin group (Tanaka, 1977). In 2014, the Nagpur mandarin attained Geographic Indication (GI) status from the Government of India due to its uniqueness (GI tag number 385). Nagpur mandarin has a unique blend of acid and sugar that does not exist in any other orange, and it can easily be peeled because of its loose skin. Its taste is very different from other mandarins in the country and it has a unique deep orange colour and a distinct aroma.

Livelihoods in the Vidarbha area of Maharashtra mainly depend on two crops – one is the Nagpur mandarin and the other cotton. Farmers are getting a good income from the sale of this mandarin in Amravati district. Demand for saplings and grafts shows an increasing trend in recent years indicating that farmers are interested in this crop and that the area under Nagpur mandarin is increasing. The livelihoods of about 200,000 families directly depend on it. However, although *Citrus* gardening contributes almost 50 per cent of the household income of farmers in this area, market linkages are poor and hence the level of income is not very high. The Vidarbha region of Maharashtra has the largest area of Nagpur mandarin production in India (150,000 ha), and more than 9,000,000 plants a year are produced and sold through 325–350 government and private nurseries. Most of the private nurseries are located in and around Shindurjana Ghat, Warud, Amravati, India, which was used for this reason as a project site in the UNEP/GEF project on 'Conservation and Sustainable Use of Cultivated and Wild Tropical Fruit Diversity: Promoting Sustainable Livelihoods, Food Security and Ecosystem Services' (see Chapter 1 for information about this project).

Propagation of *Citrus*

Citrus can be propagated through seeds or through budded plants. Propagation of Nagpur mandarin in the Amravati area shifted from seedling to budded plants (onto rootstocks) mainly following the appearance of *Phytophthora* root rot in the Azores Islands in 1842 (Singh and Ghosh, 2000). As the disease was recognized, the interest in rootstocks greatly increased because of the heavy losses experienced. The search for resistant rootstocks started and seedlings were gradually replaced so that today virtually all *Citrus* trees are propagated by budding onto rootstock seedlings (Agarwal, 1982). Initially sour orange and rough lemon dominated as resistant rootstocks in citriculture but later, because of susceptibility to *Phytophthora* and viruses of these rootstocks, screening and development of new rootstocks became a vital citriculture good practice (Arora *et al.*, 2010).

In India, more than 80 per cent of *Citrus* plants are raised on rough lemon (*C. jambhiri*), with the rest on other rootstocks such as Rangpur lime (*C. limonia*) (Plates 19 and 20). No rootstock is immune to *Phytophthora*; however, location-specific rootstock trials over the last 50 years have given good indications for regionwide use of particular rootstocks (Sonkar *et al.*, 2002; Gupta *et al.*, 2008; Singh, 2011). Much variability exists among the strains of rough lemon and Rangpur lime.

Identification of good practice

During the baseline survey of the UNEP/GEF project at the Amravati site in India, most of the farmers reported an absolute shortage of good-quality planting materials of *Citrus*. Most of the nurseries procure rootstock seeds from outside states, such as Himachal Pradesh. Many farmers suspected that the rootstock seeds they bought often contained seeds of other non-recommended *Citrus* rootstock, particularly Galgal (*C. pseudolimon*) (Plate 21). Use of rootstock from Galgal could lead to large-scale damage in the *Citrus* industry as, although saplings look vigorous initially, the rootstocks are highly undesirable and affect the quality of the plant and orchard. Erosion of an old practice of maintaining some rootstocks in every orchard or nursery has left the *Citrus* industry in bad shape. Nevertheless, some progressive growers, like Mr Udhav Futane, continue to maintain mother blocks in Warud and surrounding areas (Plates 23–25). This practice helps such farmers to produce healthy and reliable saplings and grafts for raising *Citrus* orchards. However, their capacity to produce grafts is low – they can produce only 100,000 to 200,000 plants per year against the demand of about 9,000,000 grafted saplings of Nagpur mandarin (Plate 22). The revival of the old and good practice of maintaining mother blocks of *Citrus* rootstocks by nurseries and farmers for high-quality planting materials is important for the future of the *Citrus* industry in the Vidarbha region as well in other regions of Maharashtra and in other states where the area under Nagpur mandarin cultivation is increasing. This good practice can save globally important and economically valuable Nagpur mandarin, an important *Citrus* genetic resource, and uplift mandarin-based livelihoods.

Impact on diversity

The Nagpur mandarin grown in the region is fairly uniform and does not have much diversity in terms of varieties. However, without proper rootstock, there is a danger of its cultivation being abandoned in favour of more remunerative new crops, thus resulting in its erosion as a globally important

Citrus variety. In this context, if the old practice of maintaining mother blocks is reinstated, farmers will be able to maintain at least three species (*C. jambhiri*, *C. limonia* and *C. reticulata*) in their orchards and increase orchard lifespans (Hom *et al.*, 2012).

Impact on livelihoods

The best management practices will culminate in reduced use of Galgal as rootstock. Farmers note that Nagpur mandarin on Galgal rootstock has a much shorter lifespan and it has been observed that orchards using Galgal as rootstock have declined faster. This decline and shorter lifespan of trees adversely affects the livelihoods of the farmers. With better rootstock, nurseries will be able to provide the best quality budlings (budded plants) on recommended genuine rootstock, which are more vigorous with longer lifespans and are less affected by *Phytophthora*-induced diseases. This should lead to reduced production and replanting costs, eventually generating more income and improved livelihoods. Farmers will get regular income as a result of the sustainability of orchard lifespan.

Assessment of GPD

Sustainable livelihood strategies and outcomes

Farmers can benefit by deriving additional income from raising rootstock seedlings (INR2,000–2,500 per kg seed i.e. US$30–40) in addition to what they save by avoiding purchase from commercial nurseries of the budlings required for their own use. Many nurseries that are being established require labour and this has led to the employment of local tribal inhabitants, thus creating some employment opportunities to earn livelihoods.

At the household level, human assets include the size of family (as labour), their health to work in the field and their ability to use knowledge and skills for capitalizing *Citrus* diversity. These human assets are considered a building block for achieving livelihood outcomes. Raising community awareness of the threat to the *Citrus* industry of using the wrong rootstock with the consequence of uniformity of *Citrus* diversity in the production system is essential for enhancing human capital. *Citrus* farmers and nursery experts are keen to take the initiative to: (1) identify mother plants of Rangpur lime and rough lemon and maintain blocks of 10 to 15 trees in orchards to produce enough seed to supply nursery experts; (2) enhance farmers' skills and knowledge to be able to distinguish between Nagpur mandarin budded onto Rangpur lime and that budded onto Galgal rootstocks; (3) identify three or four key nursery experts to promote these techniques through policy support or local recognition. This requires strengthened social capital such as local institutions (e.g. Maharashtra Orange Association of cooperatives and women's groups) that provide support, and monitor and regulate at the local level. These key nursery experts should be encouraged not only to maintain mother tree blocks (a small portion of an orchard) to ensure access to rootstock seed locally and supplement income, but also to train other people to develop budwood onto the appropriate rootstock. These farmers might need financial support from Community Biodiversity Management (CBM) funds (see Chapter 3) or from the National Bank for Agricultural and Rural Development (NBARD) to improve the physical facilities of nurseries. These strategies need to be refined as local teams gain experience whilst working with communities and rural institutions. Action plans should be developed using principles of CBM and mobilizing the livelihood assets of local communities based upon their comparative advantages.

Effect on vulnerability

A single cultivar, Nagpur mandarin, is the main source of income and livelihoods in Amravati communities. The uniformity of *Citrus* diversity (richness = 3 to 4, evenness = 0.02 to 0.06 and community divergence = 0.08 to 0.27) (Jarvis *et al.*, 2008), use of inappropriate rootstocks and a monoculture production system are the major threats to local livelihoods. Diversity, measured in terms of evenness and community divergence, is worryingly low and threatens to cause shocks to the livelihoods of 1,500 households in the study area alone. Orchards established using inappropriate rootstocks are vulnerable to rapid decline of the orchard caused by multiple disease complexes. In recent years the frequency of orchard decline and the sale of dead trees for fuelwood is a common sight; in their place, cotton is being planted along with other kinds of fruits and trees. One main cause of this is the unwise use of Galgal rootstocks for Nagpur mandarin.

Factors favouring or hindering successful functioning of GPD

For a farmer, it is easy to maintain a few plants of a rootstock in the periphery of his or her orchard and sell genuine rootstock seed to nurseries. The successful implementation of this practice might be hindered if the nurseries cannot provide competitive incentives to rootstock seed producers or cannot maintain quality control and supervision in a transparent manner. Supportive policies and regulatory monitoring are needed to improve access to rootstock seeds by farmers and nursery experts. Mr Vasant Madhav Rao Wankhade is a custodian farmer and self-made nursery expert who has developed this idea and is supporting citrus nurseries to introduce effective monitoring and quality control systems (Sthapit *et al.*, 2013).

Action plan for scaling up and dissemination

Any action plan drawn up for scaling up needs to be tested in farmers' fields in consultation with local service providers and farmers together. An awareness-raising programme for using genuine rootstock will help in scaling up and dissemination. Such an initiative has already been taken by the National Research Centre for Citrus (NRCC) Nagpur. Furthermore, a good practice workshop on 'Identification of *Citrus* Rootstock, Mother Block Development and Production of Quality Planting Materials of *Citrus*' was organized in May 2011 at Shendurjanaghat, Warud, Amravati (Singh *et al.*, 2014). Another one-day awareness programme on 'Accreditation and Rating of Horticulture Nurseries of Maharashtra with Special Reference to Fruit Crops' was jointly organized by NRCC Nagpur and the National Horticulture Board (NHB) in collaboration with the Directorate of Horticulture, Government of Maharashtra in August 2011 at NRCC Nagpur. With the help of the Directorate of Horticulture, Government of Maharashtra, a large-scale campaign needs to be undertaken to make it mandatory for every nursery expert to maintain genuine mother plants of rootstock. The Director of Horticulture, Government of Maharashtra has already taken an initiative in this direction to maintain mother blocks of rootstock and scions by nurseries in Maharashtra.

Because Vidarbha is famous for Nagpur mandarin production, the livelihoods of farmers depend on mandarin production on a sustained basis. The continuation and improvement of this good practice of producing reliable and recommended rootstocks will have great impact on the economy of the region and at the same time help in the continued cultivation of the globally important Nagpur mandarin.

References

Agarwal, P.K. (1982) 'Performance of different *Citrus* rootstocks in India: A review', *Agricultural Review*, vol 3, pp. 17–34

Arora, R.K., Sonkar, R.K., and Sakthivel, T. (2010) 'Promising rootstocks of *Citrus*', in *Proceedings of National Seminar on Citrus Biodiversity for Livelihood and Nutritional Security, 4–5 October 2010, Nagpur, India*, National Research Centre for *Citrus*, Indian Society of Citriculture, Nagpur, ICAR, pp. 291–314

Gupta, S.G., Srivastava, A.K., and Sonkar, R.K. (2008) 'Nagpur mandarin evaluation on commercial rootstocks in tropical central India', *Tropical Agriculture* (Trinidad), vol 85, no 4, pp. 1–5

Hom, S., Singh, I.P., Ramanatha Rao, V., Lamers, H., and Sthapit, B. (2012) 'Learning from old practices and local farmers to improve and assure rootstocks quality of Nagpur Mandarin saplings: Performance of Nagpur Mandarin buds on Ranpur lime (*C. limonia*) and Rough lemon (*C. jambhiri*) rootstock', flyer 4/12, GEF/UNEP/Bioversity, India

Jarvis, D.I., Brown, A.H.D., Cuong, P.H., Collado-Panduro, L., Latourniere-Moreno, L., Gyawali, S., Tanto, T., Sawadogo, M., Mar, I., Sadiki, M., Hue, N.T.H., Arias-Reyes, L., Balma, D., Bajracharya, J., Castillo, F., Rijal, D., Belqadi, L., Rana, R., Saidi, S., Ouedraogo, J., Zangre, R., Keltoum, R.O., Chavez, J.L., Schoen, D., Sthapit, B.R., Santis, P.D., Fadda, C., and Hodgkin, T. (2008) 'A global perspective of the richness and evenness of traditional crop genetic diversity maintained by farming communities', *Proceedings of the National Academy of Sciences PNAS* (USA), vol 105, no 14, pp. 5326–5331; www.pnas.org/cgi/doi/10.1073/pnas.0800607105

Singh, S. and Ghosh, S.P. (2000) 'High-tech *Citrus* management', Proceedings of International Symposium on Citriculture, 23–27 November 1999; NRC for Citrus, Nagpur, India

Singh, S. (2011) '*Citrus* in India: Constraints and strategies', unpublished report, NRC for Citrus, Nagpur, India

Singh, I.P., Kumar, D., and Ladaniya, M.S. (2014) *Final Report of UNEP/GEF Project "Conservation and Sustainable Use of Cultivated and Wild Tropical Fruit Diversity: Promoting Sustainable Livelihoods, Food Security and Ecosystem Services" of Amravati Site*, ICAR – NRC for Citrus, Nagpur, India, p. 77

Sonkar, R.K., Huchche, A.D., Ram, L., and Singh, S. (2002) '*Citrus* rootstock scenario with special reference to India: A review', *Agricultural Review*, vol 23, no 2, pp. 93–109

Sthapit, B.R., Lamers, H., and Ramanatha Rao, V. (2013) *Custodian Farmers of Agricultural Biodiversity: Selected profiles from South and South East Asia*, Proceedings of the Workshop on Custodian Farmers of Agricultural Biodiversity, 11–12 February, New Delhi, India; Bioversity International, New Delhi, India

Tanaka, T. (1977) 'Fundamental discussion of *Citrus* classification', *Studia Citrologia*, vol 14, pp. 1–6

14 Marcotting as a good practice for maintaining diversity of citrus in swampy lands of South Kalimantan, Indonesia

Achmad Rafieq, Muhammad Sabran, Susi Lesmayati, M. Winarno and Idha Widi Arsanti

GPD 'passport'

GPD code:	07
Focus area:	Propagation and planting material Production and crop management
Character:	Technique
Species and varieties involved:	*C. reticulata* (jeruk siam banjar) *C. sinensis* (jeruk irisan, jeruk sankis, jeruk cina) *C. grandis* (jeruk besar antalagi, jeruk bali merah, jeruk bali putih, jeruk besar sasanggan) *C. microcarpa* (jeruk sambal) *C. medica* (jeruk nipis, jeruk kuit) *C. hystrix* (jeruk purut)
Name of location:	Sungai Tuan Ilir village, Astambul subdistrict Banjar district
GIS reference of location(s):	S 03°22′05″; E 114°52′26″ Elevation: 17 masl
Name of farmer (data resource):	Mr H. Syukri, Mr H. Kalwiansyah and Mr Kasrah

Introduction

Citrus cultivation in the swampy lands of Banjar district in South Kalimantan is estimated to have been developed in the 1860s. South Kalimantan is known for diamonds and other precious stones and most probably it was diamond merchants from their trade journeys to Malaysia, China, Indochina and other countries who brought the first seeds or plant material of mandarin (*Citrus reticulata*), sweet orange (*C. sinensis*), pomelo (*C. grandis*) and lesser known species such as kaffir lime (*C. hystrix*) (Noor *et al.*, 2007). These crops are grown in both the Astambul and Cerbon sites of the project. Citrus cultivation started most likely in the swampy lands along the river Riam Kiwa, from where it spread to the surrounding villages, such as Sungai Tuan, Pingaran and Tambak Anyar. This region is now known as Astambul subdistrict. Farmers in South Kalimantan still grow citrus fruit, mostly mandarin and sweet orange, in these swampy areas. To adapt to the swampy conditions, they use marcotting for propagation instead of grafting techniques. This chapter documents why the marcotting technique is a good practice in the swampy areas of South Kalimantan that helps to maintain diversity, and it is explored whether such a practice can be scaled up in other similar areas.

Local context

Astambul is located about 45 km from Banjarmasin, the capital of South Kalimantan Province, and has a total number of 10,352 households and a total population of 34,013 (Daroini *et al.*, 2013). A total of 12 distinct varieties were found among six citrus species during the baseline survey carried out by the UNEP/GEF Project 'Conservation and Sustainable Use of Cultivated and Wild Tropical Fruit Diversity: Promoting Sustainable Livelihood, Food Security and Ecosystem Services'. The average farm size is 0.41 ha, with the average home garden being about 0.04 ha, while an orchard is about 0.09 ha. The main cash crops grown in this area are local paddy rice, mandarin, sweet orange, mango, papaya and vegetables. However, rice and fishing are the main sources of livelihoods. The average income per household is low, being Rp7,320,431 (US$790) per annum. Citrus contributes only 3.94 per cent to this but it is important in terms of nutrition and local culture.

Six citrus species are currently found in Astambul: *C. reticulata*, *C. medica*, *C. hystrix*, *C. grandis*, *C. microcarpa* and *C. sinensis*, with 12 varieties among them (see Table 14.1). Mandarin (*C. reticulata*) and sweet orange (*C. sinensis*) are commonly grown cash crops in Banjar district. Pomelo (*C. grandis*), makrut/kaffir lime (*C. hystrix*) and lemon (*C. medica*) are less common and found more often in home gardens for cultural use. The most popular species grown is mandarin, of which a unique variety is found in South Kalimantan (jeruk siam banjar) that is well adapted to swampy peatlands and grown at a commercial scale as well as in home gardens. In South Kalimantan fruit trees are traditionally planted on elevated seed beds about 1 m high with paddy growing in between the rows of trees (see diverse production system in Plates 26–29). Those seedbeds named 'tukungan' are located in swampy peatlands surrounded by canals and rivers that are sometimes flooded, particularly in the rainy season. A Four Cell Analysis of locally grown fruit trees (FCA; see Chapter 3 for details) indicated that jeruk siam banjar (*C. reticulata*), jeruk sankis (*C. sinensis*), jeruk purut (*C. hystrix*) and jeruk irisan (*C. sinensis*) were cultivated extensively as they were found in large numbers and in many households, whereas jeruk bali merah, jeruk bali putih, jeruk besar sasanggan (*C. grandis*) and jeruk cina (*C. sinensis*) are rare (in this context, as they are in fact commercial varieties common elsewhere) as few trees are found and only in few households (Figure 14.1).

For the propagation of their citrus species, farmers in South Kalimantan have traditionally used marcotting – a propagation technique invented by the Chinese about 4,000 years ago. This method is still used by most farmers and several nurseries for citrus, but also for other tropical fruits such as mango, rambutan, guava, soursop and many others. In particular, three farmers, Mr Syukri, Mr Kalwiansyah and Mr Kasrah, have perfected this technique and use it extensively as the preferred technique over other propagation methods because of its advantages in the swampy peatlands of South Kalimantan.

The demand for plant material of citrus in South Kalimantan has increased greatly during the last two decades, as many farmers have started cultivating

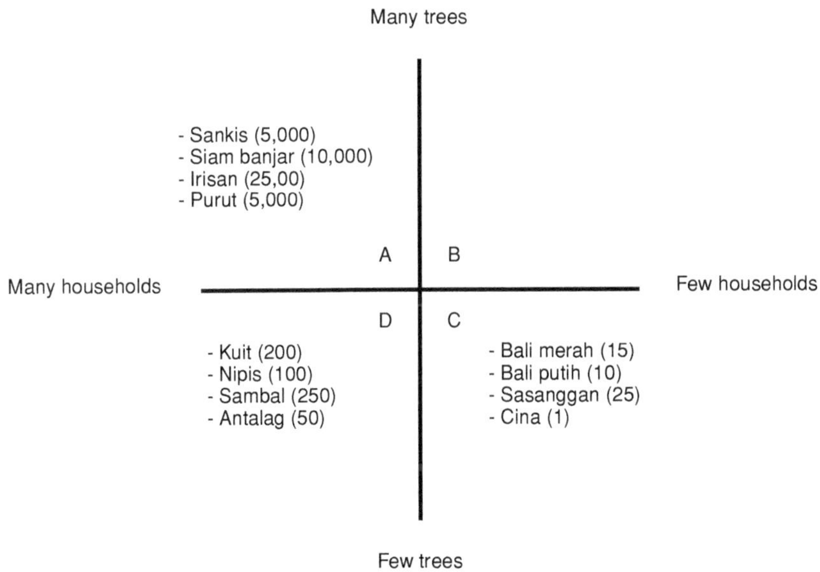

Figure 14.1 Four Cell Analysis on citrus trees in Astambul site, 2013. Cell A indicates common citrus species whereas cell C contains rare and unique varieties. The varieties in cell D are culturally important. Figures in parentheses indicate the number of trees reported.

citrus or have enlarged their area of cultivation. Farmers have planted mostly single species in these newly established orchards, especially mandarin (jeruk siam banjar) or sweet orange (jeruk sankis). To meet this demand, larger nurseries started to use stem and bud grafting techniques to be able to produce large quantities of jeruk siam banjar (*C. reticulata*), jeruk sankis (*C. sinensis*) and jeruk purut (*C. hystrix*). Those grafted saplings are generally cheaper than marcots. The increased demand for saplings has provided several farmers in Astambul with an additional livelihood option – i.e. the production of saplings through marcotting. These marcots and grafts of jeruk siam banjar, jeruk sankis and jeruk purut are mostly sold to meet the demand from other regions such as Central Kalimantan and Java. However, Mr Syukri, Mr Kalwiansyah and Mr Kasrah in Astambul target local demand for marcots of a much broader range of citrus species such as jeruk nipis and jeruk kuit (*C. medica*), jeruk irisan (*C. sinensis*), jeruk bali (*C. grandis*) and jeruk sambal (*C. microcarpa*). Fruits of these species are mostly sold in smaller quantities and mainly for the seasoning of food, use as traditional herbal medicine (jamu) and as an exotic fruit plant in the home garden. For example, jeruk sambal is traditionally used to make sambal and jeruk nipis is a traditional seasoning for fish and soups.

Methodology

The question now arises why farmers prefer marcots over grafted plants and how this contributes to the maintenance and conservation of the unique citrus diversity of this region. To answer this question, we conducted a participatory study in the Astambul subdistrict. The study began with focus group discussions involving 12 farmers, both women and men. The discussions were followed by in-depth interviews with the three key male farmers named above who seemed experienced and knowledgeable in citrus cultivation during the focus groups. The interviews with key informants focused on how they were able to maintain the diversity, in particular how they propagate the plant materials and how they pass them from generation to generation and exchange them with other farmers.

Marcotting as a good practice

The mandarins grown in Astambul are of a special type with a very thin skin that has significant demand in the local market, where it is sold mostly by women (Plates 30 and 31). This type, named jeruk siam banjar, is very well adapted to the swampy conditions of South Kalimantan. Farmers explained that the marcotting technique (a kind of air layering) has been much more successful than other propagation techniques for these swampy areas (Plate 32). Marcotting is a form of vegetative propagation that consists of inducing branches or twigs to produce roots while still attached to a tree. This is done by selecting a healthy branch (usually about pencil-thick, but thickness is not a limiting factor) and removing the bark up to the cambium tissue around the portion of stem that is to develop roots. The ring is allowed to dry for two days before it is wrapped with soil, moss or another medium to keep it moist. When sufficient roots have developed, the branch is cut from the mother tree and planted in a nursery bed to develop buds and become an independent plant. After about two months, when the new sapling looks healthy, it is planted in the field or put in polythene bags for sale.

The major advantage of marcots over grafted plants mentioned by the farmers is their longer lifespan in the swampy peat soils of South Kalimantan. Farmers reported that trees from marcots have a lifespan of 10–12 years, whereas trees from grafted saplings tend to die after four or five years. Trees grown from marcots do not have a deep taproot, as when grown naturally from seed, and instead develop many adventitious roots. These give the trees a firmer but shallow rooting in the elevated seedbeds, avoiding inundation of the root system and helping them to more quickly take up nutrients and fertilizers applied on the top soil. Under swampy conditions, a marcotted sapling usually grows faster and fruits earlier compared with a tree grown from seed. In addition, trees from marcots tend to stay smaller and can be planted in a higher density, therefore needing less space and being easier to harvest (NARI, 2004). In these swampy areas trees can also be grafted. The farmers' experience is that, although grafted trees are often more productive and faster fruiting then marcots, they are less capable of surviving prolonged periods of flooding. Further advantages of using the marcotting technique are:

- Marcotting is relatively simple to perform and requires few additional inputs or facilities to achieve high survival rates.
- Farmers who use marcots often use several genetically distinct mother plants from individual home gardens or orchards, because only a limited number of marcots can be made each time from a preferred mother plant. This leads to the use of several mother plants, which enhances the genetic diversity within the population in the field or in a community of home gardens compared with that derived from grafts taken from a single mother tree. Such a higher genetic variability within the population improves adaptation capacity of the species and lowers the susceptibility to pests and diseases.
- Marcotting can be applied successfully to a wide range of tropical fruit species, including many lesser-known and semi-wild types, whereas grafting techniques are only easily applied for a few domesticated species.

Nurseries, when they have the proper equipment to keep grafted saplings moist, in shade, well fertilized and disease free, can produce very many grafted saplings with a high survival rate from just one mother plant. This makes marcots more expensive than grafted saplings, but grafting, by using only the one mother tree, produces highly uniform progeny. Farmers stated that they buy grafted plants only in times when marcots are not available. Marcotting is an easy technique as it involves simply removing bark around the stem and packing it with soil. Many farmers find grafting more challenging as they are less familiar with it, and they obtain higher survival rates with marcots.

Citrus growers in Astambul have been practising marcotting techniques for more than a century. An essential aspect is the selection of mother plants used to make the marcots. Farmers such as Mr Syukri, Mr Kalwiansyah and Mr Kasrah have identified and maintained unique mother trees of different species and varieties using their own selection criteria. The selected mother trees are characterized and evaluated on vigorous and healthy growth, high yield and good fruit quality, and are usually free from pests such as citrus greening, foot rot, gummosis and twig blight. Farmers produce marcotted planting materials for their own use as well as for sale to other farmers, which generates additional income. Mr Syukri, Mr Kalwiansyah and Mr Kasrah know the location of the best mother plants in their village and they conduct marcotting for other farmers, sharing half of the profit earned from the sales of the marcots with the owner of the mother plant. People from the nearby city of Banjarmasin, who want to plant non-commercial citrus plants, usually look for saplings from traditional vendors at Astambul market, as the area is known for producing good-quality saplings by marcotting of a wide range of species.

Impact on diversity

Astambul subdistrict is known as the traditional district where farmers grow and produce saplings of citrus species in South Kalimantan. To put its diversity in context, we compare it below with neighbouring Cerbon district, also in South Kalimantan, an area known for its high production volume of mandarins. Though overall citrus tree populations are much higher in Cerbon, we can see that diversity indicators for citrus in Astambul are higher than in Cerbon (Table 14.1).

The larger commercial nurseries in Astambul produce cheap saplings in large quantities using grafting techniques. They focus only on mandarin (jeruk siam banjar), sweet orange (jeruk sankis) and kaffir/makrut lime (jeruk purut), using very few mother trees for scions. This results in a very homogenous population of trees for those three species with limited genetic diversity. Other citrus species or varieties like jeruk sambal, jeruk kuit or jeruk bali merah and jeruk bali putih are produced in limited quantities by farmers using the marcotting technique, for which a much broader range of mother trees are used.

Impact on ecosystem services

Half a century ago, South Kalimantan was still largely covered by lowland rainforest, which has been virtually wiped out for its timber value. Many of these swamp areas have been drained and are left idle without any green cover or are increasingly converted into agricultural land (Hendayana, 2010). South Kalimantan now has 8,109,000 ha of tidal swamp and 3,580,000 ha of lowland

Table 14.1 Genetic diversity assessment in citrus in South Kalimantan

Diversity indices of citrus	*Cerbon*	*Astambul*
Number of households interviewed	51	46
Total number of trees (*Citrus* spp.)	10,823	724
Average number of trees per household (*Citrus* spp.)	212	16
Community richness (*Citrus* species)	5	6
Community richness (*C. grandis* varieties)	1	4
Community richness (*C. reticulata* varieties)	1	1
Community varietal richness across all *Citrus* species	7	12
Community evenness *Citrus* spp. (Simpson index)	0.37	0.74
Divergence (*Citrus* spp.)	0.19	0.49

Source: Daroini *et al.* (2013).

swampy fields. The root system of marcots is more suitable than grafted planting materials for raised-bed cultivation in these marginal swampy lands. Adoption of the marcotting technique can enhance the availability of planting materials and ensure more variability within the population of citrus trees found in these areas, providing the opportunity for further adaptation to new circumstances such as salinization of drained soils. Moreover, the increase in citrus planting in the swampy lands not only makes the land more productive but also increases green cover, thus contributing to ecosystem services such as reduced loss of nutrients, reduced lowering of the water table and reduced erosion of peatlands.

Impact on livelihoods

Farmers such as Mr Syukri, Mr Kalwiansyah and Mr Kasrah, who are highly skilled in the marcotting technique, earn additional income through the sales of marcotted saplings and share the profits with the farmers whose mother plants they use. The marcotting technique has been practised for more than a century in the Astambul area and is now spreading to other parts of South Kalimantan, with several implications for the livelihoods of the region. These include: (i) increased income for marcotting experts such as Mr Syukri, Mr Kalwiansyah and Mr Kasrah through the sales of marcots; (ii) additional income for owners of mother plants from which marcots are taken; (iii) improved performance of orchards and trees leading to better productivity; and (iv) independent multiplication of citrus trees by poor farmers in swampy areas without having to purchase plant materials.

Assessment of the good practice for diversity management

Marcotting makes the traditional selection and propagation of citrus of many domesticated, semi-domesticated and wild species relatively easily. The knowledge can be transferred to new farmers by means of a short training course conducted by expert farmers. Actions that might strengthen the take-up of this good practice are: (a) conducting training on citrus mother tree management, (b) strengthening the network of mother tree collectors and planting material producers, (c) increasing farmers' awareness of the availability of elite mother trees, (d) linking to microcredit programmes and to wider markets outside the district and (e) providing marketing outlets. These actions facilitate the maintenance of unique types of citrus diversity in these harsh peat swamps where other agricultural crops are difficult to cultivate because of the high salinity.

Concluding remarks

Citrus cultivation in swampy land areas could have positive impacts on livelihoods and on the environment. It improves swampy peatlands that have been de-forested and could reduce the loss of nutrients, stop the lowering of the water table and avoid further erosion. This chapter outlines a practical case study where, remarkably, the traditional propagation technique of marcotting provides better results compared with modern grafting techniques due to its context-specific advantages that help the species to adapt to the local unique, harsh or adverse environmental conditions. The use of the marcotting technique, as proven by farmers in Astambul, helps maintain the unique mandarin species jeruk siam banjar besides several other citrus species and hence sustains the livelihoods of local communities in swampy areas.

References

Daroini, P.B., Dewi, H.A., Andri, K.B., Rafieq, A., Pribadi, Y., Fakhrina, Saderi, D.I., Lesmayati, S., and Lamers, H.A.H. (2013) *Baseline Report Indonesia for the UNEP-GEF Project Entitled: Conservation and Sustainable Use of Cultivated and Wild Tropical Fruit Diversity: Promoting Sustainable Livelihoods, Food Security and Ecosystem Services*, GEF, UNEP, Bioversity and ICHORD, Indonesia

Hendayana, R. (2010) *Assessment of Adoption Acceleration of High Yielding Varieties in Swampy Areas in South Kalimantan and Central Kalimantan* (in Indonesian), Indonesian Centre of Agricultural Technologies Assessment and Development, Bogor

NARI (2004) *Marcotting Manual KERR22E*, National Agriculture Research Institute, East New Britain, Papua New Guinea

Noor, H., Izzuddin Noor, D., Antarlina, S.S., Yanti Rina, and Noorginayuwati (2007) *Local Wisdom on Citrus Plantation in the Swampy Areas in South Kalimantan* (in Indonesian), Indonesian Wetland Research Institute, Banjarbaru, p. 49

15 Combination of side-grafting technique and informal germplasm exchange system in non-irrigated mango orchards in Thailand

Phichit Sripinta, Supattanakit Posawang, Chatchanok Noppornphan, Songpol Somsri and Bhuwon Sthapit

GPD 'passport'

GPD code:	09
Focus area:	Propagation and planting materials
Type of GPD:	Technique
Species and varieties involved:	3 species: *Mangifera indica, M. odorata* Griff, *M. duperreana* Pierre 10 commercial varieties: Kaew, Chok Anan, Namdokmai, Pimsane Mun, Kheoi Morakot, Kheoi Sawei, Namdokmai No.4, Nam Dok, Mai-Sitong, Man Khunsri 3 wild species: Khiya, JingReed and Pom
Name of location:	Mae Or-Nai subdistrict and Maena subdistrict, Chiang Dao district, Chiang Mai province, Thailand
GIS reference of location(s):	N 19°17′98″; E 99°01′76″ Elevation 527 masl
Name of farmer (data source):	Mr Suradet Tapuan and Mr Pleng Funphun

Context and introduction

Chiang Mai province is located in the mountainous northern part of Thailand. Mae Or-Nai and Maena subdistricts in Chiang Dao district are mostly hilly, rainfed areas located in forest buffer zones in the mountains (500–1000 masl). In the winter the temperature is cool enough (minimum 14°C) for the cultivation of fruit such as lychees and strawberries. Much of the fruit-based farming system in northern Thailand can be described as agroforestry or agri-silvicultural production-oriented systems used on sloping lands in the highlands (Bradford *et al.*, 2005). Tropical and subtropical fruit farming is an attractive option as it can contribute to home consumption as well as market products (Vangnai, 1996). The communities in the area have access to basic services and support systems such as sanitation, water, electricity, public health, schools, transportation, local markets and export markets to Japan. The higher mountains in this area are cut by steep river valleys and upland areas that border the central plains of Thailand. The Ping River lies in the lowlands of the Ping River watershed and is the main watershed catchment in this area. Traditionally, these natural features made several different types of agricultural cultivation possible, including wet-rice farming and fruit-tree-based home gardens in the valleys and shifting cultivation in the uplands. The forested mountains include stands of teak and other economically useful hardwoods that are dominant in the area.

Ten years ago, the Mae Or-Nai area contained marginal buffer forest lands that were not used for much fruit-based farming because of lack of irrigation. In response to growing demand for organic mango varieties for export markets, cultivation has been expanded to these dry hilltop lands in the proximity of buffer zones (Plates 33–36). Mango grown here is found to be good quality but the survival rate of commercial mango saplings is low. For this reason, some innovative farmers, such as Mr Suradet Tapuan and Mr Pleng Funphun, developed the technique of side-grafting (Plate 37) and have scaled the technique out from farmer to farmer, later supported by extension agencies of the Department of Agriculture.

Now, farmers in Mae Or-Nai subdistrict use side-grafting as a propagation technique for multiplying local and commercial mango varieties in the hilly landscape areas near forest buffer zones. This side-grafting technique, which is unique to this area, is especially popular in rainfed or non-irrigated areas where success rates of other techniques for grafting are very poor (Sripinta *et al.*, 2012). This grafting technique helps to maintain mango varietal diversity as it strengthens the local seed system in which farmers exchange scions of commercial and local varieties. This propagation technique also favourably affects the productivity and quality of the mango trees as it ensures a higher survival rate for saplings and more vigorous mature trees. This in turn positively affects the livelihoods of local farmers. In particular, small- and medium-scale farmers are directly affected as they are often more dependent on rainfed orchards, for which this technique is most commonly used. Home consumption of local fruit has also been perceived to increase, resulting in better family nutrition. This technique has been identified as a good practice for diversity management (GPD) as it is economically, socially and environmentally viable and sustainable in the long term.

Methodology

For this chapter, information was collected from the men and women of Mae Or-Nai community through participatory focus group discussions to identify and document potential good practices for diversity and the knowledge keepers of these practices. Subsequently, a checklist of questions was used to interview key informants so that they could describe the good practice and also demonstrate the steps visually.

Description of GPD: side-grafting of mango sapling for rainfed mountain areas

Side-grafting is one propagation technique used for growing mango in rainfed areas. Rootstock seeds are planted in advance in a field for two to three years, then side-grafted with local and commercial cultivars. Three types of Mangifera species – *M. indica*, *M. odorata* and *M. duperreana* – are used for rootstocks. Planting the rootstock directly in the field results in mango trees with higher survival rates compared with planting grafted saplings taken directly from a nursery. Farmers then select the variety they would like to cultivate and collect scions (local and commercial varieties) from many different mother trees in their own orchards and from other farmers or villages. This is the preferred practice because the number of different varieties sold at nurseries is low, often limited to common commercial varieties. Some farmers even side-graft two or three scions of different varieties on one tree, which they often then use as a mother branch for taking further scions (Plate 36).

The side-grafting is carried out as follows (Plate 37): First, select a good heathy branch and cut a vigorous dark green shoot about 7 cm long and containing two or three mature buds, and remove all the leaves. Prepare the stock by making a 5 cm-long slanting cut (U shape) at a height of between 30 and 50 cm above the ground. Then, make a single smooth cut on both sides at the bottom of the scion to form a wedge shape the same length as the cut on the stock. It is important that the cuts on the two pieces are as similar in size as possible in order to afford a greater chance of a successful graft union. Insert the scion into the cut of the stock and wrap the graft with plastic upwards from bottom to top to keep the cuts tight and to prevent drying. Cut the stock branch close to the graft after several weeks have passed and the scion has begun to grow. Wax the new cut with either flint coat (acidic) or red lime (alkaline). These are chemical substances that protect mango branches from stem rot fungus damage and from insects laying eggs in the exposed cut of the branches. Farmers tend to prefer red lime to flint coat because it is cheaper.

Ten farmers perform this grafting method for all the other farmers in the project area. Two men farmers, Mr Suradect Tapuan and Mr Pleng Funphun, do the side-grafting on a large scale and charge for this service. They graft about 200 to 500 trees every season and receive 10 Baht (US$0.30) per successful graft. Other farmers perform this grafting method for their neighbours or family without asking for payment for the service. Grafting is always done in May during the rainy season to ensure a better survival rate for the young grafts.

This technique has been innovated by farmers and practised for more than 10 years, and approximately 35 per cent of farmers in Mae Or-Nai village use it to varying degrees in their plantations on sloping lands. Correspondingly, local orchards contain a large number of varieties of mango species, including *M. indica*, *M. odorata* and *M. duperreana*. Farmers mostly use *M. indica*, *M. odorata* and *M. duperreana* as rootstock to graft *M. indica* or *M. odorata* varieties or landraces. For instance, Mr Suradet Tapuan and Mr Pleng Funphun prefer to graft mostly *M. indica* Namdokmai and Man Khunsri varieties. However, on occasion they also graft local varieties like Kaeo and Talap Nak and some landraces of *M. odorata* that are unavailable for purchase from commercial private nurseries or formal seed exchange programmes. Farmers prefer these landraces and varieties because they are very strong, disease- and pest-resistant and perform well without access to irrigation.

Impact on diversity

Mangifera species such as *M. indica*, *M. odorata* or *M. duperreana* are cross-pollinated and are genetically highly diverse. *M. odorata* is regarded by botanists as a cultivated hybrid of *M. indica* and *M. foetida*, as it has never been found in the wild. Species, such as *M. indica* and *M. odorata*, have several varieties or forms recognized. The availability of *M. duperreana* is limited as its status is reported to be vulnerable (Kole, 2011).

The side-grafting technique maintains intraspecific diversity at the varietal level due to the fact that it often involves local and less common varieties like Man Khunsri or less common species like *M. odorata* and *M. duperreana* in the grafting process. Likewise, private nursery experts or custodian farmers applying this grafting technique often use scions from a wider range of local mother trees or varieties that are introduced from other areas compared with nurseries, which often have a much narrower varietal portfolio. Nurseries primarily sell commercial varieties like Namdokmai, Khieo Sawoei, Maha Chanok, Chok Anan or local varieties like Kaeo and Talap Nak, neglecting rare or more indigenous species and varieties. Lastly, this technique enlarges the area planted with mango trees, as grafted saplings from nurseries often do not survive the dry season in non-irrigated lands. As a result, the interspecific and intraspecific diversity in the area is more likely to survive.

Impact on livelihoods

Because of the introduction of the side-grafting technique, land on hilltops that was not used before can now be planted with mango trees, improving marginal lands while expanding the community's cultivation capacity. This propagation technique improves the survival rate, productivity and quality of mango trees, which impacts harvest-time yields and, as a result, livelihoods derived from the sale of the fruit. As the technique is applied in rainfed areas, mostly small- and medium-scale farmers directly benefit by increasing their income through fruit tree cultivation. This grafting technique is relatively simple as well as economically, socially and environmentally viable over an extended time frame.

Both men and women farmers in the village received training or were provided with information about this technique by local experts from the government in order to improve and enlarge their mango production capacity. The increase in mango production has improved the well-being of farmer households in the target community. At the project site, local people have formed farmer groups and are selling mango fruit collected from home gardens and semi-commercial or commercial orchards to market outlets inside and outside the community. Later, women farmer groups received investment funds through a newly established cooperative dedicated to the production and distribution of several agricultural value-added products made from the fruit. Currently, through extended collaboration with various stakeholders, farmers in this community are enjoying several sustainable livelihood benefits as a result of the introduction of this grafting technique.

Factors favouring or hindering successful functioning of GPD

Farmers in the Mae Or-Nai area were in need of improved mango propagation methods, especially for rainfed areas, as many farmers in the target area do not have access to irrigation and there is a market for locally selected mango varieties. The likelihood of success was strengthened by farmers taking the opportunity to set up a village cooperative and collecting funds to invest in processing facilities to derive better economic profits from value addition processing.

The main factor that contributed to the success of this good practice was the execution of a government programme aimed at increasing mango production in this area by providing training for farmers on the application of this technique. The technique innovated by Mr Suradet Tapuan and Mr Pleng Funphun was noticed by the Department of Agriculture and scaled out through the government programme. In addition, the Export Growers Association (a group of farmers registered at the local government for export market) assisted farmers in value chain analysis assessments and worked to strengthen the capacities of value chain actors, thus empowering farmer groups socially and economically. Other factors that contributed to the success of this practice were the accumulation of financial capital for the farmer groups through the purchase of shares by participating members, and guaranteed access to markets for products through diverse outlets and channels.

Conclusion and a way forward for scaling up and dissemination

This side-grafting technique is practical, cost-effective, sustainable, easy to adopt and results in increased diversity together with improved livelihoods. The practice has potential for local, national and even regional dissemination due to the fact that side-grafting can be applied to any mango tree and does not require much investment beyond a certain set of skills and associated knowledge. This technique can be part of set of good practices that can be scaled out beyond Thailand in similar areas by sharing skills and information. However, the success of the practice was also due to the very good work of extension, cooperatives, financing and linking to markets.

References

Bradford, A., Robinson, W., and Hibbs, D.E. (2005) 'Testing an ecologically based classification tool on fruit based agroforestry in Northern Thailand', *Agrofoestry Systems*, vol 65, pp. 1223–1235

Kole, C. (2011) *Wild Crop Relatives: Genomic and Breeding Resources: Tropical and Sub-tropical Fruits*, Springer, New York

Sripinta, P., Noppornphan, C., Yoovatana, M., Somsri, S., Sthapit, B., Ramanatha Rao, V., Kaur M., and Lamers, H. (2012) 'Combination of side grafting technique and informal scion exchange system for mango diversity management in non-irrigated orchards', flyer 9/12, GEF, UNEP, Bioversity International

Vangnai, V. (1996) *Mango*, Srisombat Publishing Ltd, Bangkok

16 Propagation and pruning techniques of *Garcinia atroviridis* (asam gelugor) in Bukit Gantang, Perak, Malaysia

Zahimi Hassan, Norhayati Md Haron, Mohd Nizam Abdullah, Muhammad Shafie Md Sah, Salma Idris, Hugo A.H. Lamers and Bhuwon Sthapit

GPD 'passport'

GPD code:	08
Focus area:	Propagation and planting materials Production and crop management
Character:	Technique and practices
Species and varieties involved:	*G. atroviridis*, locally known as asam gelugor
Name of location:	Bukit Gantang, Perak, Malaysia
GIS reference of location:	N 04°45′83″; E 100°45′86″ Elevation: 40 masl
Name of farmers:	Mr Zainal Abidin Ramli, Mr Abdul Wahab Bin Ahmad, Mr Mazlan Bin Mohd Nor, Mr Mustafa Kamal and Mr Zaki

Introduction

Garcinia atroviridis l. (asam gelugor) is a large perennial fruit tree that can grow up to 30 m high and has drooping branches. The species is endemic to Peninsular Malaysia, although the tree grows throughout a large part of south-east Asia where it is valued for its culinary and medical uses. Seven of the 50

Garcinia species that exist in Peninsular Malaysia (Corner, 1988) can be found in the home gardens and orchards in the six sites chosen as part of the TFTGR project[1] (Salma *et al.*, 2012): *G. mangostana, G. atroviridis, G. forbesii, G. cowa, G. dulcis, G. hombroniana, G. prainiana* and *G. bancana*. Of these, *G. mangostana* (Manggis, mangosteen) is the most commonly cultivated species, while *G. atroviridis* (asam gelugor) is ranked the second most important species among the sites. It is also widely grown in India, Indonesia and Thailand. In the Malaysian and Indonesian tradition, the rinds of the unripe fruit are cut into pieces and dried in the sun. The dried sliced fruit is locally known as 'asam keping'. It is sold in markets or bazaars for use as a sour relish in curries in place of tamarind or for dressing fish (Plate 38). Asam gelugor is rich in vitamin C and also contains hydroxycitric acid (HCA), which can be used for reducing weight and excess fat (Khairunnisa, 2005). This has led to a growing demand for this fruit. In addition, the young leaves of asam gelugor are traditionally eaten as a salad, while lately the matured leaves have been used for making tea. The dried fruit have also been used as a dye for silk, when combined with alum as a fixative. Studies have now confirmed that the acidic nature of the fruit has anti-fungal, anti-microbial and weight-reducing properties. In traditional medicine, dried asam gelugor is soaked in hot water and then drunk to reduce high blood pressure. This range of traditional and modern uses has increased interest in this underutilized species over the last 10–15 years.

Asam gelugor trees are mostly raised from seeds planted in home gardens and orchards. These seedlings are grown from seeds selected by the farmers themselves or the seedlings are bought from private nurseries. Despite the increasing interest in the species, its cultivation and especially its propagation has been challenging for farmers. First, asam gelugor, like other *Garcinia* species, is dioecious and produces male and female trees. The farmers interviewed noted that trees raised from seedlings produce about 70 per cent male trees, which are not wanted as they do not bear fruits. Normally the farmers cut down the male trees after 6 to 10 years when the female trees should have started producing their first fruits. It would be useful for farmers not to wait so long and to be able to select female seedlings from the outset. Second, the seedling tree can grow up to 30 m tall and thus can be difficult to harvest. In addition, the species is plagiotropic; the branches grow obliquely or at an almost horizontal angle from the trunk. When new plants are produced by taking cuttings from these branches, the new sapling does not assume a normal vertical tree shape but continues to grow like a horizontal branch. This makes it difficult to propagate the species vegetatively. A few farmers from Bukit Gantang have developed some particular methods to be able to induce female trees from saplings. As well as this they have perfected some propagation and pruning techniques to keep the trees small and thus easier to harvest.

Context

Bukit Gantang is a small town located in the centre of Larut-Matang Selama District, in the state of Perak, Peninsular Malaysia. The town and its periphery cover an area of about 68,160 ha. The topography of the area is hilly in the interior western part, sloping towards the coast in the east. The area receives an average rainfall of 3,045 mm a year, with December being the wettest month while February to April is drier. The average temperature is 28°C. A few perennial rivers pass through the area including Sungai Larut, Sungai Jaha, Sungai Limau, Sungai Sepetang and Sungai Punggor. Bukit Gantang is surrounded by thick forest and the livelihoods and culture of the people are closely associated with the forest. Agriculture and horticulture are the main land use systems in the area. The main crops planted are oil palm (83 per cent), fruit trees (9 per cent), rubber (4 per cent) and vegetable crops (4 per cent). Palm oil and rubber are cultivated in large-scale estates owned by private companies, but also in small plots by farm households. In addition, farmers grow a range of fruit trees in home gardens and mixed orchards including *G. atroviridis* (Table 16.1).

Major fruit tree species grown in home gardens and mixed orchards are mango, durian, mangosteen and rambutan. Fruit trees are grown in home gardens, mixed fruit orchards or in agroforestry systems where rubber trees are used as anchor crops. Some wild fruit species such as *Parkia speciosa* (a legume fruit tree) are planted in the orchards. *Garcinia atroviridis* is planted in home gardens, mixed fruit orchards, agroforestry systems and also small-scale

Table 16.1 Comparative species richness, evenness and diversity indices of tropical fruit tree diversity in Bukit Gantang in Malaysia

System	*Richness*	*Evenness*	*Diversity Indices*
Home garden	8	0.65	0.57
Orchards	7	0.72	0.62
Forest	NA	NA	NA

Source: Baseline survey data, 2010; see Nazmi *et al.* (2013).

Table 16.2 Comparative richness, evenness and diversity indices of *Garcinia mangostana* and *Garcinia atroviridis* in Bukit Gantang in Malaysia

*Diversity parameters**	*Home garden*	*Orchards*
Total number of trees	968	1245
Average number of trees per household	18.62	23.94
Community richness	2	2
Average richness per household	1.67	1.63
Average household evenness (Simpson Index)	0.21	0.21
Community evenness (Simpson Index)	0.32	0.44
Divergence community	0.34	0.53
Number of households	52	52

* Refer to Jarvis *et al.* (2008) for definition of various parameters.
Source: Baseline survey data, 2010; see Nazmi *et al.* (2013).

monocropping. Comparative on-farm *Garcinia* diversity indices of Bukit Gantang are summarized in Table 16.2. Some community members who enjoy going to the forest, also collect fruit from the adjacent forest, such as wild mango and fruits from several *Garcinia* species and *Parkia* species among others (diversity data from forest are not available).

Table 16.2 shows that the number of *G. atroviridis* and *G. mangostana* trees are higher in orchards, suggesting farmers' interest in commercialization of these species. The project focused on farmers growing fruit trees, so those involved in oil palm activities are not included in this analysis. Among these farmers, the average monthly household income is RM750 (US$250), which is mostly derived from agricultural activities such as rubber tapping, fishing and selling of agricultural produce. The community receives strong support from government agencies such as the Department of Agriculture (DOA), the Department of Irrigation and Drainage (DID), Department of Forestry (DOF) and government-supported Farmer Associations. All the agencies play a role in developing the community infrastructure, economic development programmes, marketing and social development.

Method used for data collection and problem statement

A research team from the Malaysian Agriculture Research and Development Institute (MARDI) and officers from the Perak Horticulture Division, DOA and DOA District of Larut Matang Selama, Perak, interacted with farmers to identify, understand, evaluate and document the propagation and management of *G. atroviridis*. A semi-structured questionnaire was used, followed by a farm walk and focus group discussion with heads of households (mostly male) to understand the practice. The farming and propagation practices and other details were documented in the form of pictures, audio and video recordings and written notes.

Farmers in Bukit Gantang who cultivate asam gelugur were facing two problems: (i) having too many male trees in their home gardens and (ii) having trees that were too tall to harvest. The difficulty of obtaining female trees and the drudgery of harvesting prevented farmers from growing the species at a larger scale or in higher numbers in their home gardens and orchards. These drawbacks kept the species underutilized despite its multiple uses and increasing market value. Also, over the last three to four decades, many farmers have been steadily losing interest more generally in the diverse species grown in home gardens or mixed orchards or that are found in forests, as profits have been comparatively low and not keeping up with the high profits made in palm oil and rubber. Many farm households have kept their home gardens or mixed orchards out of tradition or merely for home use or surplus sales as maintenance costs are relatively low. As nowadays most of the younger generation have jobs outside farming and are moving to the cities, it is expected that many orchards or home gardens will be discontinued or abandoned in the near future.

The research did, however, find cases that bucked this trend: farmers who had developed techniques for inducing female trees and pruning techniques to make harvesting easier.

Description of GPD: propagation techniques and pruning practices of asam gelugor

This chapter describes four practices that could help to popularize this underutilized species:

- Inducing the growth of female trees by bending and stressing the taproot of saplings
- Traditional marcotting technique through root cuttings to multiply female trees from the forest (Plate 41)
- Patch and cleft grafting techniques to multiply female trees (Plates 39 and 40)
- Top working and pruning of trees to train tree height and reduce drudgery of harvesting (Plates 38a and 38b).

These practices have been developed and perfected by a few dedicated farmers in Bukit Gantang, who have innovated and adopted those techniques based on traditional knowledge in combination with information and workshops provided by the DOA.

Inducing the growth of female trees through root stress

Mr Mazlan Bin Mohd Nor has been growing asam gelugor for a long time and has developed in his home garden an innovative method to obtain female trees. He obtained the knowledge of how to raise female asam gelugor seedlings from the older farmers in the village. They told him to stress the taproot of the young sapling when transplanting from the seedbed into the orchard or home garden. This is done by bending the taproot several times before transplantation and placing a plank or board just below the taproot to force the taproot to grow sideward. This practice is traditionally proven to induce the growth of a female tree. Mr Zaki, another farmer, has now tested this method with 100 saplings of which about 10 seedling trees have started to bear fruits.

Traditional grafting technique by root cuttings (marcotting)

Propagation by cuttings (i.e. marcots) is the oldest known method for propagation and considered by farmers the cheapest and most convenient method to create new saplings of a preferred tree. First, one selects a female tree with good characteristics such as a tree known to produce big and good-quality fruits. In the root cutting technique, the root from this female tree is dug up, raised to the surface and a small cut is made to the root. The exposure to the air will induce the formation of a new shoot at the location of the cut and subsequently additional roots (Plate 41). This method is considered by farmers the simplest method to produce a clone from a preferred female tree.

Cleft grafting and patch budding

In cleft grafting, a shoot (i.e. scion) of the desired female tree is joined with the stem of a young seedling (stock or rootstock) of different genetic origin (Plate 39). This technique has been developed by scientists and is commonly used for popular commercial species such as mango (*Mangifera indica*) and rambutan (*Nepheleum lappaceum*), although it has proven to be more complicated for lesser known species such as asam gelugor. Mr Abdul Wahab Bin Ahmad from Bukit Gantang, however, has experimented and perfected the technique specific for asam gelugor. He is considered the most skilled farmer in the village and obtains a high survival rate (about 90 per cent) of his grafted asam gelugor plants. Another related grafting method is patch budding, where a patch of bark including a bud is carefully cut and taken from the desired female tree and placed on the stem of a rootstock where the bark has been removed. As soon as the bud starts to grow a new shoot, the stem of the rootstock above the bud is removed (Plate 40). For patch budding and cleft grafting, Mr Abdul Wahab Bin Ahmad prepares the rootstock by sowing the seed in 15 cm x 22.5 cm polybags. When the seedlings in the polybags are four to six months old they are budded or grafted using the scions selected from healthy female tree branches. Mr Abdul Wahab Bin Ahmad has been practising cleft grafting for more than 10 years in Bukit Gantang. All grafted trees planted survived according to Mr Abdul Wahab Bin Ahmad, but he prefers patch budding. He has planted in his home garden 60 healthy trees over the last 10 years using patch budding, with a survival rate of 90 per cent. Mr Mustafa Kamal learned the technique during a training course and now he has 20 new saplings of about nine months old. Patch budding develops a strong tree structure that can avoid falling and breakage in the strong winds that are common in this region. Neither grafting method is practised widely in Bukit Gantang, or Malaysia in general, as they require experience and specific skills to be successful and to obtain a high survival rate of grafts for asam gelugor.

Top working and pruning to reduce size of trees

Mr Abdul Wahab Bin Ahmad was interested in the management of asam gelugor trees to reduce their height and to optimize the tree structure to improve the quality of fruits and to make harvesting easier. He decided to carry out top working of the tree, which means cutting the trunk of a tree aged 2 to 2.5 years at 2–3 m height, maintaining three strong, healthy branches at the trunk base to form a much lower but wider canopy by regular pruning so that harvesting the fruits is easier (Plates 38a–38b). Pruning is necessary to maintain a good canopy and healthy growth. Mr Abdul Wahab Bin Ahmad has experimented and applied this technique, which is more common in mango, to asam gelugor and this has now become a popular practice in Bukit Gantang as well in neighbouring villages, especially since the Department of Agriculture has been supporting and promoting this technique as part of the TFTGR project. Farmers have shown interest in applying the technique to new trees, but often do not want to apply it to already fully grown trees as they are afraid of reducing yield potential and damaging the tree.

Benefits to farmers

A team of researchers from the DOA, along with the above mentioned innovative custodian farmers, examined the techniques closely to understand and demonstrate their use and evaluate the potential to spread the techniques to other farmers and areas with similar conditions. Which particular method of propagation to be used in a particular situation is often dependent on the experience of the farmer and the purpose of growing asam gelugor. Farmers tend to prefer traditional propagation techniques when growing the trees in home gardens on a smaller scale, but prefer the grafting techniques and top working when growing them for commercial purposes. Among the four propagation techniques that were tried and evaluated within the TFTGR project, patch budding was found to be the most suitable and practical when growing for commercial scale. The propagation and pruning practices were demonstrated and validated from 2010 up to 2014 as part of the TFTGR project. Additionally, training was provided by the Department of Agriculture in collaboration with the identified farmers, and vegetatively propagated planting materials have been supplied to other farmers in the district. The patch budding technique has now been adopted by a few innovative farmers to improve the commercialization of asam gelugor for sales to generate income. Two training courses on vegetative propagation of *G. atroviridis* were conducted for 25 men farmers, two of whom established community nurseries. When combining patch budding with top working of trees, farmers get the following benefits:

- Farmers obtain only female trees with good yields as buds have been selected from heavily fruit bearing trees with high-quality fruits
- A tree can be productive within 3–4 years compared with 10 years from a seedling raised tree
- It is easy to harvest the fruit from the dwarf trees and this therefore reduces labour costs
- The dwarf trees allow the fruit to be harvested at the right stage and ripeness to ensure the quality of the finished product
- The risk of trees falling down due to strong winds and pest damage is reduced
- The techniques are simple and inexpensive and can be practised by farmers.

Impact on intraspecific and interspecific biodiversity

These propagation and management practices have helped farmers to continue growing *G. atroviridis*, thus maintaining its diversity in home gardens and orchards. Previously in Bukit Gantang, *G. atroviridis* was mainly grown for home consumption only, but with the introduction of the vegetatively propagated female seedlings, it has become a source of income for the farmers through the sale of its products. The production of female saplings combined with the top working technique creates more interest among farmers to plant this species for commercial purposes. It has increased the population size (number of trees planted) in the area. A total of 189 new *G. atroviridis* seedlings were planted in home gardens and orchards by the farmers in the study area during the period 2010 to 2014. Through the training of farmers on the use of grafting techniques such as patch budding, cleft grafting or the use of root cuttings, farmers make better use of the available intraspecific variability of *G. atroviridis* in Bukit Gantang as they select their scions or buds from a range of source trees. Besides, the growing interest in the species asam gelugor has increased the interest in home gardens and the mixed orchard system in general, as sources of promising species that can be commercialized.

Economic impact on livelihoods

Through the adoption of the techniques described above, combined with increasing market interest and demand, growing asam gelugor has become a livelihood activity that generates income. In the study site, an average home garden has about seven trees of asam gelugor from which each household obtains an extra income of RM300–600 (US$100–200) a month through selling fresh fruit or processed asam gelugor. In addition to this, the vegetatively propagated female saplings fetch a substantially higher price compared with undefined seedlings and thus provide additional income for those farmers who have obtained the skills to graft and multiply their most productive and high-quality female trees. The combination of specific propagation techniques and pruning methods makes it possible to grow asam gelugor on a larger scale within their home gardens and mixed orchards. From 2010 to 2014, 301 grafted seedlings were produced and sold at RM15.00 (US$5) per sapling. In addition, a total of 3,120 seedlings were produced directly from seed and were sold at RM10.00 (US$ 3.3) per seedling.

Scaling up and dissemination

Based on its special advantages, cultivation of asam gelugor should be maintained and enhanced to provide an additional source of income to the households in the community while it ensures the on-farm and *in situ* conservation of this species in the country. Propagation and production practices should be combined with the improved processing and marketing of the processed product, asam keping. A set of farmer-friendly good marketing practices might include: (i) hands-on training on propagation and post-harvest techniques; (ii) specialized equipment such as a mechanized fruit slicer; (iii) good hygiene; and (iv) an all-weather drying structure that can help to improve the quality of asam keping (see Chapter 21 for more details). This good practice will be extended to other farmers in this area and also to other asam gelugor growing areas in Malaysia by introducing this technique using a 'peer to peer' group method and extension officers through a training programme organized by the DOA. The training will be on patch budding, topping, pruning and other aspects of on-farm management and good cultivation practices.

Conclusion

The difficulties in harvesting and in obtaining female trees had limited the potential of asam gelugor to become a popular, commonly grown tree species. However, the growing market demand for traditional local recipes and cuisine in Malaysia, the new application of dried gelugor as an ingredient for weight-loss products and other health-related products, combined with the successful use of improved propagation methods to multiply female trees, have increased the interest of farmers in its commercial cultivation. Farmers have found the following sets of propagation and pruning techniques useful to popularize this underutilized species:

- Inducing the growth of female trees by bending and stressing the taproot of saplings
- Patch budding and cleft grafting techniques to multiply female trees
- Traditional marcotting technique through root cuttings to multiply female trees from the forest
- Top working and pruning of trees to train tree height and reduce drudgery of harvesting.

The case of asam gelugor describes how market incentives can lead to new innovations in propagation and the continued cultivation of an underutilized species. However, to popularize a neglected species such as asam gelugor requires the concerted efforts of all stakeholders within the value chain, from farmers to retailers, including service providers such as researchers. In conclusion, this set of techniques has significant potential as it could be scaled out in other similar contexts in which this simple technique might contribute to farmers' livelihoods, food culture and income.

References

Corner, E.J.H. (1988) 'Wayside trees of Malaya', *Malayan Nature Society*, vol 1, pp. 351–359

Jarvis, D.I., Brown, A.H.D., Cuong, P.H., Collado-Panduro, L., Latourniere-Moreno, L., Gyawali, S., Tanto, T., Sawadogo, M., Mar, I., Sadiki, M., Hue, N.T.H., Arias-

Reyes, L., Balma, D., Bajracharya, J., Castillo, F., Rijal, D., Belqadi, L., Rana, R., Saidi, S., Ouedraogo, J., Zangre, R., Keltoum, R.O., Chavez, J.L., Schoen, D., Sthapit, B.R., Santis, P.D., Fadda, C., and Hodgkin, T. (2008) 'A global perspective of the richness and evenness of traditional crop genetic diversity maintained by farming communities', *Proceedings of the National Academy of Sciences PNAS (USA)*, vol 105, no 14, pp. 5326–5331; www.pnas.org/cgi/doi/10.1073/pnas.0800607105

Khairunnisa, S. (2005) 'Asam gelugor', *Berita Harian*, 27 June, pp. 8–9

Nazmi, M.S., Lamers, H.A.H., and Salma, I. (2013) *Baseline Report Malaysia for the UNEP-GEF Project Entitled: Conservation and Sustainable Use of Cultivated and Wild Tropical Fruit Diversity: Promoting Sustainable livelihoods, Food Security and Ecosystem Services*, GEF, UNEP, Bioversity and MARDI, Malaysia

Salma, I., Mohd. Shauqi, N., Shariah, U., Pearlycia, B., Wong, W.W.W., and Abd. Rahim, B. (2012) 'Distribution and diversity of *Garcinia* species on farm in six selected sites in Malaysia', poster presented during 1st Regional Conference on Agrobiodiversity Conservation and Sustainable Utilization (RAC-1), 25–27 September, Langkawi, Kedah, Malaysia

Note

1 From 2009 to 2014, Bioversity International coordinated a research-for-development project supported by the Global Environment Facility (GEF) with implementation support from the United Nations Environment Programme (UNEP) – 'Conservation and sustainable use of cultivated and wild tropical fruit tree diversity: sustainable livelihoods, food security and ecosystem services', abbreviated as 'the TFTGR Project'. This project, implemented in India, Indonesia, Malaysia and Thailand, focused on livelihood and environment benefits that people could derive from the conservation of species and varietal diversity of *Citrus*, *Garcinia*, *Mangifera* and *Nephelium*.

Case studies

Production and crop management

17 The role of a traditional festival, Chhath Puja, in the conservation and sustainable use of tropical fruits

Awtar Singh, Vishal Nath, Sanjay Kumar Singh, Bhuwon Sthapit and B.M.C. Reddy

GPD 'passport'

GPD code:	11
Focus area:	Production and crop management Collective action and social networking
Character:	System
Species and varieties involved:	Mango (*Mangifera indica*): mostly seedlings of sucking and pickle types and more than 30 commercial varieties such as Neelam, Dussehari, Alphonso, Malda *Citrus:* pomelo (*Citrus grandis* (L.) Osbeck) seedlings, mostly red/pink-fleshed; lime (*Citrus aurantifolia* (Christm) Swingle) and lemon (*Citrus limon* (L.)) Others: many other fruits (papaya, guava, gooseberry, date palm, coconut, banana, etc.) and vegetables (tomato, cabbage, coriander, radish and leafy vegetables)
Name of location:	Communities of Mahmada, Jagdishpur, Murliyachak, Dhobgama (Pusa), Bihar, India
GIS reference of location(s):	N 25°59′33″; E 85°38′46″ Elevation: 52 masl
Name of farmers (data resource):	Mr Vinod Kumar Rai (Jagdishpur), Mr Janki Raman Prasad Singh (Mahmada), Mr Kanahiya Kumar (Mahmada), Mr Rajneshwar Thakur (Dhobgama), and Mr Rajiv Kumar Pandey (Murliyachak)

Context

The communities studied are all located in the Pusa site of the UNEP/GEF project 'Conservation and Sustainable Use of Cultivated and Wild Tropical Fruit Diversity: Promoting Sustainable Livelihoods, Food Security and Ecosystem Services' in the Indian state of Bihar. This district is located close to the old Gandak River in the fertile Gangetic Plains in North India. The main fruit crops grown in this area are mango, lychee, guava, date palm, citrus, gooseberry, custard apple and papaya. Other important livelihood crops include rice, wheat, potato, tobacco, vegetables and seasonal flowers. The majority of households maintain small home gardens.

The climate in Pusa, Bihar, is humid and subtropical, with maximum and minimum average temperatures of 31°C and 19°C, respectively, and an average annual rainfall of 1200 mm, distributed over 35–40 rainy days during the monsoon season. The maximum temperature touches 46°C in the month of May and the minimum touches 4°C in January. The maximum precipitation occurs during the months from June to September. During the winter season there are occasional fogs.

This study was conducted in two villages (four communities). Mahmada village in Pusa site has a total number of 1,124 households with a population of 6,028, of whom 30 per cent are from Bhumihars and Bhramin castes and the rest are Yadavs, Muslims, Kumhars, Kushwah, Baniyas and Paswans. The overall literacy rate is 50 per cent. Dhobgama village has 744 households with a population of 3,845, of whom 12 per cent are from the Bhramin castes and rest are as reported earlier. The most predominant crops are rice, wheat and maize for staple crops and mango, lychee and pomelo for perennial fruit crops. The soils in this area are sandy loam type, rich in organic matter and with very good water-holding capacity. Farmers use mainly organic fertilizers in their fruit orchards; very few use chemical fertilizers.

The main income source in all these four communities is from agriculture, horticulture, dairy and beekeeping. Some farmers in low-lying areas also have fish ponds. Farmers sell their crops, such as rice, wheat, potato and onion, directly in local markets. Most of the fruit grown in the orchards (mango, lychee, guava, etc.) is sold to pre-harvest contractors, who take up the marketing of the produce. The fruit is generally sent to distant markets such as Kolkata or Delhi by the pre-harvest contractors. There are no functioning co-operative societies to take care of marketing. Several households have non-farm income from absentee members with jobs as maids or in construction work (INR2,000–3,000 per month) or small shops and business activities (INR5,000–8,000 per month). Some of the crops grown in home gardens are also sold and this adds to the family income. The average income for a landless agricultural labourer in this district is around INR2,500 (US$40) per month. Large-scale farmers, service workers and businesspeople earn considerably more: INR5,000 to 15,000 (US$80–250) per month.

Methodology used for data collection

Information was collected about the family setup, land holdings, types of crops cultivated, number of different fruit cultivars and seedling trees in the orchards and home gardens, and cultural practices followed for agricultural and horticultural activities by the farmers in the project communities. Information on the different types of seedling trees maintained by the farmers was used for the identification of superior clones of mango and pomelo after characterization of their fruits for different morphological and physicochemical traits. The preliminary information was collected from both primary and secondary sources that have knowledge about the local communities. The information relating to species and varieties was also validated by field visits and direct observations. Both qualitative and quantitative data were collected. Data on fruit diversity and other crop diversity were gathered from 34 households. More in-depth interaction was carried out with three custodian farmers and their families, Shri Vinod Kumar Rai in Jagdishpur and Shri Janki Raman Prasad Singh and Shri Kanahiya Kumar in Mahmada village, as they were seen to have rich traditional knowledge about these fruits and their cultural value.

In the process of identification of superior types in seedling populations of pomelo and mango, we came across a very interesting practice of home gardening whereby many fruit species are maintained for a special popular religious function practised by almost all the families in the communities of our project site, *Chhath Puja* (Singh *et al.*, 2013; 2015; Somashekhar, 2014).

Chhath Puja

Chhath Puja is the festival of cleanliness of body and soul, truth, non-violence, forgiveness and compassion. It is dedicated to the Sun God, Surya. *Chhath Puja* is one of the most important Hindu festivals in Bihar and Jharkhand during the month of *Kartik* (October/November). It is also celebrated in some parts of West Bengal, Oddisha, Assam, Uttar Pradesh and Madhya Pradesh, and even in Nepal and Mauritius (Jha, 2009). The enormous faith and belief that all the desires of the devotees who perform *Chhath Puja* will be fulfilled has made it one of the most popular festivals in this region.

This is the only festival where the function and *Puja* is led predominantly by women, thus indicating the honour and respect of women in Hindu society. The songs sung on this occasion tell about the importance of natural resources, and the conservation and use of diversity for the benefit of human welfare. This festival is celebrated for four consecutive days. The first day of *Chhath* is dedicated to cleaning, preparation and purification of items to be used in offerings. The house and surroundings are thoroughly cleaned. Meals are prepared and taken after a bath with holy water, which is why the day is called '*Nahai Khai*', meaning 'meal after bath'. *Lauki* (bottle gourd) is an important preparation on the day and the cooking and the day is also referred as '*Lauki Bhat*'.

Kharna (fasting) begins on the second day. The *Vrati* (devotee) observes the fast for the whole day until the evening, a little after sunset. After this, the family shares the *Prasad* (offerings) prepared by the *Vrati* with extreme care and devotion using rice, gur (jaggery), mild spices, fruits and holy water. Only cow's milk and *ghee* can be used to prepare the offerings. Mango twigs and dry branches are used as fuel for the preparation. From this day onwards for the next 36 hours, the *Vrati* fasts without even drinking water. The fast is broken on the morning of the fourth day. The third day is spent on preparations of *Prasad* at home. On the evening of this day, the entire family accompanies the *Vrati* to a river bank, pond or a large water body to make offerings to the setting sun.

Arghya and *Soop* are the offerings given and consist of flowers, fruit, sprouted grains, coconut, sugarcane, white radish, turmeric, ginger, sweet potato and sweets. Pomelo (*Citrus grandis*), known locally as Gagar Nimbu, is one of the main fruits specially offered to the god.[1]

Description of good practice for diversity

Bioculture

Chhath Puja is deep-rooted in folk culture and it sustains the basic concept of worship with a combination of biodiversity conservation, social integrity and local livelihood development. This is a unique Hindu cultural and ritual practice in Bihar that safeguards varieties of tropical fruits, flowers and water bodies in the villages. It has emerged as a national festival along the border of India and Nepal.

In the *Puja* ceremony, many fruits like pomelo, aonla, guava, coconut and banana, along with many root vegetables and spices, are offered to the Hindu god. Both men and women collect the fruits from their home gardens, receive them from their close relatives or buy them from the markets. Thus people plant many types of fruit trees in their home gardens so that they can harvest their fruits during this festival. Some species continue to be maintained purely due to the requirements of *Chhath Puja*. The best example of this is pomelo (*C. grandis*), which is maintained by most of the families only for this festival and not for any other purpose (Plate 42). The festival contributes strongly to the conservation of pomelo biodiversity in Bihar. A total of 13 varieties were catalogued in the Pusa site and more than 16 varieties and forms have been recorded in various parts of the state on the basis of fruit size, flesh and plant type (Dinesh *et al.*, 2014).

Another aspect of biodiversity conservation is related to the water body. *Chhath Puja* takes place on the banks of different water bodies, for example, rivers, lakes, ponds and streams, where clean water flows. The cleaning and maintenance of water bodies and other embankments is carried out for celebrations of the *Puja*, and contributes widely to diversity conservation by maintaining healthy ecosystems, especially in the wetlands (Kesari, 2009).

As long as this culture is strongly embedded in the society, the maintenance of a variety of tropical fruits and healthy water bodies will continue without economic incentives.

Home gardens

In the context of the above cultural backdrop, home gardens have multiple functions: (i) supply and supplement of subsistence requirements, (ii) goods and services for cultural needs and rituals, (iii) social interactions and exchange of materials, (iv) reservoirs of globally and locally valued crop and tree diversity, (v) platform of experimentation and innovation, and (vi) conservation of agricultural biodiversity (Eyzaguirre and Linares, 2004; Gautam *et al.*, 2008).

In the home gardens of most households in the project communities, different types of fruit – such as papaya, mango (dwarf type, sucking type, pickling type and common commercial varieties), citrus (pomelo, lime and lemon), banana, guava, pomegranate and gooseberry; seasonal vegetables (leafy vegetables, cucurbits, beans, cabbage, chilli, coriander, tomato, etc.) and some flowering plants (China rose, marigold, etc.) are planted. These provide fresh fruit, vegetables and flowers throughout the year (Gautam *et al.*, 2008: Rahman *et al.*, 2009; Singh *et al.*, 2015). Cucurbits such as bottle gourds, pumpkins and sponge gourds can be seen growing as climbers, whereas green vegetables are grown in small beds and flowering plants are planted on the periphery of the house and the home garden. The seeds used for flowering plants and vegetables are from the farmers' own saved seeds and thus are genetically highly variable. Some of the farmers are interested in maintaining unique types of fruit and vegetables as a hobby, which increases inter- and intraspecific diversity in the form of landraces of different crops.

A minimum of three different species of fruits are seen in every home garden and often a pomelo tree is planted in front of the house or in the backyard of the homestead, specifically for *Chhath Puja*. The households plant seedling types of mango and pomelo, thus providing an opportunity for the selection of improved types of these fruit crops owing to the genetic recombination and segregation in the seedling progeny. From a commercial perspective, these types do not have a high value due to the mixture of different types, but they could be potential genetic resource for selection.

Impact on diversity

Traditional home gardens maintain diversity as the house owners plant different types of fruit, flowering plants and vegetables and also different varieties of these crops. They also enhance diversity as seedling (*biju*) trees are maintained. Some house owners maintain unique types of fruits, vegetables and flowering plants as a hobby. Dissemination of the diversity is supported by the practice of farmers distributing the seeds and other planting material to their social connections through relatives, friends and neighbours. Almost all families with home gardens maintain pomelo and/or mango trees from the seed material of their choice and mostly of local origin. Thus they contribute to the maintenance of local varieties and landraces. Because mainly seedling types are planted, the trees found in different homes tend to be different from one another and variable, and this variability offers diversified uses of these fruits. Intraspecific diversity is increased as, in addition to seedlings of the targeted fruits, some improved and rare varieties of these fruits are also planted in these home gardens.

Households across the four communities maintain a total of 53 different commercial varieties of mango in their home gardens and orchards, of which Malda, Sipia, Sukul, Bathua, Bombay Green, Kishanbhog, Paharpur Sinduria, Jarda and Kanchan are the most popular. However, these four communities together maintain a large population of more than 8,400 seedlings or unnamed mango trees, making it one of the sites with highest intraspecific diversity in the tropical fruit tree project. This seedling population makes up almost 50 per cent of the total 16,916 mango trees that are maintained by the 200 households that participated in the baseline survey (Gajanana *et al.*, 2014).

The main *Citrus* species maintained in home gardens are pomelo (*C. grandis*), acid lime (*C. aurantifolia*), lemon (*C. limon*), rough lemon (*C. jambhiri*), sweet orange (*C. sinensis*) and Cleopatra mandarin/Hazari nimbu (*C. reshni*). Most varieties of these citrus trees do not have names because they are grown from seeds.

A total of 23 of farmers' best pomelo (11) and mango varieties (12) have been registered in the National Bureau of Plant Genetic Resources (NBPGR), India through the project interventions.

Impact on livelihoods

Most households obtain income from non-farm jobs and there are few full-time farming households. However, even those villagers who are not members of full-time farm households have home gardens and grow fruit and vegetables for home consumption. Diversity in home gardens means that farmers enjoy home consumption of a diversified diet of fruit and vegetables. Direct income from the home gardens is small but has a direct bearing on the health and livelihood of the communities. They can save household expenditure on food items, especially when market prices are high, and this saving in household expenditure can alternatively be used for other important purposes such as education of children and clothing. The demand for fresh fruits during the festival increases manyfold in the Indo-Gangetic plains of India and Nepal and farmers derive some income by selling to those who do not have access to home gardens.

The practice of growing food and fruit crops in home gardens enhances the human and social capital of households as it involves and supports traditional, cultural and religious activities like the exchange of fruits among relatives, friends and neighbours during the religious *Chhath Puja* celebration. Having many different fruits and a well-maintained home garden contributes to the status of the household within the village. As thc fruits and vegetables include many landraces, are mostly cross pollinated and include large seedling populations, the home gardens help enhance the natural capital of households.

Sustainability and other benefits

As most of the seed used is from farmers' own sources or sourced locally from friends and relatives, there is little dependency on outside support for seed and other inputs and the diversity in the form of landraces of plants and rare species is maintained. It also provides an opportunity for developing new types through crop selection and contributes to the process of plant domestication. Because most of the varieties grown in home gardens have been selected from pre-existing seedling trees, it can be assumed that these have evolved under natural conditions and are well adapted to local conditions. They therefore require very little spraying with pesticides, which benefits the environment in a small way. If practised in conjunction with beekeeping, the system can add further to the family economy. Beekeeping provides additional income throughout the year and also improves pollination and subsequent fruit setting, as in most home gardens the trees of a single fruit are grown in isolation and only bees ensure good pollination.

Although home gardens have persisted for generations, they have been undervalued because the home gardens make little profit and their produce is mainly used for family consumption. In recent years home gardening has come to be revalued as the fragility of modern monoculture production systems has been increasingly recognized; these modern systems are highly influenced by external price shocks and potentially reduce environmental and ecological health. Home gardens make households less susceptible to external price shocks in vegetable and fruit markets. Alternative practices, such as using these crops in pickle and jam preparation, transportation of fruit (pomelo) to distant markets, and a minimum of value-addition to the fruits have been piloted to improve the consumption of these traditionally grown fruits. Home processing of fruit could also make households less dependent on income from cash crops, which are more vulnerable to weather conditions and fluctuating sales or input prices. Strengthening and improving local methods for identification, selection, multiplication, promotion and marketing of superior types that are found in home gardens and planted in orchards could increase sales value and provide additional household income.

Driving forces for the success of the good practice for diversity

The major driving force for the success of this good practice for diversity management is the cultural tradition of celebrating *Chhath Puja* and the requirement for a wide range of fresh fruits for this religious occasion. This religious practice is a major driving force for several households to grow some unique and rare types, which increases the overall diversity in the community. Women play a key role in *Chhath Puja* and therefore women farmers favour the maintenance of fruit diversity in their gardens. Recognition of the role and knowledge of women, strengthened by scientific back-up through detailed review and verification of their knowledge, should be hastened before the knowledge that rural women have fostered through generations is lost.

In addition, increasing market prices of fruit and vegetables in recent years could create renewed interest in home gardens. Women have the opportunity to sell extra produce at local markets, which can contribute to financial empowerment within the household (Suwal *et al.*, 2008). Another driving force is raised awareness of the harmful residues of chemicals and pesticides often found on fruit and vegetables purchased from the market.

The major constraint hindering the successful adoption of the good practice is the potential scarcity of labour or increases in opportunity costs when other activities such as urban employment as maids or factory workers become much more profitable for women. Home gardening is further constrained due to the lack of clear policy support for family nutrition and health. Similarly, the small size of home gardens and lower availability of land for gardening are also constraining this practice. The planting of improved types may lead to a reduction in the population of seedling or local types and ultimately to reduction of diversity.

Conclusions and action plans for scaling up and dissemination

Home gardens are an essential element within traditional production systems and local seed systems, where farmers of both sexes identify their preferred species and varieties by natural out-crossing and natural and human selection. Home gardens provide fresh fruit, vegetables and flowers throughout the year, ensuring good nutrition, health and general well-being of farm families. The practice also reduces household expenses by providing daily requirements of fruit and vegetables. The seeds and planting material are mostly from the farmers' own seeds and thus contribute to the conservation of many vegetable, fruit and ornamental species and landraces. Because pomelo and mango are cross-pollinated, the seedling populations of these fruit trees exhibit a lot of variability in the home gardens, which offers scope for the selection of desirable types or superior cultivars. Thus the practice of home gardening leads to the conservation of diversity and to the evolution of plants in the form of new varieties, as well as keeping household members healthy.

The present case is an excellent example of a good practice (i.e. maintaining fruit tree diversity in home gardens) mainly driven by local socio-cultural needs. As long as the festival of *Chhath Puja* continues in these Hindu communities of India and Nepal, on-farm conservation and sustainable use of pomelo and mango diversity will continue to persist.

References

Dinesh, M.R., Rajan, S., Singh, I.P., Singh, A., Singh, S.K., Vasudeva R, Vinoth, S., Patil, P., Parthasarathy, V.A., Reddy, B.M.C and Sthapit, B.R. (2014) *National Fruit Catalogue of Tropical Fruit Tree Diversity (Mangifera, Citrus and Garcinia)*, GEF, UNEP, Bioversity International and ICAR, India

Eyzaguirre, P.B. and Linares, O.F. (2004) *Home Gardens and Agrobiodiversity*, Smithsonian Books, Washington, DC

Gajanana, T.M., Lamers, H.A.H., Rajan, S., Singh, A., Singh, S.K., Singh, I.P., Vasudeva R, Hegde, N., Dinesh, M.R., and Dakshinamoorthy, V. (2014) *Baseline Report India for the UNEP-GEF Project Entitled: Conservation and Sustainable Use of Cultivated and Wild Tropical Fruit Diversity: Promoting Sustainable Livelihoods, Food Security and Ecosystem Services*, GEF, UNEP, Bioversity and ICAR, India

Gautam, R., Sthapit, B.R., Subedi, A., Paudel, D., Shrestha, P.K., and Eyzaguirre, P. (2008) 'Home gardens management of key species in Nepal: A way to maximise the use of useful diversity for the well-being of poor farmers', *Plant Genetic Resources: Characterization and Utilization*, pp. 1–12 (online). DOI: 10.1017/S1479262108110930

Jha, C. (2009) 'Worship of Sun God (*Chhath Puja*) – Blessings for liberated untouchables', Biodiversity Conservation, Social Integrity and Local Livelihood, New Delhi, http://www.sulabhinternational.org/sulabh-old/admin/config/media/file-system/Chhath%20Puja.pdf

Kesari, V. (2009) '*Chhath Puja*: A combination of biodiversity conservation, social integrity and local livelihood', www.forestrynepal.org/article/627/4433

Rahman, M.L., Hasanuzzaman, M., and Islam, M.K. (2009) 'Fruit distribution and diversity in the homesteads of Southern Island, Bangladesh', *Advances in Biological Research*, vol 3, no 5–6, pp. 208–214

Singh, A., Pandey, S.D., Nath, V., Kumar, A., and Rai, R.R. (2013) *Role of Chhath Puja in Biodiversity Conservation and Social Harmony*, National Research Centre on Litchi, Mushahari, Muzaffarpur, APO Newsletter no. 60, Jan–June 2013, Bioversity International, Serdang, Malaysia

Singh, S.K., Singh, A., Vishal Nath, Parthasarathy, V.A., Sthapit, B., and Vinoth, S. (2015) 'Pummelo in homestead garden: Conservation through family farming', *Indian Journal of Plant Genetic Resources*, vol 28, no 1, pp. 132–138. DOI 10.5958/0976-1926.2015.00017.0

Somashekhar, B.S. (ed.) (2014) *Tropical Fruit Trees & Farmers' Traditional Knowledge: Community Primer on Good Practices for Diversity Management*, National Project Management Unit – India, Bioversity International, Tropical Fruit Tree Genetic Resources Project, Indian Institute of Horticultural Research, Bangalore, India

Suwal, R., Regmi, B.R., Sthapit, B.R., and Shrestha, A. (2008) 'Home gardens are within the reach of marginalized people', *LEISA*, vol 24, no 3 (September), p. 34

Notes

1 The *Chhath Puja* is performed in order to thank Surya (Sun) for sustaining life on earth. The Sun, considered as the god of energy and of the life-force, is worshiped during the Chhath festival to promote well-being, prosperity and progress. It is also celebrated for the worship of Goddess Chhathi Maiya (ancient Vedic Goddess Usha). She is believed to be the consort of Surya, the Sun God.

18 Multivarietal orchards

An age-old conservation practice in mango

T.M. Gajanana, M.R. Dinesh, Sudha Mysore, C. Vasugi, Bhuwon Sthapit, Hugo A.H. Lamers, B.M.C. Reddy, V. Ramanatha Rao and V. Dakshinamoorthy

GPD 'passport'

GPD code:	12
Focus area:	Production and crop management
Character:	System
Species and varieties:	*Mangifera indica*, including a seedling population and 28 different varieties such as Totapuri, Banganapalli, Neelum, Alphonso, Atimadhuram, Lalbaba, Gaddemar, Omelette, Rumani, Khuddus, Imam Pasand, Ali Pasand, Kalepadu, Seeri, Reddy Pasand, Dil Pasand, Chitti Bangalora, Mallika, Peter, Gadiyaram, Thorapadu, Raja Pasand, Pulira, Mulgoa, Manoranjitam and Chakkaraguttulu
Name of location:	Bangarupalyam, Talupulapalle and Polakala villages in Chittoor district, Andhra Pradesh, India
GIS reference of location(s):	Bangarupalyam: N 13°11′24″, E 78°12′37″; Talupulapalle: N 13°23′40″, E 79°03′26″; Polakala: N 13°26′14″, E 79°01′03″
Name of farmers (data resource):	Data collected from 195 male farmers (65 from each community)* and 1 woman custodian farmer

*Data were collected at the household level, with men being the primary respondents as they are considered the head of household and knowledge holders on mango cultivation.

Introduction

Mango, the king of fruits, is one of the most important fruit crops grown in India, accounting for 38 per cent of area planted and a 22 per cent share of total fruit production. Mango has been cultivated in India for at least 4,000 years and more than 1,000 varieties are recognized (Mukherjee, 1953). The large variability that is exhibited by mango is due to seed propagation and the high heterozygosity present (Iyer and Schnell, 2009).

Chittoor district in Andhra Pradesh is one of the major commercial mango production areas in India. Despite increasing commercialization and the establishment of mango pulp processing industries, many farmers in Chittoor maintain multivarietal orchards. This chapter outlines and explores this age-old practice of maintaining multiple varieties in an orchard and examines whether this practice contributes to mango varietal diversity. Although 20 to 28 varieties are grown commercially at a community level, at a household level, average household richness ranges from four to five varieties. In addition to commercial varieties, most of the orchards surveyed showed that farmers grew indigenous seedling types (*naati* type, often unnamed). Farmers explained that they maintain local varieties because they are the most suited to local home use and consumption (in the form of pickles, juice, spice ingredients or fresh), but also because they improve pollination services by combining both early and late varieties, thus spreading nectar availability for pollinators. They also get better prices and higher yields.

Context

Chittoor has developed into one of the main mango growing belts in South India. Chittoor district in Andhra Pradesh has a hot, dry climate with low temperatures of 12–18°C in winter, high temperatures of 38–46°C during summer and an average rainfall of 918 mm per year. The region receives rainfall from the southwest monsoon from June to September and the northeast monsoon from October to December. Land is mostly covered by red sandy soils. The mango cultivation is rainfed, but irrigated by a tube well irrigation system at critical stages. The water table has dropped to dangerously low levels in the last two decades and bore well irrigation has become a necessity, although it is also failing due to dwindling groundwater resources. Mango cultivation is less water-demanding than other water-intensive crops like sugarcane and paddy.

The average farm size in Chittoor is 2.6 ha and the main crops grown are mango, sugarcane and groundnut. During the last two decades, several, both small and large, processing plants of mango pulp have been established in Chittoor due to the increased demand for exports. Hence, over the last two decades farmers have slowly replaced traditional varieties with commercial types that are in demand from pulp processing plants. As a result many new plantations have been initiated with commercial types. A survey revealed that more than 95 per cent of the total number of trees belong to the four most popular commercial varieties, namely Totapuri (local name Bangalora), Neelum, Banganapalli (Baneshan) and Alphonso (Khadar). The varieties Alphonso and Totapuri are in demand from the processing plants and Alphonso is also used as a table variety. The varieties Neelum and Banganapalli are mainly sold as a fresh fruit for table purposes in regional markets. In spite of the dominance of these four commercial varieties, many other mango varieties are still maintained in Chittoor (Table 18.1). The tail of the tail of these (rare and unique types) have potential scope for diverse use and commercialization.

Identification of the good practice for diversity

A number of tools were used to identify good practices for diversity management in three communities (Bangarupalyam, Talupulapalle and Polakala) in the Chittoor site as part of the TFTGR project.[1] They included: (i) consultations with local experts, (ii) participatory rural appraisal (PRA) including Four Cell Analysis (FCA) and Diversity Fairs, (iii) Focus Group Discussions and (iv) a baseline survey. For the baseline survey, a random multistage sampling methodology was used to select respondents from each community. In each community of about 500 households (HH), a sample of 10 per cent (approximately 50 households) was selected. Further, another 15 HH were selected as a control group in each community. In total, 65 households (50 samples + 15 controls) were selected for each community, thereby making a total sample size of 195 respondents from the three communities. In addition, one custodian farmer (female) was identified and her views were also added to the analysis. The sampling structure was designed in such a way as to give representation to small (<1 ha), medium (1–2 ha) and large (>2 ha) growers. A pre-tested questionnaire was administered by interviewing the head of the sample households for collection of relevant information. Genetic diversity analysis was performed using the Simpson Index (SI) (Meng *et al.*, 1998; Kruijssen and Mysore, 2010) and the Margalef Index (MI) (Nagarajan *et al.*, 2007).

Description of good practice for diversity

During the PRA, FCA and baseline survey, researchers observed that Chittoor farmers have increasingly adopted the practice of keeping multiple varieties of mango in current and new commercial and semi-commercial orchards and home gardens. A multivarietal orchard is defined as an orchard that has more than five varieties. This definition was considered based upon the baseline data of the average number of varieties (i.e. richness of 4.7) per orchard found in the community. It is interesting to note that this practice was observed in a very commercially oriented mango production system. Then the question is, why do farmers do this? The multivarietal orchard, containing seedling and traditional types, is considered a good practice for diversity management in commercial production areas because the system serves the purpose of multiple home uses and also helps maintain indigenous seedling varieties and lesser-known varieties along with established commercial varieties. Moreover, a multivarietal orchard often includes a combination of rare or local varieties in addition to the most common commercial types, in this case Totapuri (Bangalora), Neelum, Banganapalli (Baneshan) and Alphonso (Khadar). The general characteristics of good practice for diversity management are a combination of many varieties – whether commercial or local – in one orchard. Based on expert consultation there are basically five key reasons why farmers have multivarietal orchards:

- Commercial interest
- Pollinator effect to improve yields
- Minimize risk of losing income from monopoly markets
- Including early and late bearing varieties or regular bearers to lengthen harvest season
- Home use and personal interest.

This multivarietal orchard system in Chittoor tends to consist of South Indian varieties that are mostly regular bearers that fruit every year and usually includes some or all of the four most popular varieties – Totapuri (Bangalora), Neelum, Banganapalli (Baneshan) and Alphonso (Khadar) – for income generation, in addition to a few seedling (*naati*) types or traditional varieties such as Imam Pasand, Atimadhuram and Lalbaba.

These seedling and traditional types flower profusely, thus attracting bees and other insects, which improve the pollination and fruit setting of the commercial varieties. According to Mr Chitti Reddy, one of the farmers interviewed, this practice improves yields by 10 per cent to 25 per cent. Some farmers, like Mr Chandrasekhar Reddy (one of the custodian farmers having >10 varieties of mango), keep beehives inside the orchards to further improve pollination services and also obtain additional income by selling honey. Honey production improves if orchards have many varieties flowering at different times. Traditional and *naati* seedling saplings tend to be taller, so birds are attracted to them and eat their fruit, thus minimizing fruit loss of the most popular commercial varieties.

The choice of varieties in multivarietal orchards is driven by the farmers' livelihood strategy. For example, to avoid the low prices during the production glut of the harvest season, farmers often combine several trees of an early or late variety at the time of establishment that fruit outside the main season. This extends the period of income for farming households. In such cases, Pulira (Sendura) is often used as an early variety and Neelum, Rumani and Mulgoa as late maturing types. Alphonso (Khadar) and Banganapalli (Baneshan) are used as mid-season varieties, while Totapuri (Bangalora) is rather late with a guaranteed market for the processing industries.

Some farmers choose to integrate traditional landraces or seedling types in their orchards because of their potential robustness against adverse weather conditions (dry monsoon, unexpected heat or sudden rains during flowering) and tolerance against pests (mango hopper, anthracnose). Other reasons worthy of note given by some farmers are that they maintain diversity as a form of hobby or because they did not want to cut down older trees planted by their ancestors.

Many multivarietal orchards maintain a composition of one or two trees of traditional types mostly used for different home consumption and food culture uses: sucking (e.g. Chakkaraguttulu), pickling (e.g. Ali Pasand, Gaddemar, Omelette) or table varieties such as Atimadhuram (shy bearer). Fruits from these trees are often used for traditional recipes like 'amchoor' (powder made from green unripe mangoes that is used as a flavouring condiment), 'mango leather' (sweet dried mango paste used as candy), several types of mango pickle (made from tender or slightly mature green mangoes), chutney (slightly mature green mango ground along with chillies and other spices) or 'panna' (a juice made of mango and several spices, mostly consumed during summer). These different uses reflect consumer preferences. Families sometimes maintain in their yards older trees of rare and lesser-known varieties or seedling selections for fresh consumption that have excellent taste or skin colour such as Atimadhuram, Imam Pasand or Lalbaba. High-value fruits and products of some of the varieties (Ali Pasand, Gaddemar, Imam Pasand, Atimadhuram, Kalepadu) are often shared with neighbours and relatives during special family occasions or celebrations. These examples clearly illustrate that farmers are maintaining multivarietal orchards for their multiple uses and, in most cases, having five varieties addresses the majority of farmers' household needs.

Table 18.1 illustrates a comparative overview of all 28 varieties found in Chittoor. From this table it becomes clear that, depending on their livelihood strategy (e.g. self-sufficiency, risk minimization or income maximization), farmers choose and combine a set of popular commercial varieties that generate income (Category 1) with varieties from one or more of the other categories: early or late ripening (Category 2), pollinator services (Category 3) or home consumption (Category 4). Commercial and short-term gain-oriented farmers plant their orchards predominantly with commercial (Category 1) varieties, which is often more risky, while more risk-averse or long-term sustainability-oriented farmers seem to use a larger share of their orchards for other types of varieties such as early varieties, late varieties, sturdy varieties or pollinator-friendly varieties (e.g. *naati*), and tend to maintain more trees of traditional pickle, table or sucking type that are predominantly used for home consumption.

Based on qualitative and quantitative information, one can identify three types of farmers:

- Type 1 (monocropping) maintains only one to four varieties (81 farmers), mostly because they have a strong market orientation, can afford to take risks or are not aware of the advantages of multivarietal orchards and plant commercial varieties only.

- Type 2 (multivarietal) maintains from five to nine varieties (practised by 68 farmers): they include some *naati* or local varieties for improved pollination or maintain a few pickling, table or sucking type of trees for home consumption.
- Type 3 (custodian) maintains 10 or more varieties (seven farmers): they do it for one or more reasons – as a hobby, because of genuine cultural or historical interest in diversity, for improved pollination, for home consumption and/or to explore multiple uses and characteristics.

Effect on crop genetic diversity

Multivarietal orchards impact significantly and positively on intraspecific diversity of mango. This results from the diverse needs of farmers and their special preferences for specific uses. The 195 male farmers interviewed, representing about 12 per cent of all the families with fruit trees in the three communities, have on average 327.4 trees per household. An average of 4.7 varieties are maintained per household and in total 28 different varieties are known within the three communities (Table 18.1). Indigenous varieties such as Gaddemar, Omelette, Chakkaraguttulu, Reddy Pasand, Atimadhuram, Raja Pasand, Kalepadu, Torapadu, Imam Pasand, Dil Pasand, Khuddus and Gadiyaram are combined with commercial varieties such as Totapuri (Bangalora), Neelum, Banganapalli (Baneshan), Alphonso (Khadar), Mallika, Mulgoa, Rumani and Pulira (Sendhura). Of the total number of trees, 95 per cent belong to the four most popular commercial varieties: Totapuri (Bangalora), Neelum, Banganapalli (Baneshan) and Alphonso (Khadar).

About 52 per cent of the households maintain between one and four varieties, 44 per cent maintain five to nine varieties and about 4 per cent of the households maintain more than nine varieties on their farms. About 98 per cent of the households have the most popular variety, Totapuri, and 24.6 per cent of the farmers maintain local varieties called '*naati*' (i.e. seedling origin or unnamed varieties). Within this group of 48 households with *naati* trees, the average number of *naati* trees or seedlings per household is 5.3 trees. Pickling types such as Gaddemar, Ali Pasand, Omelette and Reddy Pasand are maintained by 23.7 per cent of the households and sucking types such as Chakkaraguttulu are maintained in about 11 per cent of the households.

When looking at the average orchard age, households with one to four varieties have an average orchard age of 16 years, households with five to nine varieties have an average orchard age of 14.3 years and households with ten or more varieties have an average orchard age of 31.1 years (Table 18.2). This corresponds with a trend showing that the oldest orchards harbour the most diversity. However, after a swift reduction in the level of diversity about 10 to 25 years ago, plantations established in the last decade show again a slightly higher level of diversity.

The 4 per cent of farmers who have more than nine varieties on their farm maintain 89 per cent of the richness in varieties (25 of the 28 varieties found).

Table 18.1 Comparative overview of available diversity at the Chittoor site with their main characteristics

Serial number	*Landraces of mango – Chittoor*	*FCA result*	*Botanical and agronomic traits*	*Uses*	*Morphological and market traits*	*Other interesting features or characteristics*	*Category*
1	Ali Pasand	Few HH, Few trees	Off-season variety	Pickling type	Acidic and fibrous	Off season	Cat 4
2	Alphonso/Khadar	Many HH, Many trees	Mid (May)	Table	High price High juice content	Excellent table variety	Cat 1
3	Amini/Omelette	Few HH, Few trees	Early	Pickling	Acidic and big	Fibrous	Cat 4
4	Atimadhuram	Few HH, Few trees	Pollinator, shy bearer	Table	Excellent taste/sweet Yellow colour		Cat 4
5	Banganapalli/ Baneshan	Many HH, Many trees	Mid	Table	Large fruits Yellow colour High yield		Cat 1
6	Chakkaraguttulu	Few HH, Few trees	Mid to late	Sucking type	Excellent taste	Size very small preferred for cooking	Cat 4
7	Chitti Bangalora	Few HH, Few trees	Mid	Table	Taste is good	Small type of Totapuri	Cat 1
8	Dil Pasand	Few HH, Few trees	Mid	Table	Taste is good		Cat 4
9	Gaddemar	Few HH, Few trees	Early to Mid	Pickling type	Big size fruits		Cat 4
10	Gadiyaram	Few HH, Few trees	Mid	Table	Good taste and skin colour	Regular bearer	Cat 2
11	Imam Pasand	Few HH, Few trees	Pollinator, shy bearer	Table	Excellent taste	Fetches highest price, shy bearer, need to improve productivity	Cat 3
12	Kalepadu	Many HH, Few trees	Mid–late, shy bearer	Table	Sweet taste	Good keeping quality	Cat 1
13	Khuddus	Few HH, Few trees	Mid–late	Table	Good colour	Coloured variety	Cat 2
14	Lalbaba	Few HH, Few trees	Pollinator	Table	Red colour Good taste	Good keeping quality	Cat 2

15	Mallika	Few HH, Few trees	Mid	Table	Good taste	Hybrid between Neelum and Dashehari	Cat 1
16	Manoranjitam	Few HH, Few trees	Late, shy bearer	Table	Long shelf life (more than 20 days), High price	Good skin colour	Cat 4
17	Mulgoa	Many HH, Few trees	Mid/late, shy bearer	Table	Shy bearer	Farmers need to improve productivity	Cat 1
18	Naati (seedling)	Many HH, Few trees	Pollinator, disease resistant	Table/ pickling type	Good skin colour	Regular bearer	Cat 2
19	Naati Baneshan	Few HH, Few trees	Mid	Table	Green skin	Regular bearer	Cat 2
20	Neelum	Many HH, Many trees	Late	Table	Medium size, High yield	Regular bearer	Cat 2
21	Pither (Peter)	Few HH, Few trees	Mid	Table	Medium size	Coloured, regular bearing	
22	Pulira/ Sendura	Many HH, Many trees	Very early	Table Pulp	Red colour, Good price	Regular bearer, good substitute for Khadar (Alphonso)	Cat 2
23	Raja Pasand	Few HH, Few trees	Mid	Table	Good taste, Round shape fruits		Cat 4
24	Reddy Pasand	Few HH, Few trees	Mid	Pickling type	Heavy bearing		
25	Rumani	Many HH, Many trees	Mid/late	Table	Good taste, Round shape fruits		Cat 1
26	Seeri	Few HH, Few trees	Mid	Table	Good taste, Oblong fruits		Cat 4
27	Thorapadu	Few HH, Few trees	Mid	Table	Large-sized		Cat 4
28	Totapuri/ Bangalora	Many HH, Many trees	Late	Mostly for pulp, but also for table	High yield, Large fruit with good pulp recovery, Parrot beak shape	Regular bearer, Originates from Chittoor area	Cat 2

Note: Category 1 (Cat 1) = commercial; Cat 2= early or late or regular; Cat 3 = pollinators; Cat 4 = home consumption. FCA = Four Cell Analysis; HH = households.

Source: Baseline survey, 2010; see Gajanana *et al.* (2014).

Table 18.2 Orchard age per household diversity group

Household richness	*Average orchard age*
1–4 varieties	16.0
5–9 varieties	14.3
> 9 varieties	31.1

This constitutes a significant amount of intraspecific diversity in *Mangifera indica* that is maintained by very few farmers in Chittoor district. It is a rather risky long-term strategy leaving the maintenance of this significant diversity subject to so few custodian farmers.

Environmental, social and economic benefits

Multivarietal orchards ensure benefits by promoting ecosystem functions and services and thereby increasing better fruit set, creating easy access to a wide range of potentially interesting varieties and allowing for income diversification. They also contribute to economic improvement as some of the indigenous varieties command better prices and are used for market heterogeneity in roadside stalls. The mango trees serve a social cause through exchange of the diverse fruits among friends and relatives at different times to enhance social well-being. Multivarietal orchards are good sources of elite materials for breeding programmes; in fact, a total of 13 *naati* varieties with unique traits from these orchards in Chittoor have already been sent for registration with the Protection of Plant Varieties and Farmers' Rights Authority (PPV&FRA) (Dinesh *et al.*, 2014; Plate 43).

Farmers' knowledge about some of the indigenous varieties that are likely to be resistant or tolerant to several types of abiotic and biotic stress is maintained, updated and adapted to changing environmental conditions. For example, farmers maintain rare varieties like Omelette and Manoranjitam, which are reported to perform well under high temperatures by some of the farmers.

Improved livelihoods

The analysis of baseline data indicated that all respondents found fruit trees to be a very important source of income for their household (very important = 96.1 per cent) and as a source of nutrition (42 per cent very important and 48.7 per cent a little important). Mr Ravindranath (one of the custodian farmers) opined that fruit trees are also important for their medicinal, cultural or natural value. However, just 15.2 per cent, 6.1 per cent and 13.4 per cent respectively, of the farmers interviewed in the three communities think these values have some importance. The majority of the respondents (75–95 per cent) were in favour of multivarietal orchards for maintenance of mango varietal diversity to reduce yield risk (95–100 per cent), provide income security to the household (95–100 per cent) and ensure minimum price risk (75–95 per cent). The main driving force behind this practice mentioned by farmers is the improved fruit set and enhanced productivity by enhancing pollinator services and livelihood security. The practice also has a direct bearing on livelihood security by reducing the risk of dependence on one variety.

From the analysis of the baseline data, it was observed that mango contributes more than 75 per cent of total household income. The analysis of data from the three communities in the Chittoor site indicated a positive (although not significant) association between income and mango diversity (number of varieties) maintained by the farmers in two of the three communities (Bangarupalyam and Talupulapalle). Furthermore, it was also observed in these communities that the income from mango was higher per hectare (INR 53,073–61,264) (US$9,000–10,000) for those who maintained more than six varieties compared with those who maintained fewer than three varieties (INR 39,460–47,437) (US$6,000–8,000). Thus, there is an indication that the maintenance of multiple varieties in an orchard by the farmers may be justified as a means of livelihood.

Conclusion

These multivarietal orchards help farmers to obtain better income as well as to reduce risks – both yield and price risks – and to ensure livelihood security. Major gains from multivarietal orchards are or could be:

- Extended harvesting season and lower dependency on single markets and varieties
- Reduction of harvest risks through improved pollination and sustainable yields
- Reduction of risk of failure of one or two components (varieties) in the orchard
- Wider base of varieties for varied uses
- Source of elite material identification
- Allowance for on-farm experimentation and sharing of information with other farmers and communities about important traits and best varietal combinations
- Potential for new entrepreneurial ventures for the marketing of best indigenous fruits and varieties for targeted consumer groups or markets.

Improved understanding of the different types of benefits connected to multivarietal orchards and the distinct varieties that are connected to these benefits can help to target and tailor on-farm conservation programmes. Multivarietal orchard designs with beneficial combinations of certain types of varieties could be popularized among 'monocropping' farmers and the respective varieties could be made available among nurseries.

References

Dinesh, M.R., Gajanana, T.M., Reddy, B.M.C., Parthasarathy, V.A., and Sthapit, B. (2014) *Elite indigenous (Naati) Mango Varieties of Chittoor*, Indian Institute of Horticultural Research, Bangalore, India

Gajanana, T.M., Lamers, H.A.H., Rajan, S., Singh, A., Singh, S.K., Singh, I.P., Vasudeva R, Hegde, N., Dinesh, M.R., and Dakshinamoorthy, V. (2014) *Baseline Report India for the UNEP-GEF Project Entitled: Conservation and Sustainable Use of Cultivated and Wild Tropical Fruit Diversity: Promoting Sustainable Livelihoods, Food Security and Ecosystem Services*, GEF, UNEP, Bioversity and ICAR, India

Iyer, C.P.A. and Schnell, R.J. (2009) 'Breeding and genetics', in R.E. Litz (ed.), *The Mango: Botany, Production and Uses*, 2nd edn., CABI, Wallingford, UK, pp. 67–96

Kruijssen F. and Mysore, S. (2010) *Integrating Biodiversity Conservation and Livelihood Improvement: The Role of Markets for Mango Varieties and Garcinia Species in India*, Bioversity International, India

Meng, E.C.H., Smale, M., Bellon, M.R., and Grimanelli, D. (1998) 'Definition and measurement of crop diversity for economic analysis', in M. Smale (ed.), *Farmers, Gene Banks and Crop Breeding*, Kluwar, Boston, MA, pp. 19–31

Mukherjee, S.K. (1953) 'The mango – its botany, cultivation, uses and future improvements, especially as observed in India', *Economic Botany*, vol 7, pp. 130–162

Nagarajan, L., Smale, M., and Glewwe, P. (2007) 'Determinants of millet diversity at the household-farm and village-community levels in the dry lands of India: the role of local seed systems', *Agricultural Economics*, vol 36, pp. 157–167

Notes

1 From 2009 to 2014, Bioversity International coordinated a research-for-development project supported by the Global Environment Facility (GEF) with implementation support from the United Nations Environment Programme (UNEP), 'Conservation and sustainable use of cultivated and wild tropical fruit tree diversity; sustainable livelihoods, food security and ecosystem services', abbreviated as 'the TFTGR Project'. This project, implemented in India, Indonesia, Malaysia and Thailand, focused on livelihood and environment benefits that people could derive from the conservation of species and varietal diversity of *Citrus*, *Garcinia*, *Mangifera* and *Nephelium*.

19 Integrated home gardens for maintaining mango and *citrus* diversity and for family well-being in East Java

Kuntoro Boga Andri, Putu Bagus Daroini, M. Winarno, Prama Yufdy, Nono Sutrisno and Idha Widi Arsanti

GPD 'passport'

GPD code:	14
Focus area:	Production and crop management
Character:	System
Species and varieties involved:	4 Mango species (30 varieties): *Mangifera indica*, *M. odorata*, *M. foetida* Lour, *M. lalijiwa* 7 Citrus species (16 varieties): *Citrus reticulata*, *C. maxima*, *C. medica*, *C. hystrix*, *C. aurantiifolia*, *C.sinensis*, *C. limon*
Name of location:	Kediri and Magetan
GIS reference of location(s):	Kediri: S 06°48′47″; E 107°36′52″ Elevation: 500–600 masl Magetan: S 07°39′50″; E 111°11′43″ Elevation: 100–110 masl
Name of farmer (data resource):	Mr Jemu, Mr Mustari, Mr Tumini, Mr Yasuladi (Kediri) Mr Pardi, Mr Subarno, Mr Sudarman, Mr Sadino (Magetan)

Introduction

Home gardens are traditional time-tested multi-crop systems (similar to agroforestry systems) that may harbour globally and locally important tropical fruit genetic resources. They are characteristic production systems in Indonesian rural landscapes (Soemarwoto, 1987; Abdoellah *et al.*, 2001; Seneviratne and Kuruppiuarachchi, 2006). This production system is valuable for the livelihoods of local people throughout the country as a source of both dietary diversity and income. In other words, home gardens are a general agroforestry concept for a land management system combining trees and agricultural crops or tree gardening (Weersum, 1982).

Tropical home gardens are generally found around the homestead or backyard. In a home garden system a diverse mixture of annual and perennial crops is practised on private lands. Indonesian home gardens include: (i) a small area of mixed cropping around the homestead; (ii) rich, unique, rare and family preferred crops and fruits; (iii) multistoreyed combination of underground crops, shrubs, trees for maximizing light interception and efficient use of soil, water and nutrients; (iv) a mixture of annuals and perennials; and (v) production primarily for family consumption using family labour (Weersum, 1982). Home gardens in Kediri and Magetan range from 200 m^2 to 400 m^2 and are planted with a mixture of fruit trees, tuber crops, vegetables, ornamentals, spices and medicinal crops. Farmers plant mostly local varieties of a wide range of crops. Besides crops, farmers commonly keep poultry in the home garden. Some wealthier farmers also keep livestock. Household men are commonly the custodians of fruit trees, tuber crops and livestock, and women are particularly the custodians of vegetables, ornamentals, spices, medicinal crops and poultry. Both work together at household level.

Despite the number of classification schemes proposed for tropical home gardens, none has been universally accepted (Kehlenbeck and Maass, 2004; Jacobi *et al.*, 2009) because they evolve in specific conditions. Traditional home gardens have received special research attention in Indonesia since the 1970s (Abdoellah *et al.*, 2001). Home gardens, particularly those in Java, have been investigated in some depth (Pamungkas *et al.*, 2013).

Under the UNEP/GEF project 'Conservation and Sustainable Use of Cultivated and Wild Tropical Fruit Diversity: Promoting Sustainable Livelihoods, Food Security and Ecosystem Services', we identified the home gardens as a good practice for maintaining tropical fruit tree diversity (GPD) while addressing family well-being. The GPD was identified in two communities in East Java by the Assessment Institute for Agricultural Technology (BPTP) Java Timur jointly implemented with the Indonesian Center of Horticulture Research (ICHORD) and Bioversity International, focusing on mango and citrus fruit tree species. These home garden practices are, however, slowly disappearing from East Java because of rapid commercialization of agriculture (Abdoellah *et al.*, 2006) and lack of policy support for family farming. Because the practice is shrinking, the unique and rare varieties of fruits with their associated knowledge are also disappearing (Arifin, 2012). This will directly affect access to unique germplasm and availability of nutrition dense food for poor smallholder farmers and consumers as supply will be limited.

Methodology

This study is based on three sources of information. First, we drew much of the preliminary information from secondary sources such as publications, reports and local expert knowledge. Second, information was obtained through conducting participatory rural appraisal using tools such as a transect walk, resource mapping and Four Cell Analysis to gain insights about local crop and fruit tree diversity in the communities. Third, a baseline survey, key informant interviews and focus group discussions (FGDs) with men and women separately were conducted in the Kediri and Magetan communities. Kediri and Magetan were selected because they are the main production centres in Indonesia of mango and citrus, respectively. We visited farmers' fields in these two communities, in which 85 to 100 per cent of households have agro-ecosystems located close to the homestead. In East Java, there is a long tradition of home gardens that use family labour to produce a diverse range of food and fruits preferred by the family members.

Community FGDs were conducted to understand the dynamics of these systems and how they result in management of local fruit tree diversity. Participatory Market Chain Approach (PMCA) methodology was used to intervene to boost the value of some of the underutilized varieties through value-addition activities with women's groups.

The local context

Home gardens in Kediri and Magetan sites in East Java Province were included in this study. General characteristics of households between the two sites are summarized in Table 19.1. The majority of fruit farmers have more or less the same area.

Magetan village is mostly commercially oriented whereas Kediri village is subsistence-oriented farming. Kaligayam village in the Kediri site covers a lowland area of about 26,700 ha, at an elevation of 500 masl, located in the centre of East Java Province. The home gardens are small (0.01–0.5 ha) with an average annual income generated from tropical fruits of IDR1,500,000 (US$150) per household. The mango fruits are usually sold directly by farmers to intermediaries or in the local market. Sale of mango fruits contributes 15–20 per cent of family income (Daroini *et al.*, 2013; Table 19.2).

The Bibis community in the Magetan site has citrus as the main crop in their home gardens. The site covers a lowland area of about 247,830 ha, at an altitude of 105 masl, located in the mid-west of East Java Province. The size of a home garden ranges from 0.04 to 1 ha, with an average annual income

Table 19.1 Information of community household (HH) characteristics and sample surveyed

Site characteristics	*Tiron, Kediri*★	*Bibis, Magetan*
Total number of households in the study site	643	647
Total households with fruit trees	559	647
Household head age (years)	52.6	56.7
Education (% > high school)	0.00	2.67
Illiterate (%)	26	5.33
Female head (%)	7.5	7.7
Family size (no.)	4.0	3.0
Land size (ha)	0.18	0.19
Average home garden size (ha)	0.01	0.04
Number of respondents (*n*) as % of fruit growers	54	65

★In Tiron village, the study site selected only one hamlet, namely Kaligayam hamlet.
Source: Daroini *et al.* (2013).

Table 19.2 Average income contribution for households from targeted tropical fruit crops (%)

Community	*From tropical fruit based home gardens*	*Other agricultural activities*	*Non agriculture*
Tiron, Kediri	12	28	60
Bibis, Magetan	32	16	52

from pomelo fruits of IDR 1,500,000 (US$150) per farm family, which is the main household income. The total population of pomelo is more than 50,000 trees, with an average yield of 100–200 kg/tree. The fruits are usually sold directly by each farmer to intermediaries or to the local market by the farmers' group (Daroini *et al.*, 2013).

On-farm fruit tree diversity

Most of the lands are in the form of home gardens and orchards, which are managed intensively by their owners and planted with diverse tropical fruit tree species (Table 19.3).

In this study the distribution of genetic diversity of mango and citrus was assessed in both home gardens and orchards (Table 19.4). Wild fruit tree diversity is reported in the forest (Idris *et al.*, 2015; Sthapit *et al.*, 2015) but the data were not collected from the forest in the proximity of these two sites.

The sites were chosen because they are located in diversity rich regions where there is already a rich diversity of mango and citrus species and varieties and this is reflected in their home gardens. Four species of *Mangifera* with 29 varieties were found in the Tiron, Kediri community, while six species of *Citrus* with 21 varieties were found in the Bibis, Magetan community (Table 19.5).

Table 19.3 Comparative richness and evenness of tropical fruit tree species diversity in two villages of East Java, Indonesia

Site	*Richness*		*Evenness*	
	Home garden	*Orchard*	*Home garden*	*Orchard*
Tiron, Kediri	8	5	0.709	0.698
Bibis, Magetan	6	8	0.942	0.799

Table 19.4 *Mangifera* and *Citrus* genetic diversity assessment in two sites of East Java

Diversity indices	*Tiron, Kediri (Mangifera spp)*	*Bibis, Magetan (Citrus spp)*
Number of households interviewed	54	65
Total number of trees	1105	2747
Average number of trees per household	20	42
Community richness	13	8
Average richness per household	4	6
Community evenness (Simpson)	0.234	0.734
Divergence	0.175	0.367
Average age of fruit trees (years)	20	15

Source: Daroini *et al.* (2013).

Table 19.5 *Mangifera* and *Citrus* diversity found in two sites

Species	*Varieties*
Mango in Kediri	
M. indica	Podang Urang, Podang Lumut, Golek, Gadung, Madu, Santok, Manalagi, Lanang, Santok Kapur, Santok Buto, Bader, Jempol, Kopyor, Dodonilo, Beruk, Empok, Sengir, Ireng, Dasa Muko, Cantek, Gajih, Gurih, Lulang, Apel, and Cantrik (25)
M. odorata	Kweni (1)
M. foetida	Jaran, Pakel (2)
M. lalijiwa	Lali Jiwo (1)
Citrus in Magetan	
C. reticulata	Keprok Siem, Keprok Pulung (2)
C. grandis	Adas, Adas Duku, Sri Nyonya, Bali Putih, Adas Nambangan, Jeruk Gulung, Jeruk Jowo, Bali Merah, Pomelo Magetan, Jowo Besar, Jowo Kecil, Java Pomelo (12)
C. medica	Jeruk Sitrun (1)
C. hystrix	Jeruk Purut (1)
C. aurantiifolia	Jeruk Nipis, Pecel (2)
C. sinensis	Jeruk Manis, Sunkis, Keprok Manis (3)

Very few mango varieties were found in Magetan and, similarly, very few varieties of citrus species were found in Kediri. The home gardens of Magetan maintain unique pomelo varieties such as Jeruk Gulung, Jowo Besar and Jowo Kecil and these are currently exported to Taiwan for their highly valued qualities such as their red flesh and longer shelf life of around 6 months compared with the normal 3–4 months. In Kediri, despite the high number of total varieties, evenness is low, with 89 per cent of the total tree population dominated by two varieties, Podang Urang (72 per cent) and Gadung (17 per cent). The remaining 27 varieties each represent 11 per cent.

Of the 29 mango varieties, 13 have been identified as having readily marketable traits. A women's group, locally known as 'Budidaya', with support from a private company called Resource Exchange International (REI) and the TFT project collaborated to improve the economic situation by introducing diverse value-added products of local mango varieties in Kediri and pomelo in Magetan. Existing women's groups were initially processing only a few varieties of mango (Podang Urang/*Mangifera indica*) and citrus (Adas Nambangan/*Citrus grandis*). Currently, the groups process a variety of products such as dried mango (from Podang Urang, Podang Lumut, Madu), dodol[1] mango (from Gadung, Golek, Dodonilo), mango juice (from Kweni), mango sweets (from Pakel, Jaran), mango leather (from Podang Urang, Podang Lumut, Madu). Products derived from Citrus – sweet pomelo rind (from Java Pomelo, Adas, Bali Merah), jelly (from Jeruk Gulung, Pomelo Magetan), juice (from Jeruk Manis, Keprok Manis, Sunkis), as well as local uses as medicine and spices (Jeruk Purut, Jeruk Sitrun, Pecel and Jeruk Nipis) – help to main local diversity *in situ*.

Home garden integrated with animals and beekeeping

Home gardens in East Java are found to be integrated with the objectives of sustainable intensification, which reduces dependency on external outputs, maximizes synergies between farm activities and generates better harvests and thus income as well as a range of non-monetary benefits (Figure 19.1). Networks of diversity rich home gardens in the community contribute multiple ecosystem benefits that are still poorly quantified.

The integration of livestock and domestic poultry with diverse crops and fruits reinforces food and nutritional security for families and saves on the cost of food expenses (Figure 19.1). Cattle or goat and chicken farming provides organic manure for tropical fruit trees in addition to providing food and income. The harvest of safe and healthy fresh vegetables, fruits and crops with home produced eggs, meat and milk are of special value to the family. Galhena *et al.* (2013) reported the perceived key benefits of such an integrated approach:

- Improves family food security
- Increases the availability of safe, healthy and nutritionally rich and fresh foods for the family

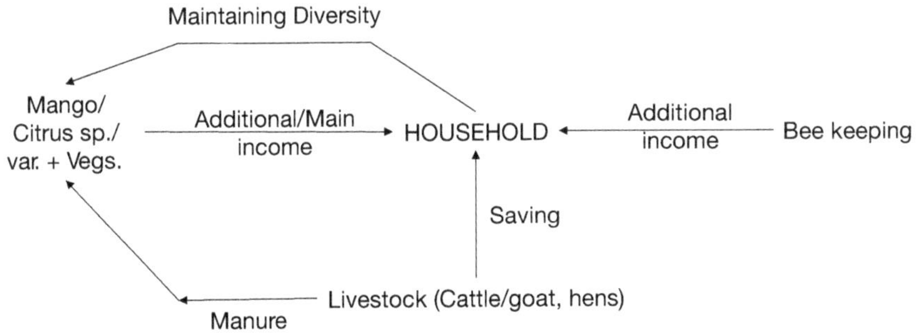

Figure 19.1 Component diagram of integrated home garden management.

- Maintains unique and rare species and intraspecific diversity (biodiversity) as family heritage
- Reduces the risk of crop failure by crop diversification
- Continuous supply of small but regular fresh supply to family kitchen
- Environmental benefits by recycling water and nutrients
- Harbours pollinators
- Reduces family expenditure on food
- Platform for first biodiversity education and skill transfer from elder generation to children and social learning.

Integrated home gardens (Plate 44) capitalize various natural capitals for sustainable livelihoods, income and sustainable management of land, water and ecosystem on a small scale.

The actual composition of the home garden varies depending on the community. In Kediri, tubers, yams and aroid crops are intercropped under the mango, whereas under the citrus trees in Magetan farmers grow mixtures of peanuts, leafy vegetables and aroids. Yams, taro, cassava and aroids are particularly valued during floods and typhoons as food security crops. A few farmers also combine crops and animals with beekeeping, which ensures pollination services that increase crop productivity and provides direct income by selling organic honey and honey products (Table 19.6). Debris and dried plant parts found in the home garden are usually burnt; the ash is used as minerals for nurseries and the smoke may induce early fruit ripening.

Impact on crop genetic diversity

These two sites were selected because of the presence of integrated home gardens and rich biodiversity and traditional food culture. The integrated home gardens of these communities have continued to maintain inter- and intraspecific mango and citrus diversity (Table 19.4). High intraspecific diversity is found in home gardens of custodian households both in Tiron, Kediri and

Table 19.6 Percentage of households (HH) managing livestock and other crops in the studied areas

Site	*Goat (%)*	*Beef cattle (%)*	*Poultry (%)*	*Honey bee (Total HH)*	*Major crops grown under tropical fruit tree diversity*
Kediri	75	23	95	4	Cassava, taro, yam, aroids, vegetables such as Labu Siam (*Sechium edule*), eggplant, tomato, chilies, amaranth
Magetan	30	16	74	1	Groundnut, many kinds of leafy vegetables, sweet potato, cassava, taro or yam

in Bibis, Magetan, and some of them have elite traits that have been multiplied in community fruit nurseries and distributed to other community members. Social capital building of local community helps to link various service providers and research stations, and such connection could reap more economic benefits from the diversity that farmers are already conserving *in situ*. Such diverse uses of varieties and products help to maintain and enhance inter- and intracrop diversity as different varieties are suitable for different uses and products. Furthermore, to support source of raw material farmers expand areas to planting many fruits in their home gardens and managed forest.

Impact on livelihoods

The fruits of some species and varieties have low commercial value but provide good nutrition for family consumption and have higher value as raw materials for the small household processing units that exist in the village. In Bibis, almost all varieties grown in farmers' home gardens are commercial varieties. Nevertheless, part of the produce is retained for home consumption thus adding to food and nutrition security. All these benefits contribute to improved household livelihoods and well-being.

Most of the mango or citrus fruits of different varieties produced in the home garden are used for sale as well as home consumption. Surplus produce is sold fresh by farmers on the spot or through a wholesale system ('*Tebasan*' system in Bahasa) to intermediaries, who harvest the fruits. In some instances, tropical fruit trees in the home garden provide the main income for the households. Income from tropical fruits is important for the households in the studied area. Income from mango fruits in Kediri contributes 12 per cent on average to family income, whereas in Magetan income from citrus is as high as 32 per cent of the total income (Table 19.2). Because the system provides women with wages, a source of food and nutrition for family well-being and generates additional income for households, home gardens can be an entry point for women's economic empowerment.

The practice of establishing and managing fruit species in home gardens helps maintain diversity of tropical fruits and provides raw materials for women's groups to produce value-added products from different varieties of local mango and citrus through processing at home and group marketing.

Conclusion

The practice of keeping integrated home gardens has been found to maintain local mango and citrus species and variety diversity for home use and marketing fresh and surplus products. With increased commercialization, the structure and composition of home gardens in Magetan has been changing whereas the home gardens of Kediri still remain stable and subsistence-oriented. It is interesting to note that Magetan farmers have maintained rich diversity of pomelo despite commercialization and have better income than Kediri farmers. The conditions favouring success for this good practice are the availability of supportive government policy and related institutions for integrated family farming.

References

Abdoellah, O. S., Parikesit, Gunawan, B., and Hadikusumah, H.Y. (2001) 'Home gardens in the Upper Citarum Watershed, West Java: A challenge for *in situ* conservation of plant genetic resources', paper presented at the international workshop on 'Contribution of home gardens to *in situ* conservation of plant genetic resources in farming systems', 17–19 July, 2001, Witzenhausen, Germany

Abdoellah, O.S., Hadikusumah, H.Y., Takeuchi, K., Okubo, S., and Parikesit (2006) 'Commercialization of homegardens in an Indonesian village: vegetation composition and functional changes', *Agrofrestry Systems*, vol 68, pp. 1–13. DOI 10.1007/s10457-005-7475-x

Arifin, H.S. (2012) Pekarangan, Workshop Program Lestari, Solo, Jawa Tengah (in Indonesian), [Home garden, Workshop on Sustainable Program, Solo, Central Java]

Daroini, P.B., Dewi, H.A., Andri, K.B., Rafieq, A., Pribadi, Y., Fakhrina, Saderi, D.I., Lesmayati, S., and Lamers, H.A.H. (2013) *Baseline Report Indonesia for the UNEP-GEF Project Entitled: Conservation and Sustainable Use of Cultivated and Wild Tropical Fruit Diversity: Promoting Sustainable Livelihoods, Food Security and Ecosystem Services*, GEF, UNEP, Bioversity and ICHORD, Indonesia

Galhena, D.H., Freed, R., and Maredia, K.M. (2013) 'Home gardens: A promising approach to enhance household food security and wellbeing', *Agriculture & Food Security*, vol 2, no 8; http://www.agricultureandfoodsecurity.com/content/2/1/8

Idris, S., Changtragoon, S., Chatchanok, C., Idha, W.A., Elina, M., Nataya, D., Songpol, S., Winarno, W., Syauqi, N., Shariah, U., Shukor, A., Noorhayati, M.H., Wong, W.W.W., Brooke, P., Shafie, M.S.M., Gerten, A.H.A., Lhumphet, S., Sthapit, B.R., and Ramanatha Rao, V. (2015) Diversity in Cultivated and Wild Nephelium species in South-East Asia, *Indian J. Plant Genet. Resource*, vol 28, no 1, pp. 55–61. DOI 10.5958/0976-1926.2015.00008.X

Jacobi, J., Drescher, A.W., Amerasinghe, P.H., and Weckenbrock, P. (2009) 'Agricultural Biodiversity Strengthening Livelihoods in Peri Urban Hyderabad, India', *Urban Agriculture* magazine, no 22, June

Kehlenbeck, K. and Maass, B.L. (2004) 'Crop diversity and classification of home gardens in Central Sulawesi, Indonesia', *Agroforestry Systems*, vol 63, pp. 53–62

Pamungkas, R.N., Indriyani, S., and Hakim, L. (2013) 'The ethnobotany of homegardens along rural corridors as a basis for ecotourism planning: a case study of Rajegwesi village, Banyuwangi, Indonesia', *Journal of Biodiversity and Environmental Sciences (JBES)*, vol 3, no 8, pp. 60–69

Seneviratne, G. and Kuruppiuarachchi, K.A.J.M. (2006) 'Nutrient cycling and safety-net mechanism in the tropical home gardens', *International Journal of Agricultural Research*, vol 1, no 2, pp. 169–182

Soemarwoto, O. (1987) 'Home gardens: A traditional agroforestry system with a promising future', in H.A. Steepler and P.K.R. Nair (eds), *Agroforestry: A Decade of Development*, International Centre for Research in Agroforestry, Nairobi, Kenya, pp. 157–170

Sthapit, B.R., Vasudeva R, Idris, S., Changtragoon, S., Winarno, Roongrattanaku, P., Limpiyaprapan, S., and Ramanatha Rao, V. (2015) 'Harnessing the diversity of wild relatives of tropical fruit tree species for sustainable livelihoods', *Acta Horticulture*, 1101. DOI 10.17660/ActaHortic.2015.1101.31

Weersum, K.F. (1982) 'Tree gardening and taungya on Java: Examples of agroforestry techniques in the humid tropics', *Agroforestry Systems*, vol 1, no 1, pp. 53–70

Notes

1 Dodol is a sweet toffee-like confection, popular in Indonesia, Malaysia and the Philippines. Dodol is made with coconut milk, mango pulp, jaggery and rice flour, and is sticky, thick and sweet. In Muslim majority countries, such as Indonesia and Malaysia, dodol is commonly served during festivals such as Eid ul-Fitr and Eid al-Adha as sweet treats for children.

20 Management of *Garcinia fusca* for sustainable use

Tawatchai Nimkingrat, Ratchanee Siriyan, Auttapon Rukkaphan, Margaret C. Yoovatana and Songpol Somsri

GPD 'passport'

GPD code:	20
Focus area:	(a) Propagation and planting material (b) Production and crop management (c) Collective action and social networking
Character:	Process, method, technique and institutional arrangement for the management of *Garcinia fusca*
Species and varieties involved:	*Garcinia fusca*
Name of location:	Ise subdistrict, Pho Si Suwan district; Huai Thap Than subdistrict, Huai Thap Than district, Sisaket province
GIS reference of location(s):	Ise subdistrict: N 15°26′64″; E 104°01′91″ Huai Thap Than subdistrict: N 15°05′21″; E 104°03′08″ Altitude of both sites: 120-150 masl
Name of farmer (data resource):	Mr Peerachai Vonglert, Mrs Prayong Chaisawang

Introduction

Wild madan (*Garcinia fusca* Pierre) is one of the wild relatives of the more popular fruit mangosteen (*G. mangostana* L., family Clusiaceae). The tree, locally known as wild madan, is a medium-sized tree or shrub which grows abundantly and naturally along the Thap Than River and other riverbanks (Plate 45a–45c), streams and swamps (Shu *et al.*, 2007; Subhadrabandhu, 2001). The young leaves are traditionally used as an ingredient in soup, giving it a sour taste, and the fruit is used in chilli paste. While *Garcinia* generally is used as a spice in India and as a flavouring agent in soup in Malaysia, the Thai people from Sisaket have discovered a unique way of using the madan stem. The villagers in Huai Thap Than subdistrict in Sisaket use the branches of the wild madan tree as a skewer for a grilled chicken recipe called 'Kai Yang Mai Madan' (Siriyan *et al.*, 2015b). When chicken meat is impaled on a madan skewer and grilled over a fire, gum secreted from the green stick blends with the meat, giving it a distinctive flavour and aroma that is found nowhere else and making it a favourite among local consumers (Hom *et al.*, 2010). Huai Thap Than subdistrict is well-known in Thailand for this unique product (Plate 45f). This recipe has been popularly known for more than 50 years.

This particular traditional dish has now grown into a large-scale business. The grilled chicken produced in the area totals approximately 2,000 chickens per day. About 10,000 individual wild madan skewers a year are used in the several roadside stalls and restaurants in Sisaket Province. One labourer can earn about 600 Baht per day (US$20/day) by cutting these wild madan skewers. The grilled chicken is priced at 35 Baht/portion package (US$1.17/portion package). There are about 53 chicken grill stalls in Huai Thap Than, hence the combined sales turnover amounts to about 350,000 Baht/day (US$11,660/day). The production and management of madan is illustrated in Plate 45.

With the great demand for these wild madan skewers, the plant species is now vulnerable to overexploitation if proper conservation efforts are not made. During the last few years, the population of wild madan bushes in its habitat along the river has been severely damaged and its population has started to dwindle. Hence efforts are being made by the community to sustainably manage and use *G. fusca*. In order to conserve this plant species, community members have started efforts that can constitute the domestication process of *G. fusca*. A complete cultivation system covering propagation and pruning techniques, nursery management, replanting, sustainable harvesting practices and commercially viable practices of *G. fusca* has been developed, demonstrated and mainstreamed. This domestication process contributes to the conservation of *G. fusca* in the wild by taking pressure off it. The efforts have been implemented and managed by the community itself, with support from local government and the Sisaket Horticultural Research Centre.

This chapter describes the combined efforts and activities of the Huai Thap Than and Ise communities to avoid the overexploitation of this unique native wild species that has become victim of its own commercial success. The chapter focuses on propagation techniques for *G. fusca* to explore the multiplication and domestication of this wild species. It also describes the participatory research activities conducted with the environmental group of the Ise community to explore additional products and cultivation techniques for wild madan.

Methodology used for data collection

The area of the study covers Ise, Pho Sri Suwan and Huai Thap Than districts, Sisaket Province, located in the lower part of northeast Thailand. Data were collected through survey questionnaires, interviews and community group meetings. Transect walks were carried out by a team of experts to observe various home gardens, plantation boundaries and riverbanks in order to understand local management practices. Four Cell Analysis was used to assess the amount and distribution of wild madan diversity in the community and get a deeper socio-economic and biological understanding of this distribution. Community nurseries were established, in which evaluation trials were conducted to compare different propagation techniques and cultivation regimes. In addition, research was conducted to explore potential by-products from *G. fusca* through the development of prototype products from its fruit, bark and leaves.

Description of GPD

A complete cultivation system, covering propagation, nursery management, replanting, sustainable harvesting practices and commercially viable practices of *G. fusca* has been developed, demonstrated and mainstreamed in the Huai Thap Than watershed. In order to protect the local grilled chicken industry and conserve the madan plants, the community has made an effort to manage madan for long-term, sustainable use. This good practice developed gradually as a concerted effort that involves many stakeholders in the village. The first step was the establishment of a community forest of around 350 hectares in 1995 to protect the leftover area of fertile wild madan natural habitat. Since then, the community has developed its own propagation techniques whereby seeds and seedlings are collected from the river forest, grown in nurseries and then replanted in the forest and also in home gardens. Regulations for community forestry have been agreed in the community, such as zoning the area and assigning each harvest zone to a group. The community facilitates management and ensures that individuals have an invested responsibility. Up to now, seed collection from riverbanks is still the proper technique for wild madan propagation. This technique produces regular and numerous saplings and a vigorous rooting system and it is convenient for transportation. The Ise and Huai Thap Than communities established two nurseries in 2012, one located at Ban Ise Kururadwitthaya School and one located at Huai Thap Than Witthayakom School. The nurseries were trained and encouraged by Sisaket Horticultural Research Centre, the schools themselves and local government. The nurseries are managed by Mr Peerachai Vonglert and Mrs Prayong Chaisawang. Local conservation groups, the schools and the community engage in activities such as public awareness, collection of seed and propagation, restoration of wild madan along fences and degraded lands, and domestication of wild madan. The nurseries have so far produced at least 4,000 saplings of wild madan (Plate 45g–h).

The relatively low value of the madan skewers limits the potential of commercial cultivation, thus to increase its market value, additional products have been explored from the waste materials. For example, the bark of wild madan sticks was used to extract natural dyes that have been used to dye clothes for increased income. New products have also been developed from the wild madan fruit and leaves (Plate 46).

Impact on diversity

Until recently there has been indiscriminate exploitation of *G. fusca* in the wild. Such exploitation had become a serious threat to the diversity of madan species. However, wild madan is now being domesticated from forests into home gardens. The sustained adoption of seed propagation techniques has helped to reduce the human pressure on the wild madan population along the Thap Than River and facilitate the restoration of degraded land areas. The propagation techniques from seeds have also taken up the multiplication of other species such as a domesticated madan (*G. schomburgkiana*) and Cha muang (*G. cowa*). Similar efforts are being made to explore their potential market value and the most appropriate propagation and cultivation techniques. The replanting and restoration efforts of the conservation group are supporting the richness of species found along the riverbanks and could lead to the diversification of home gardens and farmers' fields. This case illustrates how delaying the rate of genetic erosion of genetic resources also maintains agricultural biodiversity (refer to Chapter 24).

Impact on livelihoods

So far the conservation efforts have not generated direct income; the saplings from the nursery are handed out for free to community members. However, the number of requests for saplings has increased and the saplings may soon have a monetary value attached. At present, the majority of the saplings are replanted as part of the restoration efforts, supported by local government, of the Non Yai community forest conservation group in the Ise community. To internalize the costs of conservation of wild madan, the hope is that the restaurant and grill stand owners will support the initiative taken by Ise community and financially support the nursery and replanting efforts. Moreover, the Ise community hopes to sell the newly developed products such as the naturally dyed clothing and wild madan juice to the same grill stand owners and tourists (Siriyan *et al.*, 2015b; Hom *et al.*, 2010). These multifaceted benefits and income opportunities connected to wild madan have encouraged the local Ise community and rural institutions to protect the natural habitat, the environment, *G. fusca* and its genetic diversity.

Additional benefits and ecosystem services

This good practice could also protect the Thap Than River banks from soil erosion phenomena. Wild madan saplings have been replanted along the river. Conserving biodiversity along the riverbanks could help shield waterways against nitrogen pollution, such as that released from agricultural fertilizers and waste, human sewage and fossil fuel burning. Recent research reported by Cardinale (2011) showed how streams with more species are better at removing excess nutrients from water. The findings imply that developing countries that keep rivers and lakes species-rich could save money on water treatment and provide benefits to downstream communities. This could be a case of how biodiversity provides ecosystem services and functions. The communities' participation through knowledge exchange and sharing within and between communities could strengthen the communities' unity towards a common goal of conserving the *G. fusca* species and its ecosystem.

Contribution to social and human capital, strategies and impact

In 1995, Ise community established a conservation group that initially had 60 members consisting of teachers and students. The group is connected and led by the local biology teacher and also 'agent of change', Mr Peerachai Vonglert, who was concerned about the degradation of the Thap Than River area. He learned that the overcollection of wild madan sticks was one of the major causes for this degradation. The group's first activity was to conserve wild madan in its habitat by monitoring the cutting of wild madan branches and collaborating to create the forest community rules. Another activity was to control the adoption of creepers for handicrafts and replanting in the forest.

The conservation group currently has a total of about 230 members. They have been empowered through training on wild madan seedling propagation organized by the Ise subdistrict authority. Training followed by public awareness activities on watershed management enhanced the community's awareness and participation in the sustainable use and management of wild madan sticks. The community has in this way been empowered for self-directed decision making on the management of the *G. fusca* resource for the benefit of people. The formation of *G. fusca* Conservation Groups and the establishment of community nurseries for the propagation of *G. fusca* species has enhanced the social capital to conserve the ecosystem and unique diversity of the *G. fusca* species.

The community group conducted trials into how to use wild madan more beneficially. They learned to make use of various parts of the wild madan tree besides the sticks, and finally they found that the bark can be used as a natural dye for many type of clothes (Siriyan *et al.*, 2015a). Based on these research results, four on-site training courses on dyeing cotton cloth with wild madan dyes were set up for the people in Ise subdistrict, Sisaket Province (Plate 46c–h). The idea for this research was stimulated by visits by the site coordinator and farmers to Kiriwong community in Thailand, where similar techniques are used with locally available materials.

Driving forces for the success of the GPD

The increasing demand for wild madan skewers for grilled chicken has resulted in overuse of *G. fusca*, which hence has become vulnerable to destruction and local extinction. Moreover, wild madan's natural habitats are now also being encroached upon by the cultivation of new economic crops. However, the strong commitments of the community and local leaders to conserve the *G. fusca* species and to protect the Thap Than watershed are the major driving forces ensuring the success of the good practice. The local leadership of Mr Peerachai Vonglert has been instrumental in convincing the community to take part in collective actions to start up the nursery and to explore the identification of diverse products. The project played a key role in bringing together all key actors to build social capital and consolidate combined efforts.

Constraints for scaling up or dissemination of GPD

Initially, many activities for conserving wild madan were started in Huai Thap Than community, but responses from community people have been muted. Instead, it was observed that Ise community was much more receptive to this idea. This was mostly because Ise community has strong leadership of the conservation group. The group has created awareness for the conservation of wild madan to both adults and children in the community. Activities for conserving wild madan are part of the study course in the school, which increases the knowledge of how to manage it among the children. To educate students and adults, the teacher used brainstorming, experimenting, conducting special projects, practising in the real field and other activities. This way of tackling the problems represents a group creativity technique designed to generate a large number of ideas for its solution.

Free and easy access to wild madan sticks from riverbanks has resulted in a general phenomenon of 'the tragedy of the commons'. There are still some members of the community who hold negative attitudes towards conservation. Lack of awareness and limits to promotion campaigns are constraints to successful implementation. Transitions in local governance could be a threat to continued policies as well.

Action plan for scaling up and dissemination

One of the action plans for scaling out the good practice is to make extra income from using the by-products from making madan skewers to dye cotton cloths with dyes made from wild madan bark. The bark yields a pleasant brown/brownish-yellow colour depending on the mordant used. Experiments to check the colour fastness to washing and rubbing of these dyed cotton clothes have found it mostly good. Dyeing cotton cloth with dyes extracted from wild madan bark is a new finding and has sparked new interest. Training and setting up of women's groups as small business enterprises selling these clothes might generate income and stimulate the community to conserve their wild madan trees, as has happened in the village of Kiriwong. The success in the scaling up and dissemination of the good practice needs strong commitment from the community groups and local government to propagate and multiply *G. fusca* to sustain the supply of raw materials for grilled chicken. Also, the community's understanding and enhanced awareness of the risks to the *G. fusca* population, and strong policy support from the local government, contribute to the successful scaling up and dissemination of this good practice.

References

Cardinale, B.J. (2011) 'Biodiversity improves water quality through niche partitioning', *Nature*, vol 472, pp. 86–89

Hom, S., Lamers, H., Yoovatana, M., Nimkingrat, T., Somsri, S., and Sthapit, B.R. (2010) 'Local foods and markets support conservation of *Garicinia fusca* in Sisaket Province, Thailand', a flyer of GEF, UNEP and Bioversity, Rome

Shu, T.H., Xiwen, L., Jie, L., and Stevens, P.F. (2007) '*Garcinia Linnaeus*, sp.', *Flora of China*, vol 13, pp. 40–47

Siriyan, R., Rukkaphan, A., and Nimkingrat, T. (2015a) 'Wild madan tree, from grilled chicken to clothdyes' (in Thai), *Kasikorn*, vol 87, no 4, pp. 67–76

Siriyan, R., Rukkaphan, A., Nimkingrat, T., Noppornphan, C., and Somsri, S. (2015b) 'Value addition of wild madan tree (*Garcinia fusca. L*) conservation: From famous grilled chicken skewer to natural cloth dyes in Sisaket, Thailand', *Indian Journal of Plant Genetic Resources*, vol 28, no 1, pp. 39–43. DOI 10.5958/0976-1926.2015.00005.4

Subhadrabandhu, S. (2001) *Under-Utilized Tropical Fruits of Thailand,* Food and Agriculture Organization of the United Nation, Regional Office for Asia and the Pacific, Bangkok, Thailand, p. 70

21 Production and management of an underutilized fruit

Aroi aroi (*Garcinia forbesii* King) in home gardens and orchards

William W.W. Wong, Jamaluddin Lani and Hugo A.H. Lamers

GPD 'passport'

GPD code:	17
Focus area:	Production and crop management
Character:	Process
Species and varieties involved:	*Garcinia forbesii* King. Varieties: 1. Aroi Tulen/Biasa, 2. Aroi Jambu and 3. Aroi Batu
Name of location:	1. Kg. Kelatuan, Papar, Sabah, Malaysia 2. Kg. Penampang Baru, Papar, Sabah, Malaysia
GIS reference of location(s):	1. N 5°38′46″; E 115°57′3″. Elevation: 9–10 masl 2. N 5°36′40″; E 115°56′11″. Elevation: 5–9 masl
Name of farmer (data resource):	1. Mr Salleh bin Sawal 2. Mr Meon bin Gimbul

Local context

Garcinia consists of roughly 300 species of trees and shrubs distributed across South America, Africa, Madagascar and southeast Asia. Most of the species diversity in the genus is concentrated in Malaysia, where more than two thirds of the species in the genus are found. *Garcinia forbesii* King is a lesser-known mangosteen that can be found in the wild in the lowland forests of the foothills of the Crocker Range, and as a semi-wild or cultivated fruit tree in home gardens and orchards in the district of Papar in the state of Sabah, Malaysia. Usually, the trees can grow up to 10 m tall and are conical to broad shaped, with characteristic rose-coloured cherry-like fruits (Plate 47a).

The district of Papar (Sabah, Malaysia) covers an area of 124,320 ha, of which 48.7 per cent is undulating and hilly terrain and 51.3 per cent is coastal plain. The climate is a monsoonal tropical type, with an average rainfall of 3,186 mm per year, the wettest month being November and the driest month being February. The average temperature ranges from 27°C to 32°C. The soil of the site is of the Brantian Association (terrace alluvium) on undulating land. The local population numbers 107,000 people, mostly rice farmers, fisher folk and traders. This district was chosen for study of its tropical fruit trees and good practices for diversity because of the high levels of diversity of fruit trees that can be found here, combined with a wide diversity of plant habitats or ecosystems for various species of plants (including fruit trees) to flourish.

Aroi aroi (*G. forbesii* King) as a species is popular mainly among the Kadasandusun community of Papar and throughout the west coast of Sabah. It is not used by the general population and, in recent times, even the younger generations of this community are increasingly abandoning use of this species. It is now mainly used only by the older generation for culinary and traditional medicinal purposes. For this reason, there are no marketing facilities or cooperatives for the target product, i.e. fresh fruits and dried rinds of *G. forbesii*, and as a result farmers arrange their own sales.

Methodology used for data collection

Potential good practices for diversity were identified through a baseline survey using a set of questionnaires developed by the Malaysian Agriculture Research and Development Institute (MARDI) in 2007. The preliminary questionnaire-driven baseline survey was conducted in 24 households in the form of structured questions. The results of the survey were reported under Salma *et al.* (2008). The results of this survey formed the basis for identifying potential households that could be included in the present full-scale project. The key areas are the diversity of fruit species each household manages and how these fruit species contribute to their daily dietary intake to supplement their nutrition.

The present study is to find out how fruit trees and their species diversity have played a role in the local community towards contributing to their food, income and traditional uses and to know whether these species are being threatened due to the introduction of modern varieties. We also wanted to find out how the elderly population of the community keep records of their traditional knowledge in order to maintain the cultivation of these species. Good practices for diversity (GPD) are one methodology of this project for us to capture the uses and practices of these people so as to enhance or even promote these GPDs for conservation of tropical fruit species.

In the present study, the following methods were used for data collection:

1. A baseline and impact assessment survey was carried out on 52 respondent households (the participating households of the project) and 12 control households before the project commenced. A set of comprehensive survey questions were used to gather information from these participants in these main areas:
   a. Household details
   b. Household income
   c. Fruit species diversity in home garden (diversity, area, tree numbers)
   d. Uses (any GPDs), e.g. food, medicine, handicraft, etc.
   e. How much these species contribute to the household income.

   The data collected were used for the action plans to be implemented as the project progressed and to gauge the impact of the project upon completion, to see if the project intervention had a positive or negative impact for the community.
2. A Four Cell Analysis (FCA) was conducted to understand the local context of the community and the diversity of fruit species in their home gardens (Sthapit *et al.*, 2006; refer to Chapter 3 of this book). In particular we were able to understand what the common fruit species are, what fruit species are cultivated for generating income, what fruit species are for household uses and what fruit species are rare and threatened and needing conservation.
3. Genetic Diversity Analysis (GDA) was also conducted to gauge the species diversity, evenness and richness in the home gardens or orchards of the respondents and the control households. Farm walks were conducted to identify species and numbers of individual trees. The data collected were analysed and the results presented as the Simpson's Diversity Index.
4. Traditional knowledge and GPDs were collected through interviews with the farmers and documented in the form of written records such as the Community Biodiversity Register (CBR), as audio recordings in the form of CDs and also as video.

Culinary, medical and other traditional uses of dried Aroi aroi rind (*G. forbesii* King)

Aroi aroi rind has been used for generations for traditional culinary and medicinal purposes, such as a means to help women recover after childbirth and as a remedy for coughs and minor stomach ailments. Over the years it has become less popular due to modern remedies, and nowadays Aroi aroi trees are mainly planted by the older generation who still maintain the diversity of the species and the traditional knowledge associated with its cultivation and varied uses. Younger demographic groups tend to rely more heavily on modern medicines than on customary herbal remedies.

There are three varieties or forms of Aroi aroi that most farmers can describe: Aroi Batu ('batu' means stone in the Malay language, referring to the hard texture of the fruit rind), Aroi Tulen ('tulen' means original) and

Aroi Jambu ('jambu' is the local word for guava, which this variety resembles) (Plate 47b). The three varieties have preferred uses. All three are used as a spice. Aroi Batu is the most preferred as a spice, due to the ease of removing its hard rind for drying. However, it is quite rare and therefore more valuable than the other two varieties. Aroi Tulen and Aroi Jambu are more acidic in nature and used as cleaning agents, while all three varieties are also used for medicinal purposes. Aroi aroi rind can also be used as a replacement for asam gelugor or asam keping (*G. atroviridis*) in the flavouring of laksa (a popular, spicy noodle dish). The rind of Aroi aroi is dried and sold locally in the market as a spice (Plate 47f).

Processing and use

The traditional method of processing Aroi aroi involves first washing the fruits and letting them air-dry before removing the rind by cutting the skin around the equator of the fruit with a sharp knife, and removing the pulp and seeds. The remaining rind is spread out on traditional woven bamboo mats for sun-drying (Plate 47c). Farmers still use this simple sun-drying method that has not changed over time, despite modern technological advances in the mainstream fruit processing industry. The local communities were not aware of modern methods for sun-drying until the intervention of this project. This method of drying is highly dependent on local weather patterns, as an afternoon shower can ruin a whole morning's work.

Once dried, the rinds are packed in plastic bags or containers. The rinds are then predominantly used as a condiment together with fish or other ingredients in local dishes like curries (similar to *asam gelugor* curry fish dish) or soups to attain a sought-after sour flavour, which is made by cooking the rinds with the other ingredients, grinding the rinds into powder or crushing them into smaller particles mixed into the dish. The rinds are seen as a cheap substitute for tamarind and are used together with spices such as ginger and turmeric.

When used as a cleaning agent, the fresh pulp and seeds, after being separated from the rind, are soaked in water and used to clean bronze by soaking the metal in the mixture. As a medicine, rinds are consumed by slowly boiling them in water so the decoction can be consumed for its medicinal properties.

Solar drying: an emerging good practice for sustainable livelihoods

The search for good practices for diversity brought to light two farmers who, in conjunction with the Department of Agriculture in Papar, in 2012, had tested a prototype solar-drying cabinet aimed at improving the processing technique for the rinds to attain higher-quality products for their variety of uses (Plate 47d). The results indicated that the drying time was reduced by 50 per cent, from three days to one and a half days, with the prototype portable solar-drying cabinet. Temperatures inside the solar-drying cabinet can reach

45°C versus the ambient temperature of 31°C. The dried product is very similar to those dried under open sun drying and both farmers found it acceptable.

The Agriculture Department has also helped the local community to develop proper packaging and labelling of the product to add value so as to gain wider market acceptability. The products are packed in 50g size plastic containers and labelled with instructions as well as nutritional value (Plate 47e). The income gained through improved processing and sales of rinds for culinary, medicinal and cleaning purposes has generated interest amongst other farmers in cultivating this particular species.

Impact on diversity

Mangosteen (*Garcinia mangostana* L.) is the preferred commercial species of *Garcinia* that is widely grown in home gardens and orchards of the community. Interspecies diversity has been maintained and increased through the continued use and planting of new Aroi aroi seedling trees (*G. forbesii* King) alongside mangosteen. The Agriculture Department has so far multiplied more than 200 seedlings and supplied to the local community for supplementary planting in their home gardens. Most *Garcinia* spp. can be planted as a mixed crop due to their tolerance to some degree of shading (Dela Cruz, 2001). This practice has improved interspecific richness and evenness in the area and strengthened the use of multiple species in the home gardens to cope with stress conditions and adversity.

From this study, it was found that there is a current research gap regarding the genetic diversity of *G. forbesii*. This species is dioecious, having male and female trees that are normally characterized by high levels of genetic diversity. However, almost all the Aroi aroi trees found locally here are female trees. Seeds from Aroi aroi are apomictic and consequently produce trees that are clones of the female parent, limiting the genetic diversity within its population. However, farmers have distinguished three distinct varieties or forms of Aroi aroi, which may indicate that despite their apomictic character, some genetic diversity is apparent within this species. Further research into the botany of this species is recommended.

This study also found that the Aroi aroi trees cultivated are propagated mainly through seeds. Farmers obtain planting materials by taking seeds from their own trees as well as through the informal exchange of seeds or saplings with other farmers and neighbours. Intraspecies diversity is maintained through the specific uses of the three varieties/forms of Aroi aroi mandated by local custom and culture. As such, farmers are expected to have an interest in maintaining each of the three forms for cultivation, thus helping to maintain richness and evenness within the population of *G. forbesii*.

Impact on livelihoods

On average, home gardens in this area have three to five Aroi aroi trees, each producing 200 to 300 kg of fruit per mature tree and 40 to 50 kg of fruit per younger tree. The prices for dried rind range from RM15–25 per kg (i.e. US$5–8). Farmers that maintain mature trees can earn an annual income of approximately RM3,000–4,500 even in years of lower yield or sub-optimal harvest.

The hydroxycitric acid (HCA) content of the fruit rinds of Aroi aroi is valued as a souring agent for local cuisines and for other purposes. The introduction of technologies such as solar-dryers has helped improve value-added product quality and reduce potential losses incurred during the drying process due to unfavourable weather. The improved product is drier and less prone to mould, and as a result can be kept for a much longer period of time. The livelihood benefits gained from the introduction of this new technology are lower post-harvest losses and increased value of the final product. Apart from domestic use, households have benefited by generating a secure source of income that is not bound to the commercial fruit season and is more evenly spread out over the course of the year. Further scientific research, particularly regarding the nutritional and health properties of the fruit, is required to support and expand further value-addition initiatives and undertakings. As an example, Rejab *et al.* (2008) used dried rind powder to make an ointment to treat eczema and other skin-related diseases. Similar alternative options may be explored for diversifying livelihood options.

The establishment of these market channels for Aroi aroi products affects the livelihoods of the community as a whole through the social capital accrued from the exchange of planting materials and the enhancement of human capital through the exchange of traditional and cultural knowledge, as well as the development of processing methods and market outlets. Malaysia has seen a surge in 're-discovered and re-invented' products over the last decades that are based on traditional knowledge, especially those products related to health and nutrition. Given this ongoing consumer trend, a better quality of life for rural communities is eminently achievable, and will result from the increase in income and the availability of locally preferred food and products in the target community and beyond.

Additional benefits of this good practice for biodiversity: enhanced ecosystem services

This practice will support the use and benefits derived from multispecies home gardens that were traditionally common in this region but are currently being replaced with cash crops. Multispecies home gardens are an important feature in the local human and natural landscape and provide ecosystem services such as habitats and food for animals and canopy cover to retain shade beneficial for other crops, all the while conserving genetic resources and diversity (Plate 47h).

Nowadays, Aroi aroi cultivation is very much dependent on domestic use, mainly for food and traditional medicine. The older population demographic, through acquired traditions and cultural practices, has played a vital role in keeping this crop in cultivation. This knowledge has been transmitted to the younger generations in some families that have been cultivating the crop. Families who have the economic incentive of additional income derived through sales are displaying a renewed interest in wanting to continue to plant Aroi aroi in their fruit orchards. Some farmers have started to raise seedlings for replanting as well as for giving to other families who want to plant the crop. In doing so, as well as contributing to families' livelihoods, the replanting of Aroi aroi may also contribute to the ecosystem services generated by multispecies home gardens.

Factors favouring or hindering successful functioning of the GPD

Local products of Aroi aroi that are available in local markets at present call for value-addition through better processing techniques, presentation and quality enhancement in order to be able to penetrate a wider market. Inroads have already been made by the Agriculture Department to introduce and test solar dryers and a solar drying house as well as packaging and labelling to add value to the products (Plate 47g).

These value-added traits may serve as a driving force for further scaling up and lead to an increase in the volume of national and regional trade. The production of a higher-quality product has been achieved through the use of an 'all-weather' drying facility for the community (i.e. the solar-drying cabinets and solar-drying house), allowing processing to continue regardless of weather conditions during the drying stage. Establishing reliable market links or outlets is also crucial to the long-term success of this GPD, which can help distribute the product to an even wider consumer base. The Agriculture Department is working towards involving other agencies such as the Federal Agriculture Marketing Authority (FAMA) and tourism facilities to further favour the success of this practice.

References

Dela Cruz Jr., F.S. (2001) *Status Report on Genetic Resources of Mangosteen (G. mangostana L.) in Southeast Asia*, IPGRI Office of South Asia, New Delhi, p. 24

Rejab, M., Teck, C.S., Zain, K.M., and Muhamad, M. (2008) 'Mangosteen', , in C.Y. Kwok, T.S. Lian, and S.H. Jamaluddin (eds), *Breeding Horticultural Crops @ MARDI*, MARDI, Malaysia, pp. 155–174

Salma, I., Radziah, M.L., Awang, K., Amron, A., and Rahman Milan, A. (2008) *Conservation and Use of Rare Tropical Fruit Species Diversity with Potential for Enhanced Use in Malaysia*, unpublished report LOA No.: APO/07/019

Sthapit, B.R., Rana, R.B., Subedi, A., Gyawali, S., Bajracharya, J., Chaudhary, P., Joshi, B.K., Sthapit, S., Joshi, K.D., and Upadhyay, M.P. (2006) 'Participatory four cell analysis (FCA) for local crop diversity', in B.R. Sthapit, P.K. Shrestha, and M.P. Upadhyay (eds), *Good Practices: On-farm Management of Agricultural Biodiversity in Nepal*, NARC, LI-BIRD, IPGRI and IDRC, Nepal

Case studies

Linking farmers with markets (commercialization that supports diversity maintenance and livelihoods of the poor)

22 How can markets contribute to the conservation of agricultural biodiversity on farms?

From theory to practice

Hugo A.H. Lamers, Froukje Kruijssen, Bhuwon Sthapit and V. Ramanatha Rao

Introduction

Although the role of cultivated and wild biodiversity in agricultural systems and rural livelihoods is widely recognized and understood, little focus has been put on how agricultural biodiversity contributes to economic well-being and how it can be conserved and promoted through market strategies (Lockie and Carpenter, 2010). The relationship between agricultural biodiversity and markets is not one-sided. In the past, markets have had a negative impact on agricultural biodiversity, and market pressures have most probably been one of the major causes for the decline in crop and varietal diversity found on farms and in fields (Pimbert, 1999; Lenzen *et al.*, 2012; Rao *et al.*, 2005; Van Dusen and Taylor, 2005). This is because agricultural markets have long favoured monocultural systems specializing in few crops with high-yielding varieties that are more uniform in terms of quality and form (Prescott-Allen and Prescott-Allen, 1990; Gruère *et al.*, 2006). These systems, however, lead to a narrowing of the genetic base. In more recent years, interest among high-end and middle-income consumers has been growing in local food culture, origin and exclusivity of food products, and concern has been raised about food quality, authenticity and long-term sustainability (Gruère *et al.*, 2006). These interests are reflected in a growing market for organic or natural products, products branded by area of origin or nutritional value and the revival of products based on traditional recipes (Willer *et al.*, 2008). This attention represents an opportunity to reverse the usual negative impacts of markets on diversity, leveraging consumer interest to conserve and promote neglected or underutilized species (NUS) and landraces.

It should be recognized that a substantial part of agricultural biodiversity has limited market value and there are often only a few varieties within a species (or species within a genera) that attract strong demand from large consumer groups and generate revenues for many farmers. Furthermore, a value chain approach based on unique species or varieties is not automatically agricultural biodiversity-friendly. A market strategy for a single lesser-known species or variety runs the risk of 'replacement' or the 'crowding-out effect', in which the farmers decide to replace the remaining diversity with the newly promoted 'superior' species or variety. In this scenario, the successful promotion of a single underutilized species or variety could lead to the replacement of many other less successful crops or landraces. Examples are the replacement of a wide range of quinoa landraces by smallholder farmers in Bolivia with the now internationally very popular white and red types (Bioversity International, 2013; Drucker *et al.*, 2013) or the replacement of a wide range of old mango varieties by mango farmers in Uttar Pradesh in India with the popular and geographic indication-protected variety Dashehari (see Chapter 12). To avoid such a scenario, value chain strategies should be integrated with community-based conservation strategies for those species and landraces that may have little market (or use) value today but could generate market (or use) value tomorrow (see Chapter 3).

This chapter explores market strategies that contribute to community biodiversity management, for both improved livelihoods and conservation of agricultural biodiversity. The chapter reflects on existing value chain methods and approaches that have guided the formulation of the 12 tools that have been used in a UNEP/GEF regional project titled 'Conservation and sustainable use of cultivated and wild tropical fruit diversity: Promoting sustainable livelihoods, food security and ecosystem services' (TFT project) to create markets for neglected fruit species and landraces and generate income for custodians of fruit tree diversity. The chapter describes 16 case studies from the TFT project that showcase how farming communities can use markets to generate income from local, unique fruit tree species and varieties as an integral part of a community-based on-farm or *in situ* conservation effort. The final section reflects on lessons learned about different types of market strategies to consider when designing value chain development based on natural capital such as agricultural biodiversity and introduces a tool to assess the level of agricultural biodiversity in a value chain, which can guide market interventions and monitor their impact on agricultural biodiversity.

Value chain approaches, methods and tools for agricultural biodiversity

Several value chain development approaches and methods have been applied for biodiversity-based value chains, such as the Marketing Approach to Conserve Agricultural Biodiversity (MACAB), the Participatory Market Chain Approach (PMCA) and Value Chain Development for NUS (VCD-NUS). When designing and implementing value chain strategies for farming communities within the TFT project we used and refined tools, concepts and methods of those three approaches and ensured that market strategies generating income were complemented with conservation efforts of communities to safeguard the remaining diversity locally available.

Marketing Approach to Conserve Agricultural Biodiversity (MACAB)

Bernet *et al.* (2004) formulated nine steps of a MACAB – the first value chain approach designed specifically for agricultural biodiversity. This intervention strategy was developed based on experiences with potato and yacon diversity in Peru and involves: (i) discovery of promising crop attributes, (ii) development of a potential new product, (iii) analysis of the economic feasibility of the product, (iv) elaboration of a sound marketing concept, (v) testing of the marketing concept with consumers, (vi) protection of brand name and concept, (vii) defining criteria for selecting private enterprises, (viii) transparent transfer of the marketing package to the private enterprise and (ix) examination of enterprise behaviour and social impact (Bernet *et al.*, 2004).

Participatory Market Chain Approach (PMCA)

The PMCA is a three-stage facilitated process that promotes technical and social innovation by strengthening trust and constructive interactions among value chain actors to facilitate the exploration of market value from agricultural biodiversity (Bernet *et al.*, 2006). The approach brings together value chain actors in a multi-stakeholder platform to share the costs of innovation. The approach has been developed and applied for potato diversity in Bolivia, Peru and Ecuador (Thiele *et al.*, 2011; Horton *et al.*, 2011; Cavatassi *et al.*, 2011), chili diversity in Bolivia and Peru (Jaeger *et al.*, 2015) and in Uganda and Indonesia (Devaux, 2014).

Value Chain Development for NUS (VCD-NUS)

Will (2008) provided a guidance document with good practices for VCD for NUS. The book formulates five steps in VCD design and discusses guiding principles and preconditions when VCD-NUS can contribute to agricultural biodiversity conservation and poverty elevation, drawing upon lessons learnt and good practices described in eight case studies and additional literature. Detailed case studies on capers (Giuliani *et al.*, 2005), emmer (Giuliani *et al.*, 2009), coffee and potatoes (Nill and Bohnert, 2006) and several case studies on tropical fruit tree diversity (Kruijssen, 2008; Kruijssen *et al.*, 2009; Kruijssen and Mysore, 2010) highlight the exploration of niche markets, need for collective action (horizontal and vertical) and active community participation for market-based approaches that support on-farm agrobiodiversity management and livelihood improvement. Padulosi *et al.* (2014) describe VCD-NUS as an holistic approach for the promotion of NUS based on experiences gained over the last 15 years in different contexts (Andean grains, minor millets in India). The approach takes into account all aspects along the value chain from genetic diversity and seed supply to final use and consumption (see Figure 22.1) with the goal of contributing to better incomes, improved nutrition, enhanced

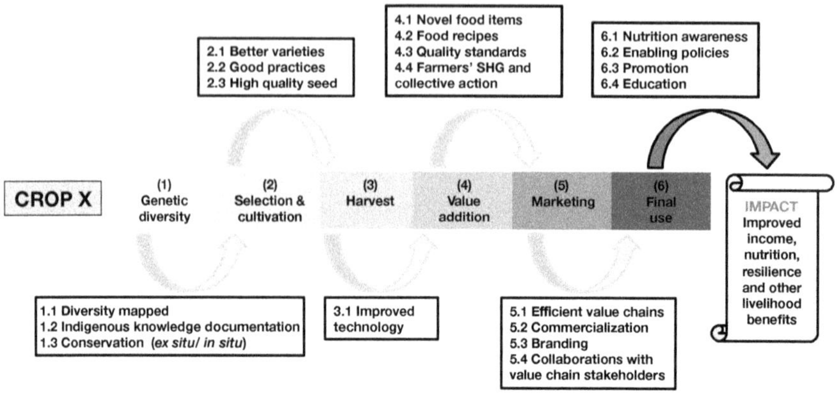

Figure 22.1 An holistic approach to promoting NUS (neglected and underutilized species).

Source: Padulosi (2014 *et al.*).

livelihood resilience and the conservation of NUS. The figure above shows the different stages of a value chain from genetic diversity to final use and all elements that need to be addressed to ensure the envisioned impact on livelihoods, including improved nutrition, income and resilience of the farming community.

We borrowed insights and methods from the approaches and case studies listed above to open the treasure box of local fruit tree diversity and generate income for the 36 villages in the 22 project sites. We used the sequence of steps as formulated in MACAB and developed a specific tool to identify crop attributes. We also used and refined tools from PMCA such as theatre play, impact filter and rapid market appraisal. We used the value chain map as described in the guidelines on VCD-NUS and a participatory focus to build local collective action as highlighted in the case studies. Based on the above, the following participatory tools were applied across the 22 sites and 36 communities in different combinations and sequences depending on the local needs and contexts.

1. Four Cell Analysis (see Chapter 3) to identify common, threatened, rare and unique fruit species and varieties
2. Participatory identification of crop attributes based on traditional recipes and home uses to evaluate promising market traits and to identify potential products (MACAB)
3. Joint assessments by farmers and traders of potential impacts by using an impact filter (PMCA) to evaluate and select best products and markets
4. Theatre play or sketches to facilitate discussion among stakeholders and explain the concept of a value chain, demand orientation and the importance of collaboration (PMCA)

5. Participatory value chain mapping to provide insights into the value chain such as constraints, opportunities or knowledge gaps and monitor the increase in market knowledge of participants (VCD-NUS)
6. Participatory rapid market appraisal to collect market information and identify market trends, consumer preferences, competitive products, niche markets and product requirements, differentiate consumer groups and estimate market potential (PMCA)
7. Identification of collaborating entrepreneurs or potential buyers who are interested in advising or jointly developing and testing novel products (MACAB)
8. Stakeholder meetings and workshops to foster collaborations, build trust and establish a shared vision between traders and farmers (PMCA)
9. Action plan to develop sample products, design the brand and label, develop prototype packaging and test improved processing equipment by research partners or community groups (PMCA, VCD-NUS)
10. Product evaluation by laboratory analysis of biochemical components such as micro-nutrients or vitamins, or during tasting events at trade fairs, workshops or diversity fairs (PMCA)
11. Participatory assessment of the level of agricultural biodiversity in a certain value chain using the market pyramid to assess and monitor how far species or varieties have entered the value chain
12. Community-based conservation strategies such as diversity blocks where a selection of all local varieties and species will be maintained, the marking of superior source trees that need to be protected, distribution of saplings or seeds from promising heirloom varieties or the promotion of sustainable harvesting practices in forests (VCD-NUS).

Results of market interventions to create income from fruit tree diversity

A high level of fruit tree diversity was found in the project communities (see Table 22.1) of the TFT project. In total, 43 distinct species of the genera *Citrus* (13 spp.), *Garcinia* (12 spp.), *Mangifera* (13 spp.) and *Nephelium* (5 spp.) were identified in the 36 project communities across four countries, of which orange, mangosteen, mango and rambutan are the most commonly known species for each respective genera. In addition, within the mango species (*Mangifera indica*) a total of 211 distinct named varieties were identified across the four countries. After initial assessments conducted by partners, a total of 67 potential products derived from 21 species were initially identified and selected for value chain development activities. After further assessments regarding their market potential, 35 products from 14 species were selected and developed into novel or improved prototypes that were tested and promoted by project partners with processing groups or cooperatives in all four countries. Finally, 18 products from 10 species were taken up by cooperatives, processing groups or entrepreneurs and generated income or resulted in increased turnover.

The other case studies are still being developed or encountered value chain constraints that hampered successful adaptation by communities. These market constraints include lack of an entrepreneurial manager or leader, lack of collective action, difficulties in finding appropriate buyers with interest in diversity-based products, lack of trust between farmers, traders and support organizations, lack of skills in obtaining appropriate market intelligence and lack of physical or financial assets to invest in enterprise activities.

Different types of market strategies can be deployed by entrepreneurs to generate income from local agricultural biodiversity. The Ansoff matrix (see Figure 22.2) is an assessment tool that identifies four types of growth strategies for an enterprise based on new or existing products and for new or existing markets (customers): (a) market penetration, (b) market development, (c) product development and (d) diversification. The level of market risk is lowest with a penetration strategy, increases when entering new markets or engaging in new products, and is highest when trying both at the same time.

We used the Ansoff market growth strategies to describe 16 most representative market case studies of the TFT project that both create income and contribute to the conservation of fruit tree diversity. Each case study is described by explaining the product, entrepreneur and target market, which species are involved, type of growth strategy, economic success (or relevant constraints) and contribution to conservation (Table 22.2).

Table 22.1 Overview of market activities in TFT project

	India	*Indonesia*	*Malaysia*	*Thailand*	*Total*
Number of project communities	18	6	6	6	36
Number of targeted households	5,681	3,405	1,328	3,931	14,345
Combined richness of four genera (*Citrus*, *Garcinia*, *Mangifera* and *Nephelium*) encountered in project communities	13	16	13	17	43*
Number of initial potential products identified	23	18	10	16	67
Number of prototype products developed, tested and evaluated	12	8	5	10	35
Number of products promoted and adopted that generate revenue	7	3	3	5	18
Number of most representative products that contributed to both income generation and conservation	6	3	3	4	16

*Numbers do not add up to the total as some species occur in more than one country.

	Existing Product	New Product
Existing Market	**(a) Market Penetration**	**(c) Product Development**
New Markets	**(b) Market Development**	**(d) Diversification**

Figure 22.2 Four distinct growth strategies for an enterprise.
Source: Ansoff (1957).

Analysis of market case studies

In most cases entrepreneurs, producer groups (including women, men and mixed groups) or cooperatives engaged for the first time in processing activities or marketing activities targeting buyers beyond the traditional channel of farm-gate sales of fresh or preliminary dried fruits to collecting traders. The majority of the case studies (8 out of 16) are characterized as a market development strategy while seven describe a market diversification strategy and one a product development strategy. None represents a market penetration strategy, which focuses on improving the marketing of an existing product within an existing market. This supports the idea that pursuing a value chain development strategy for agricultural biodiversity often entails a diversification strategy simultaneously exploring new products and engaging with a new type of customers or market channels. This results in higher potential profits but also increases risks and thus the chance of failure of the enterprise. Pursuing such a diversification or market development strategy requires substantial market intelligence and a minimum level of market skills and experience of the respective entrepreneur to succeed. Though dedicated value chain tools such as street theatre, value chain map and rapid market appraisal increase the knowledge about markets of individual farmers, women's groups or cooperative/association members, the lack of an experienced entrepreneurial, skilled leader or manager has proven to be a major barrier to developing a viable and profit-making enterprise (as experienced in cases 2, 4, 8, 11 and 12 in Table 22.2), whereas the entrepreneurial skills of leaders as demonstrated in cases 1, 3, 13, 14, 15 and 16 in Table 22.2 have contributed substantially to an enterprise's success.

Table 22.2 Overview of 16 market case studies from TFT project

	Description of product, entrepreneur and target market	*Species or varieties involved*	*Market growth strategy (Ansoff)*	*Indicators of economic success*	*Contribution to conservation*
	India				
1	Sales of heirloom varieties of mango in single and combination cartons branded 'Chittoor origin' at roadside stall during mango season by two farmers in Chittoor, India – targeting city dwellers travelling by bus and car from Bangalore to Tirupati.	*Mangifera indica* – range of lesser-known heirloom varieties such as Kudadat, Kalepadu, Green Baneshan, Dil Pasand, Rumani, Pulira, Lal Baba, Athimaduram.	Market development by direct sales to consumers through improved and branded cartons for heirloom varieties.	Prices obtained are generally higher than sold in local *mandi* (government-controlled wholesale market) or to pre-harvest contractors who prefer commercial varieties only. The two farmers bought additional landraces from other farmers who sold out in 2014 and plan a bigger stall for next year.	Loss of diversity reduced as interest in lesser-known landraces has increased among farmers and consumers in this commercially oriented mango production belt, which generally focuses on common varieties such as Totapuri, Banganapalli and Alphonso only. Diversity block established in private orchards of two farmers.
2	Sales of organic fresh mangoes of indigenous varieties by a farmer group in Chittoor, India, to Eosta, the Netherlands (www.eosta.com) packed in mango diversity boxes targeting customers of organic retail shops in Germany.	*Mangifera indica* – common varieties such as Banganapalli, Totapuri and Alphonso combined with lesser-known varieties such as Lal Baba, Athimaduram, Rumani, Pulira.	Market development targeting consumers in the EU using a novel idea of combining several mango varieties in one carton.	Initial sample of 300kg delivered to Eosta received positive feedback from organic retailers; however, a ban in 2014 and 2015 on mango exports from India to the EU disrupted trade. Only a few farmers changed over to organic due to persistent pest problems and lack of assured market demand and trust among farmers, traders and support organisations.	Eosta is willing to support conservation activities of supplying farmers when trade is re-established, such as agronomic training, the maintenance of diversity blocks, support for community nurseries and the multiplication and distribution of lesser-known varieties.

3	Sales of branded and Geographic Indication protected 'Satpura Hills Mandarins' packed in cartons to major cities like Nagpur, Mumbai and New Delhi by farmer cooperative MOGA in Amravati district, India.	*Citrus reticulata* – local adapted type and rootstock species such as Rough lemon and Rangpur lime.	Market development by targeting urban consumers by improving the packaging and promoting the GI status of Satpura Hills, which are known for the quality of fruits.	GI status obtained from Indian Government allowed use of the GI logo. Price and profit obtained per kg increased substantially compared with sales to local *mandi* (government-controlled wholesale market) and pre-harvest contractors.	Two nurseries established to conserve rootstock types in mother blocks and to provide high-quality rootstock material to local farmers.
4	Display and sales of indigenous mango varieties at mango *melas* (fairs) in New Delhi and Lucknow in single variety and combination boxes by the Society For Conservation of Mango Diversity (SCMD) in Malihabad targeting urban consumers.	*Mangifera indica* – popular varieties such as Dashehari, Langra, Chausa and various lesser-known traditional varieties such as Malihabad Safeda, Gola, Husnara, Fazli, Ramkela, Gilas, Khasul Khas, Nawab Pasand.	Market development targeting consumers in bigger cities through mango fairs.	Price and profit obtained per kg increased substantially compared with sales to local *mandi* (government-controlled wholesale market) and pre-harvest contractors. However, sales volume is limited.	Local diversity of more than 200 mango varieties maintained in diversity blocks and hedgerows of a few dedicated farmer members. Local nurseries established by SCMD to promote multiplication and distribution of traditional varieties.
5	Mango pickle made from specific wild aromatic mango types collected from forests by one women's group in Sirsi, India sold in plastic jars branded 'appe midi' or the specific variety names targeting consumers in Uttara Kannada and neighbouring districts (Chapter 28).	*Mangifera indica* 'appe midi' types Milanji, Haldota and Nandagara among others that are collected from forests.	Diversification by developing a new product based on unique crop attributes 'distinct taste, texture and flavour' of preferred varieties.	Sales turnover increased from 30 to 450 kg in 3 years with a total value of US$2,000 in 2014. Now at least five other women's groups have initiated production of similar products.	Several community nurseries were established and the number of saplings sold of appe midi types increased sharply in the last few years. Women's group members have planted saplings of appe midi types in their home

Table 22.2 continued

	Description of product, entrepreneur and target market	*Species or varieties involved*	*Market growth strategy (Ansoff)*	*Indicators of economic success*	*Contribution to conservation*
					gardens. Superior source trees have been demarcated in the forest by grafting experts and seven diversity blocks with at least 30–40 varieties each have been established in farmers' fields and village forests as well as at the premises of the College of Forestry, University of Dharwad. Interest in following sustainable harvesting practices increased among collecting households after awareness training and workshops.
6	Higher quality fruit rinds of *G. gummi-gatta* sold by collecting households in Uttara Kannada district, India through improved dryers for direct sales to Himalaya Drug Company (www. himalayawellness.com) as ingredient for Ayurslim, a weight loss product sold in capsules to consumers in India, Asia and USA (Chapter 11).	*Garcinia gummi-gatta* collected from forests.	Market development by direct sales to major processor.	First trade to be established but several meetings have taken place. Price expected to be higher compared with sales to collecting traders. In process of developing producer company and obtaining organic certification.	Himalaya Drug Company agreed to support supplying farm households with obtaining organic certification, technical training in the cultivation of medicinal plants. They will support local nurseries to distribute saplings of *G. gummi-gatta* and other native types and support sustainable harvesting campaigns financed through their Corporate Social Responsibility (CSR) programme.

	Indonesia				
7	Dried mango from varieties Podang Urang, Madu, Podang Lumut and Santok, branded 'Java bite' by farmer groups in Tiron, Kediri, Indonesia, for domestic market (Jakarta) and export to USA in collaboration with NGO Resource Exchange International (REI) and subsidiary company Sun Rei.	*Mangifera indica* – local heirloom varieties, e.g. Podang Urang, which is most preferred by consumers and originates from this area.	Market development by targeting domestic and overseas consumers in USA and Jakarta with a branded product.	Established own brand name 'Java bite' and managed to increase turnover. Exploring diversification of product range with dried pineapple and papaya.	Diversity block including 24 varieties established in Tiron village on the mountainside.
8	Sales of indigenous fresh fruits (Hampalam, Kasturi, Kweni and Banana, Orange, Mandarin, Kafir lime or Guava) by women's group in Sungai Tabuk to Swiss Bell Hotel (www.swiss-belhotel.com) in Banjarmasin in South Kalimantan, Indonesia.	*Mangifera casturi, M. odorata, M. indica, C. reticulata, C. sinensis, C. hystrix.*	Market development by targeting hotels for native species.	First delivery took place during International Environment Day in 2014, volume of repeat orders is still small.	Swiss Bell Hotel planted several indigenous trees at hotel premises as conservation strategy during Environment Day 2014.
9	Dodol (a sweet, sticky toffee-like confection popular in southeast Asia) made by women's group from Telaga Langsat subdistrict from indigenous species Kasturi for sales in tourist or gift shops in Kandagan, Banjarmasin, Banjarbaru and Martapura in South Kalimantan, Indonesia.	*Mangifera casturi*	Diversification based on new product with unique taste of Kasturi, a native species emblematic of South Kalimantan.	Prototype product has been made, but large scale production hampered by lack of appropriate equipment.	Diversity block established including 22 species and varieties of *Mangifera* on the edge of the village.
	Malaysia				
10	Production of higher quality dried rinds of aroi-aroi (*G. forbesii*) by two entrepreneurial farmers in Papar, Sabah, Malaysia through improved sun dryers, to be sold as ingredient for spice condiment in curries and	*Garcinia forbesii*	Product development by improving quality of dried rinds for various uses.	Sun dryers installed and used by two entrepreneurial farmers which increased quality of the dried rinds. Proper packaging developed for	Interest by local farmers in planting aroi-aroi trees increased and substantial number of saplings have been planted across households to increase crop area.

Table 22.2 continued

Description of product, entrepreneur and target market	*Species or varieties involved*	*Market growth strategy (Ansoff)*	*Indicators of economic success*	*Contribution to conservation*
soups (laksa, fish), as traditional medicine (for stomach ailments, cough, skin problems or eczema and for recovery of women after birth) and cleaning agent for bronze antiques (Chapter 21).			the sales of this raw product, which attracts increasing demand for its use as food ingredient, traditional medicine and as cleaning agent. Powdering and capsulation of the product is being explored with a businessman from Penampang near Kota Kinabalu and funding from Ministry of Science, Technology and Innovation (MOSTI).	
11 Sweet and sour mango pickle from Mampalam (*M. pentandra*) made by women's group in Bungai for sale to local market and shops in Sibuti and Miri, Sarawak, Malaysia.	*Mangifera pentandra*	Diversification based on new product using unique local species that looks like mango.	Processing location established and prototype product developed with positive feedback from several shopkeepers.	Interest in maintaining Mampalam trees increased and a diversity block for local fruit tree species established at Bungai village (see Chapter 23).
12 Agro-tourism package including jungle trekking with a visit to a fruit diversity garden and traditional Bidayuh lunch with fresh fruits offered to domestic and foreign tourists by villagers of Kakeng, Malaysia. Other tourism activities are a welcome dance, sales of handicrafts, demonstration of how to make a parang (jungle knife), paddy planting or harvesting activities, pepper and fruit harvesting in the forest (Chapter 23).	Range of *Garcinia*, *Mangifera* and *Nephelium* species (12 in total).	Diversification into a range of new tourism products based on unique environmental biodiversity and cultural practices.	No tourists attracted yet, but a few tourist agents from Kuching have visited the village and shown interest.	Wide range of species found in the area conserved in the fruit diversity garden. Awareness about the uniqueness of the village increased among villagers.

Thailand				
13 Home stay for visiting guests (48 single rooms with families and 10 mini resorts for group accommodation) in Kiriwong, Thailand, combined with tourism activities such as jungle trekking, demonstration of soap making (see 14 below) or tie-dying technique (see 15 below), cycling, fishing and rafting (Chapter 24).	Various fruit and non-fruiting tree species.	Diversification into a range of new tourism products based on unique environmental biodiversity and cultural practices.	Substantial increase in number of domestic tourists over recent years. Being selected as exemplary eco-tourism site by tourism department of the Thai government.	Strong sense of responsibility for the conservation of the traditional biodiverse orchards, forests and farming system established over last decades due to a combination of group efforts into eco-tourism, reforestation and rehabilitation of traditional multi-species orchards.
14 Herbal and natural soap and shampoo branded as 'Mr Mankud' made by herbal processing group in Kiriwong, Thailand which is sold to hotels, shops and consumers in Kiriwong, Nakorn Sri Thammarat state, Bangkok and overseas (Chapter 24).	*Garcinia mangostana*, *G. atroviridis*	Diversification based on new product with anti-bacterial and anti-inflammatory attributes.	Increased turnover, successful exports achieved and awarded by Thai government OTOP prize for best high quality product.	
15 Several colours for tie-dye technique for apparel (shirts, dress, bags, shawls) using fruit parts: rambutan peel or *Parkia spesiosa* (grey), mangosteen leaf (orange and pink), jackfruit bark (yellow), tropical almond (yellowish green) by processing group in Kiriwong, Thailand for sales in local village shops and sales to shops in Bangkok (Chapter 24).	*Nephelium lappaceum* and *Garcinia mangostana* among other fruit or tree species.	Diversification based on unique dying technique and natural colours made from local fruit tree species.	Increased turnover of items sold, especially higher-quality items such as bags, shirts and shawls.	
16 Spicy pork curry 'moo chamuang' using *Garcinia cowa* leaves sold in cans and plastic packs produced by one women's group in Chantaburi, Thailand, targeting consumers in Chantaburi district and Bangkok (Chapter 25).	*Garcinia cowa*	Diversification based on unique crop attributes sour taste of chamuang leaves.	Sales turnover increased from 600 packs to 1400 cans and 600 packs per month representing a value of US$1330 per month in 2014.	Population of *Garcinia cowa* trees increased over last 5 years as interest in the species among farmers increased.

The most promising and successful market case studies can be grouped into six broad categories of market strategies:

1. Products based on unique crop attributes
2. Products that display a unique diversity of types or forms
3. Tourism based on local agricultural or natural biodiversity
4. Geographic Indication or certification to obtain premiums from consumers to compensate for conservation efforts and related costs
5. Farmers' group or enterprise saves revenues to finance conservation efforts
6. Supply to larger processors or buyers that are willing to support conservation efforts through Corporate Social Responsibility (CSR) programmes.

Some case studies are a combination of two or more strategies.

Products based on unique crop attributes of minor crops and landraces

This strategy entails the exploration and identification of unique or distinctive attributes of a product based on traits, characteristics and uses of the minor crop or landrace that can be linked to the interests of specific consumer groups. European consumers, for example, prefer red mangoes with a slightly sour aftertaste whereas Asian consumers often prefer sweet mangoes with less emphasis on the colour. Another example is the fast-growing market emerging over the last ten years in the Western Ghats of India for home-made mango 'appe' pickle made from a unique aromatic sub type of mango (appe midi) collected from forests, which is locally preferred above the conventional mango pickle. These socio-cultural associations provide opportunities for the marketing of products derived from specific native species and varieties. However, they can also entail barriers. In a different project in India, marketing of minor millets in India was initially negatively influenced by social factors ascribing a 'poor' connotation to them, as they were traditionally produced and consumed by poor households, even though they have exceptionally high nutritional values compared with the more prestigious wheat (Padulosi *et al.*, 2013). Additional barriers are more intensive processing requirements (drudgery) or a shorter shelf life, such as the case of juice made from Kuini (*Mangifera odorata*) compared with mango (*Mangifera indica*).

Products displaying a diversity of types or forms

A market strategy or product development strategy can focus on products or services that 'celebrate' or promote the range of types, distinct forms, varieties or species that are available. For example, Eosta recently introduced heirloom tomatoes in German and Dutch supermarkets – a mixture of different coloured and shaped traditional tomato varieties combined in one package (personal communication Volkert Engelsman, 2014). Similar products exist for heirloom potato varieties in supermarkets in the EU or include gift packages with a range of spices targeting tourists in India. Often such types of products are bought by consumers for special occasions and celebrations or entail the targeting of tourists (one-time buyers).

Agro-tourism or ecotourism

This type of market strategy entails ecotourism or agro-tourism activities in which the visitor or tourist enjoys the natural and agricultural ecosystem and all its intrinsic diversity. Agro-tourism is of growing interest especially in areas of rapid urbanization where urban dwellers seek re-connection with nature, countryside and socio-cultural traditions. A prerequisite is, however, that the environment should be of a unique quality and enable the generation of a range of interesting tourism activities to attract tourists.

Certification and Geographic Indication

Certification strategies entail a certification or product branding strategy in which consumers pay a premium for the product to finance agricultural biodiversity conservation activities. This can be achieved through third-party certification in which existing or new labels include conservation targets in the label requirements, such as set by the International Federation of Organic Agriculture Movements (IFOAM) for organic labels or the Forest Stewardship Council (FSC) programme for forests (Mutersbaugh and Klooster, 2010). Geographic Indication (GI) registration guarantees that a good originates from the locality or region where a given quality, reputation or other characteristic is attributable to its geographic origin (WTO, 2012). For example, in India partners managed to register 'Satpuri Hills mandarin' as a GI, as this hilly region is known to provide an ideal climate for mandarins. Similarly, a GI strategy was followed and approved in 2009 for 'Malihabadi Dashehari' from Malihabad district and 'appe midi' types of mangoes from the Western Ghats. However, in both cases farmers have had difficulty translating the GI registration into a quality brand and thus a premium price for their mangoes or derived products. The social, economic and environmental impacts of GI registration are debated (Thevenod-Mottet, 2010; Bowen and Zapata, 2009) and achieving benefits from GI requires substantial investment and time. A GI strategy is less suitable for a novel product, but protects a product with a certain history and reputation against counterfeit or competitive products produced outside the GI region which claim the same qualities or characteristics (Ngo Bagal and Vittori, 2011; Jaeger and Padulosi, 2012).

Voluntary conservation fund by producer groups and enterprises

This entails a market strategy in which the farmers' group or enterprise decides to save funds voluntarily to finance conservation efforts. This can be facilitated through a Community Biodiversity Management Fund (Subedi *et al.*, 2006; Shrestha *et al.*, 2013) in which a percentage of the revenues earned from commercial activities are saved to finance conservation activities (see Chapter 2 and Chapter 29). This is sometimes referred to as a self-declaration strategy or participatory guarantee system (PGS) in which producers agree to maintain the traditional agro-ecosystem or safeguard local biodiversity through a certificate or seal on the product label, company website or in advertisements (May, 2008). This strategy differs from a certification strategy as no external agency is involved to monitor and verify the claims made and the monitoring is based on trust, peer-review, social learning and direct linkages between consumers and producers. Often such a strategy is combined with obtaining funds from government or other agencies to enlarge the conservation fund.

Linking CSR programmes to conservation efforts of suppliers in farming communities

This market strategy entails a larger processor or retailer that supports conservation efforts by its suppliers through its CSR programme and budget. Examples of such a strategy are the supply of mangoes by Chittoor farmers to Eosta, the largest organic importer of exotic fruits in the EU, who are willing to support local conservation activities such as supporting the nursery and the distribution of rare species and varieties, contributing to the maintenance costs of the diversity block and the training of farmers in agronomic practices. Similarly, the Himalaya Drug Company is committed to assisting collecting households in Uttara Kannada district in the Western Ghats to obtain organic certification, and is willing to finance the supply of saplings and awareness-raising activities for sustainable harvesting practices. Barriers of such a strategy are often the more sophisticated product requirements and quality standards of such large-scale processors or traders, and that CSR-based financial support is only provided when a trade relation is established.

Tool to assess agricultural biodiversity in value chains

When designing and implementing value chain development activities as part of an agricultural biodiversity conservation programme, it is useful to have insights into how far locally available diversity has entered markets; i.e. which species or varieties already have market value and which do not. In addition, a way to measure impact is needed; that is, whether more diversity reached the market after the interventions. As no clear tools existed for these two purposes at the beginning of project design, the TFT project designed the 'market pyramid' which helps to assess and monitor the level of agricultural biodiversity at species or varietal level in a certain value chain. The pyramid in Figure 22.3 depicts the varietal diversity (intraspecific) of mango (*Mangifera indica*) available in markets in India. The x-axis represents the number of varieties and the y-axis shows the distance or distinct levels of the value chain

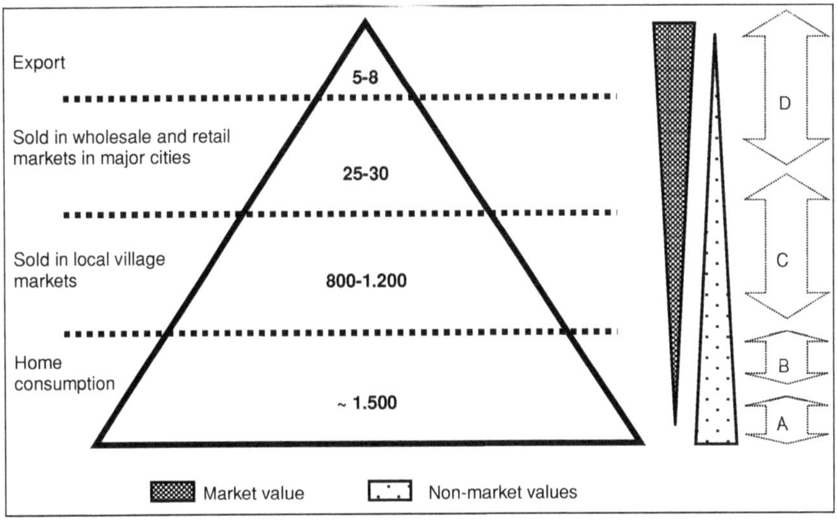

Figure 22.3 Market pyramid to assess the level of diversity within the mango value chain in India. The species and varieties at the top of the pyramid can be characterized as those with high market value and limited non-market values, while those at the bottom of the pyramid comprise limited market value but high non-market values. Based on this we categorized locally found diversity into species or varieties with:
(i) high market value targeting distant urban consumers (category D);
(ii) medium market value targeting local rural consumers (category C);
(iii) low market value but high direct use value at home (category B);
(iv) low market and use value but high indirect, option or non-use values requiring a safeguarding strategy (category A).

including home consumption, local village markets, distant wholesale markets and exports. For example, in India, about 1,500 distinct named varieties of mango are grown and found on farms across the country including 1,000 commercial varieties (Mukherjee, 1953; National Horticultural Board of India, 2015), which form the base of the pyramid. Based on consultation with traders and researchers it was estimated that only between five and eight varieties are exported, about 25 to 30 varieties are traded across states and found in wholesale and retail markets in major cities such as Delhi, Mumbai and Chennai, and an estimated 800 to 1,200 varieties are sold in local village markets across the country.

The market pyramid can be developed for any species, genus or functional group for a given geographical area. For example, one could depict the level of diversity found for the functional group of fruit trees in a particular village and list how many of the locally found species are used at home within the village, are sold in the local village market, in the nearby district markets and reach up to more distant wholesale markets. A market strategy for agricultural biodiversity in this context means bringing more species or varieties to a higher level in the value chain pyramid; that is, introducing varieties and products used for home consumption to local village markets or bringing products based on species sold only in local village markets to more distant wholesale and retail markets or exports.

Applying this pyramid and categories to the range of species or varieties as encountered in the field helps the identification of market traits based on unique crop attributes (e.g. colour, taste, storability, etc.) and the design of market interventions and safeguarding strategies. The exercise, when carried out with farming communities at the start of a value chain development project, provides a first insight into the level of market penetration of the wide range of locally available species and varieties. When repeating the exercise after the value chain interventions have taken place, it gives an estimate of the impact of activities on the level of agricultural biodiversity in the market. The categorization of species or varieties is not static, but can change when novel uses or market traits are being explored and discovered.

Conclusion

This chapter explored market strategies that contribute to community biodiversity management that simultaneously improves livelihoods and promote conservation of agricultural biodiversity, building on existing value chain methods and approaches, and drawing on lessons learned from 16 case studies from the TFT project. The case studies have shown that successful market strategies that contribute to the conservation of agricultural biodiversity in these case studies often focus on the exploration of niche markets and entail a market or product diversification strategy with higher prospected profits but also higher risks. However, farming communities in biodiversity-rich regions often have limited exposure to markets and thus limited entrepreneurial capacities and market skills. Other value chain barriers that were encountered across the case studies include lack of collective action, difficulties in finding appropriate buyers with interest in diversity-based products, lack of trust between farmers, traders and support organizations, lack of skills in obtaining appropriate market intelligence and lack of physical or financial assets to invest in enterprise activities. A tailored set of tools, presented in this chapter, and support activities can enable these communities to engage in value chains and create added value for local agricultural biodiversity.

Farmers do not maintain a wide range of diverse varieties or species at farm level for a single reason, but for a combination of several distinct benefits such as income, nutrition, managing harvest or market risks, socio-cultural values or beneficial ecosystem services. A value chain approach based on agricultural biodiversity should take these multiple aspects into account. When marketing biodiversity, the focus should not be on a single value chain for a single purpose (economic gains) but to create income based on the range of unique species and landraces that have market potential while ensuring maintenance of the other benefits of diversity for farming households through on-farm conservation strategies.

To be able to manage the above risks associated with exploring new markets or products while dealing with the barriers and interests of smallholder farmers, it is required to have a tailored set of tools and support activities that enable these communities to engage in value chains based on local agricultural biodiversity. The TFT project helped to identify, test and refine 12 value chain tools that can guide market interventions for biodiversity-based value chains, and identified six different marketing strategies that could help others when pursuing a value chain development strategy that generates income and conserves agricultural biodiversity.

Acknowledgements

We would like to thank all men and women farmers that collaborated in the TFT project and the field staff that supported them. This chapter is based on their experiences and knowledge. The following people have directly contributed to this chapter through the implementation of field activities and the documentation of market case studies: V Dakshinamoorthy, T.M. Gajanana, Dinesh Kumar, Shailendra Rajan, M.R. Dinesh, Sanjay Singh, Awtar Singh, I.P. Singh, Vasudeva R, Narasimha Hegde, V.A. Parthasarathy, B.M.C. Reddy, Kuntoro Boga Andri, M. Winarno, Achmed Rafieq, Idha Widi Arsanti, Faridah Aini Muhammad, William W.W. Wong, Shariah Umar, Pearlycia Brooke, Adam Haris Gerten, Zahimi Hassan, Jamaluddin Lani, Muhammad Shafie Md Sah, Salma Idris, Chatchanok Noppornphan, Kasemsak Palakorn, Samroeng Changprasert, Pichit Sripinta, Montree Issarakraisila, Tawatchai Nimkingrat, Margaret C. Yoovatana and Songpol Somsri.

References

Ansoff, I. (1957) 'Strategies for diversification', *Harvard Business Review*, vol 35, Issue 5, Sep–Oct 1957, pp. 113–124

Bernet, T., Hibon, A., Bonier, M., and Hermann, M. (2004) 'Marketing approach to conserve agricultural biodiversity', *BeraterInnen News*, pp. 21–24

Bernet, T., Thiele, G., and Zschocke, T. (2006) 'Participatory Market Chain Approach (PMCA) user guide,' CIP, June; http://tools4valuechains.org/tool/participatory-market-chain-approach-0

Bioversity International (2013) 'Harvesting quinoa diversity with payment for agrobiodiversity conservation services', Bioversity International; http://www.bioversityinternational.org/uploads/tx_news/Harvesting_quinoa_diversity_with_Payment_for_Agrobiodiversity_Conservation_Services_1664_03.pdf

Bowen, S. and Zapata, A.V. (2009) 'Geographical Indications, terroir, and socio-economic and ecological sustainability: The case of tequila', *Journal of Rural Studies*, vol 25, no 1, pp. 108–119; doi:10.1016/j.jrurstud.2008.07.003; http://linkinghub.elsevier.com/retrieve/pii/S0743016708000454

Cavatassi, R., González-flores, M., Winters, P., Andrade-Piedra, J., Espinosa, P., and Thiele, G. (2011) 'Linking smallholders to the new agricultural economy: The case of the plataformas de concertación in Ecuador', *Journal of Development Studies*, vol 47, no 10, pp. 1545–1573; doi:10.1080/00220388.2010.536221; http://www.tandfonline.com/doi/abs/10.1080/00220388.2010.536221

Devaux, A. (2014) 'The Participatory Market Chain Approach: From the Andes to Africa and Asia'; http://www.apip-apec.com/pe/goodpractices/1_10/devaux_2014_the_participatory_market_chain_approach_from_the_andes_to_africa_and_asia.html

Drucker, A., Padulosi, S., and Jaeger, M. (2013) 'No free lunches: PES and the funding of agricultural biodiversity conservation insights from a competitive tender for quinoa-related conservation services in Bolivia and Peru', Bioversity International, Rome, Italy; http://www.fao.org/fileadmin/user_upload/pes-project/docs/FAO_RPE-PES_Bioversity_BoliviaPeru.pdf

Giuliani, A., Adbulkarim, N., and Buerli, M. (2005) 'Linking biodiversity products to markets to improve the livelihoods of the resource poor: Case study on the market chain of capers in Syria'; https://www.researchgate.net/profile/Alessandra_Giuliani2/publication/228477721_Linking_biodiversity_products_to_markets_to_improve_the_livelihoods_of_the_resource_poor_Case_study_on_the_market_chain_of_capers_in_Syria/links/00b49514c99643a259000000.pdf

Giuliani, A., Karagöz A., and Zencirci, N. (2009) 'Emmer (Triticum Dicoccon) production and market potential in marginal mountainous areas of Turkey', *Mountain Research and Development*, vol 29, no 3, pp. 220–229, International Mountain Society; doi:10.1659/mrd.00016; http://dx.doi.org/10.1659/mrd.00016

Gruère, G., Giuliani, A., and Smale, M. (2006) 'Marketing underutilized plant species for the benefit of the poor: A conceptual framework', IFPRI discussion paper; http://www.ifpri.org/sites/default/files/publications/eptdp154.pdf

Horton, D., Oros, R., Ybarnegaray, R.P, Lopez, G., Velasco, C., Rodriguez, F., Escobar, E., Rotondo, E., Hareau, G., and Thiele, G. (2011) 'The Participatory Market Chain Approach: Experiences and results in four Andean cases', Social and Health Sciences Working Paper 2011-1, International Potato Center (CIP), Lima, Peru

Jaeger, M. and Padulosi, S. (2012) 'Reflections on market-based strategies for enhancing the use of agrobiodiversity', in S. Padulosi, N. Bergamini, and T. Lawrence (eds), *On-farm Conservation of Neglected and Underutilized Species: Status, Trends and Novel Approaches to Cope with Climate Change*, Proceedings of the International Conference, Friedrichsdorf, Frankfurt, 14–16 June, 2011; Bioversity International, Rome, Italy, pp. 241–248

Jaeger, M., Amaya, K., Oros, R., Pena, K., and Bejarano, C. (2015) 'Linking genebanks and small farmers to high-value markets through collective action: The case of chili peppers in Bolivia and Peru', unpublished, Cali, Colombia

Kruijssen, F. (2008) *Compilation of Publications on Marketing of Tropical Fruits in South- and Southeast Asia*, Bioversity International, Serdang, Malaysia

Kruijssen, F., Keizer, M., and Giuliani, A. (2009) 'Collective action for small-scale producers of agricultural biodiversity products', *Food Policy*, vol 34, no 1, pp. 46–52; doi:10.1016/j.foodpol.2008.10.008; http://linkinghub.elsevier.com/retrieve/pii/S0306919208000808

Kruijssen, F. and Mysore, S. (2010) *Integrating Biodiversity Conservation and Livelihood Improvement: The Role of Markets for Mango Varieties and Garcinia Species in India*, Bioversity International, India

Lenzen, M., Moran, D., Kanemoto, K., Foran, B., Lobefaro, L., and Geschke, A. (2012) 'International trade drives biodiversity threats in developing nations', *Nature*, vol 486, no 7401, pp. 109–112; doi:10.1038/nature11145;http://www.nature.com/nature/journal/v486/n7401/full/nature11145.html#affil-auth

Lockie, S. and Carpenter, D. (2010) 'Agriculture, biodiversity and markets. Part I: Agrobiodiversity in context', in S. Lockie and D. Carpenter (eds), *Agriculture, Biodiversity and Markets; Livelihoods and Agroecology in Comparative Perspective*, Earthscan, London

May, C. (2008) *PGS Guidelines; How Participatory Guarantee Systems Can Develop and Function*, IFOAM; http://www.ifoam.bio/sites/default/files/page/files/pgs_guidelines_en_web.pdf

Mukherjee, S.K. (1953) 'The mango – its botany, cultivation, uses and future improvement, especially as observed in India', *Economic Botany*, vol 7, no 2, pp. 130–162; doi:10.1007/BF02863059; http://link.springer.com/10.1007/BF02863059

Mutersbaugh, T. and Klooster, D. (2010) 'Environmental certification: Standardization for diversity', in S. Lockie and D. Carpenter (eds), *Agriculture, Biodiversity and Markets; Livelihoods and Agroecology in Comparative Perspective*, Earthscan, London, pp. 155–174

National Horticultural Board of India (2015) 'Mango' ; http://nhb.gov.in/report_files/mango/MANGO.htm

Ngo Bagal, M. and Vittori, M. (2011) *Practical Manual on Geographic Indications for ACP Countries*, Geneva, Switzerland; http://www.origin-gi.com/images/stories/PDFs/English/OriGIn_publications/manual_acpcomplet.pdf

Nill, D. and Bohnert, E. (2006) *Value Chains for the Conservation of Biological Diversity for Food and Agriculture: Potatoes in the Andes, Ethiopian Coffee, Argan Oil from Morocco and Grasscutters in West Africa Global*, GTZ, Germany

Padulosi, S., Amaya, K., Gotor, E., Rojas, W., and Valdivia, R. (2014) 'A holistic approach to enhance the use of neglected and underutilized species: The case of Andean grains in Bolivia and Peru', *Sustainability*, vol 6, pp. 1283–1312; doi:10.3390/su6031283

Padulosi, S., Thompson, J., and Rudebjer, P. (2013) *Fighting Poverty, Hunger and Malnutrition with Neglected and Underutilized Species (NUS): Needs, Challenges and the Way Forward*, Bioversity International, Rome, Italy; http://www.bioversityinternational.org/uploads/tx_news/Fighting_poverty__hunger_and_malnutrition_with_neglected_and_underutilized_species__NUS__1671_03.pdf

Pimbert, M. (1999) *Sustaining the Multiple Functions of Agricultural Biodiversity*, IIED, London

Prescott-Allen, R. and Prescott-Allen, C. (1990) 'How many plants feed the world?' *Conservation Biology*, vol 4, no 4, pp. 365–374

Rao, R., Kruijssen, F., and Keyu, B. (2005) 'Socioeconomic factors affecting genetic diversity and genetic erosion on farm', unpublished, Hydrabad, India

Shrestha, P., Sthapit, S., Subedi, A., and Sthapit, B.R. (2013) 'Community biodiversity management fund: promoting conservation through livelihood development in Nepal', in W.S. de Boef, A. Subedi, N. Peroni, and M. Thijssen (eds), *Community biodiversity management: promoting resilience and the conservation of plant genetic resources*, Earthscan, London

Subedi, A., Shrestha, P., Shrestha, P.K., Gautam, R., Upadhyay, M.P., Rana, R., Eyzaguirre, P., and Sthapit, B. (2006) 'Community-based biodiversity management: Empowering community to manage agricultural biodiversity', in B.R. Sthapit, P.

Shrestha, and M.P. Upadhyay (eds), *On-farm Management of Agricultural Biodiversity in Nepal: Good Practices*, NARC/LI-BIRD/ Bioversity International, Nepal

Thevenod-Mottet, E. (2010) 'Geographical Indications and biodiversity', in S. Lockie and D. Carpenter (eds), *Agriculture, Biodiversity and Markets: Livelihoods and Agroecology in Comparative Perspective*, Earthscan, London, pp. 201–212

Thiele, G., Devaux, A., Reinoso, I., Pico, H., Montesdeoca, F., Pumisacho, M., and Andrade-Piedra, J. (2011) 'Multi-stakeholder platforms for linking small farmers to value chains: Evidence from the Andes', *International Journal of Agricultural Sustainability*, vol 9, no 3, pp. 423–433; doi:10.1080/14735903.2011.589206;http://www.tandfonline.com/doi/abs/10.1080/14735903.2011.589206

Van Dusen, E.M. and Taylor, J.E. (2005) 'Missing markets and crop diversity: Evidence from Mexico', *Environment and Development Economics*, vol 10, no 4, pp. 513–531; doi:10.1017/S1355770X05002317;http://journals.cambridge.org/abstract_S1355770X05002317

Will, M. (2008) *Promoting Value Chains of Neglected and Underutilized Species for Pro-Poor Growth and Biodiversity Conservation*, Global Facilitation Unit for Underutilized Species, Rome, Italy; http://www.bioversityinternational.org/uploads/tx_news/Promoting_value_chains_of_neglected_and_underutilized_species_for_pro-poor_growth_and_biodiversity_conservation_1294.pdf

Willer, H., Yussefi-Menzler, M., and Sorensen, N. (2008) *The World of Organic Agriculture, Statistics and Emerging Trends*, International Federation of Organic Agriculture Movements (IFOAM), Bonn, Germany

WTO (2012) *Uruguay Round Agreement: TRIPS. Part II – Standards Concerning the Availability, Scope and Use of Intellectual Property Rights. Sections 3 and 4*, World Trade Organization, Geneva, Switzerland; https://www.wto.org/english/docs_e/legal_e/27-trips_01_e.htm

23 Agrotourism in Kampung Kakeng, Serian

Development and challenges

Pearlycia Brooke, Salma Idris, Rateng Girid, Lau Cheng Yuon, Muhammad Shafie Md Sah and Hugo A.H. Lamers

GPD 'passport'

GPD code:	33
Focus area:	Commercialization and home use Collective action and social networking
Character:	A process to use agrotourism to secure monetary benefits from conserving tropical fruit trees *in situ*
Species and varieties:	*Mangifera indica*, *M. odorata*, *Nephelium lappaceum*, *N. ramboutan-ake*, *Garcinia mangostana*, *Lansium domesticum*, *Psidium guajava*, *Artocarpus integer*, *Baccaurea macrocarpa* and many others
Name of location:	Kampung Kakeng, Malaysia
GIS reference of location(s):	N 01°17′32″; E 110°56′65″ Elevation: 100–200 masl
Name of farmer (data resource):	Mr Juny ak Nyaud Mr Sindon ak Nuong

Introduction

Natural attractions have always drawn crowds, but recent years have seen a boom in 'agrotourism' in Malaysia as more tourists seek alternatives to traditional vacations and a deeper understanding of natural and agricultural environments. This is reinforced by a trend of urbanization in which urban dwellers and younger generations are losing touch with nature and the realities of the origins of their food culture and related traditional production systems. In Malaysia, Kampung Kakeng in the district of Serian, located 60 km north of Kuching, Sarawak, is a village rich in natural resources. Because of its uniqueness in terms of agricultural resources, such as its tropical fruit tree diversity, paddy fields and pepper farms, combined with an attractive landscape of forested and steep mountains, and its cultural and historical attractions, it has potential to be developed into an agrotourism destination. Despite the growth in agrotourism in Malaysia generally, it is relatively new in Sarawak province. The most crucial phase in developing agrotourism in Kampung Kakeng is identifying existing assets for tourism and converting them into profitable businesses. Through focus groups and unstructured interviews with the community and tourism stakeholders, ten potential tourism products were identified in this study. Over the years, several efforts have been made to develop and promote some of these agrotourism products, such as jungle treks and fruit diversity gardens. However, the development of an agrotourism industry in the village has been slow due to numerous obstacles such as a lack of entrepreneurial skills and knowledge in key individuals about how to develop and promote agrotourism products, a lack of tourism infrastructure and low participation from the wider community, especially young people. Besides looking into the potential agrotourism products, the interventions that have been implemented are discussed in this chapter along with the challenges faced by the villagers in developing Kampung Kakeng as an agrotourism destination.

Description of good practice for diversity maintenance

Agrotourism can, in theory, create conditions and incentives for local biodiversity conservation by increasing the local benefits of tourism in rural areas. Agricultural tourism is the process of attracting visitors to agricultural areas, generally for educational and recreational purposes (Veeck *et al.*, 2006). Agrotourism has been widely promoted as one strategy for the conservation of agricultural resources (Ceballos-Lascurain, 1996). In agrotourism, a local farmer offers tours to their agricultural farms to allow visitors to view local people growing, harvesting and processing locally grown crops the visitors would not come across in their own areas or countries. Often, the farmers also provide farm-stay opportunities including educational programmes and recreational activities (Nilsson, 2002; Weaver and Fennel, 1997).

Over the centuries, farmers have developed a range of crops that can generate income and sustain their day-to-day lives. These natural assets, such as agricultural biodiversity (agrobiodiversity), can be developed into tourist attractions if found to be interesting and viable. Indirectly, this is expected to encourage farmers to maintain and conserve the agricultural biodiversity in their area due to the economic benefits generated. By providing economic incentives to protect agrobiodiversity, agrotourism makes conservation efforts possible, as well as providing revenue to continue supporting conservation efforts (CBD and UNEP, 2007).

Kampung Kakeng is a typical Bidayuh village in the south of Sarawak province in Borneo. The Bidayuh (or land Dayaks), who are the native inhabitants of this part of Borneo, live among Iban, Malay and Chinese communities. Besides the spectacular natural landscape, it is also an area abundant in fruit trees. There are five *Mangifera* species and at least ten landraces of rambutan (*Nephelium lappaceum*) cultivated and maintained in Kampung Kakeng. Fruit trees are cultivated in home gardens and orchards situated from lowland to highland areas. The fruit trees in the orchards were domesticated from the forest in an agroforestry system. The rich diversity of fruit trees accompanied by the beautiful topography of the area makes Kampung Kakeng very suitable for agrotourism. In addition, the relatively easy accessibility of the village and cultural and historical attractions in the village add value to the agrotourism products.

Identification of good practice for diversity maintenance

Department of Agriculture (DoA) staff and members of the local community met several times to explore agrotourism as a good practice to increase the local benefits of natural resources and biodiversity. Given the exploratory nature of this study, focus groups and unstructured interviews served as the primary method of data collection and were conducted alongside a sequence of interventions to establish agrotourism activities. The focus group and interview participants included men and women community members of Kampung Kakeng and tourism stakeholders such as tour and travel agencies, representatives from the Ministry of Tourism Sarawak and the Sarawak Tourism Board. From these activities, interviews and discussions, several potential agrotourism products were identified.

Potential agrotourism products in Kampung Kakeng

The most crucial phase in developing agrotourism in Kampung Kakeng is identifying existing tourism assets (natural, cultural assets and local products) and assessing their potential. Several discussions took place between villagers and stakeholders to identify potential agrotourism products in the village. From the multiple discussions, ten tourism assets in Kampung Kakeng were shortlisted. As shown in Table 23.1, the assets can be categorized into three groups: (1) nature, (2) history, tradition and culture and (3) agriculture.

Table 23.1 Current tourism assets in Kampung Kakeng, Serian and their potential

Tourism assets	*Category*	*How to turn these assets into a potential tourism business?*
Traditional Bidayuh Serian dance with traditional music	History, tradition and culture	Tourists can watch cultural shows performed by the villagers. Access fees will be charged.
Jungle trekking	Nature	Tourists can trek in the jungle while enjoying nature and scenery. Fee will be charged for hiring of guides.
Fruit diversity garden	Agriculture	Tourists can harvest and enjoy tropical fruit during the fruit season and learn about the different types and uses. Access fees will be charged.
Gua Antu (limestone cave)	History, tradition and culture/ nature	Introduce to tourists the history and legend of ghosts living in the cave. Villagers also can make postcards as souvenirs and sell to the tourists. Fee chargeable for hiring of guides.
Traditional paddy farming	Agriculture	Tourists can participate in paddy farming activities ranging from planting and weeding to harvesting and processing, depending on the farmers' current progress at their farms. Access fee chargeable.
Traditional pepper farming	Agriculture	Tourists can participate in pepper farming activities ranging from planting and weeding to harvesting and processing, depending on the farmers' current progress at their farms. Access fee chargeable.
Traditional handicrafts and costume	History, tradition and culture	Tourists can watch how villagers make their traditional clothes. Access fee will be charged and tourists also can buy costumes directly from the villagers.
Traditional 'parang' (machete) making	History, tradition and culture	Tourists can watch how villagers make machete using traditional equipment. Access fee chargeable.
Rumah padi (paddy seed storage house)	Agriculture/ history, tradition and culture	Introduce the history and show the uniqueness of the 'rumah padi'. Exhibition of artefacts or old farm tools used in paddy farming can be shown. Access fee chargeable.
Traditional food	Agriculture/ history, tradition and culture	Cooking demonstration of traditional Bidayuh food and durian buffet during fruiting season. Tourists can harvest and enjoy tropical fruit whilst in jungle trekking. Access fee chargeable.

From the table above, it is clear that a number of potential tourism products were identified in Kampung Kakeng that might bring economic benefit to the community. Based on suggestions from the tourist agents, the villagers assigned tasks among themselves; they defined a fee for each activity and a group of people that would be responsible for managing the activity. The fee is planned to be charged by the tourist agent for each tourist that opts for that activity. The community decided to save a percentage of the income earned in a common fund and distribute the rest between the people conducting the activity. Within the activity groups, people take turns to ensure that all families participate and benefit. The tourist agent promotes the activities as a full-day or half-day package to tourists.

Development of agrotourism in Kampung Kakeng

Several activities and interventions were implemented to develop Kampung Kakeng as an agrotourism destination (see Table 23.2). These efforts have revolved around three primary objectives: (1) to increase intra- and interspecific fruit species diversity; (2) to develop key agrotourism products using existing natural resources that function as unique selling points of the village and (3) to develop additional agricultural, cultural or heritage-based products in order to generate additional activities and benefits.

Intervention 1: Increasing intra- and interspecific fruit species diversity for demonstration and conservation

A community fruit nursery was built in Kampung Kakeng in 2012 to propagate fruit planting materials and distribute them to home gardens and a fruit garden that had been established as part of the same project (Figure 23.1: Fruit diversity garden). Some commercial varieties of rambutan were provided by the DoA Sarawak as source trees. Prior to the establishment of the nursery, in 2011, project participants were given training in nursery management and grafting techniques of fruit trees at the Agriculture Research Centre in Semengok. Subsequently, seven follow-up training courses were held *in situ* in 2013–2014 to improve the farmers' propagation skills. The fruit nursery currently relies on dedicated members of the community who work alongside DoA staff to manage and propagate various types of fruit trees for the project.

Intervention 2: Developing key agrotourism products using existing natural resources as the unique selling points of the village

A 5-hectare fruit tree garden was established in 2012 in which fruit trees from the genera *Mangifera*, *Nephelium*, *Garcinia*, *Durio*, *Artocarpus* and many more were planted. The layout of the fruit diversity garden is provided in Figure 23.1. Currently, more than 100 plants of different fruit tree species are growing in the fruit garden. The garden includes 10 fruit tree species, including

Table 23.2 Overview and sequence of interventions conducted in Kampung Kakeng

Type of activities	*Timeframe*	*Method/tools used*	*Result*
Develop broad ideas about potential agrotourism activities	Nov 2011	Focus group discussion (both men and women)	Homestay, agrofruit park, trekking identified as potential activities
Consultation meetings with Ministry of Tourism Sarawak	May 2012	Meetings	Feedback on potential activities and related regulations or procedures
Evaluation meeting to select activities	Feb 2013	Focus group discussion (both men and women)	Decision not to focus on homestay and large agrofruit park but jungle trekking and fruit diversity garden instead
Establish trekking route	2014	Tender contract and gotong royong (voluntary labour)	Improved paths with safety measures such as steps, handrails, bamboo bridge, mapped routes of 3 km and 5 km
Develop fruit diversity garden at start/finish point of the trekking route	2014	Tender contract and gotong royong (voluntary labour)	Cleared 1.5 hectare of forest and planted 10 fruit species, placed resting benches and signboard, improved road access
Assess and understand value chain actors, constraints and opportunities for agrotourism	June 2014	Participatory theatre play and value chain map by men and women	Identified knowledge gaps such as lack of information about 'competitors' and downstream value chain actors
Collect market information from two tourist agents in Kuching (Planet Borneo and Borneo Exploration)	June 2014	Participatory rapid market appraisal and value chain map	Learned that tourist agents take care of promotion, only need to assign tasks, people and fees per activity
Tourist guide training conducted by Mr Edward Mansel from Sarawak Tourist Guide Association (SkTGA)	Nov 2014	Training workshop	Villagers (both genders) learned how to manage tourists, develop anecdotal and informative stories connected to items or places along the trail.
Visit to Matthew Ngau's Homestay and Tebakang Homestay	Nov 2014	Exposure visit	Villagers understand concept of homestay and requirements
Visit to to Matang Wildlife Centre and Matang Family Park and briefing by officer in charge Mr AubyIlias	2014	Exposure visit	Saw nature trail at the Centre and learned about design, planning, construction, maintenance and visitor safety
Develop simple operational and business plan	2014	Meetings	Appointed respective groups, people responsible and agreed fees to charge
Invited tourist agent and tourism department officials to learn about all agrotourism activities in Kampung Kakeng	2014	Familiarization trip	Ministry of Tourism, Sarawak Economic Development Corporation (SEDC), Sarawak Tourism Board and Planet Borneo were sensitized and provided feedback for further improvements

56 plants of *Mangifera*, 23 of *Nephelium* and 7 of *Garcinia* species. The fruit diversity garden also retained useful forest trees such as tapang (*Koompassia excelsa*), which harbours wild bees. A community programme to raise awareness of the value of these tropical fruit trees and the importance of their conservation was organized with 30 school children from neighbouring schools, SK Parun Suan and SMK Taee (Plates 48 and 49). All students involved in this programme were tasked to take care of the trees they planted and to monitor the trees' growth. New seedlings are also planted on the occasion of visits from very important people to the community.

A 3.8 km jungle path used by the villagers to collect forest products was identified and upgraded into a jungle trek in early 2014 (Figure 23.2). The trek (Plates 50 and 51) was clearly marked with signboards, and hand railings made of rope and wood were installed at strategic points along the jungle trek to ensure the safety and comfort of the visitors.

A tourist guide course was conducted by DoA Sarawak to train villagers to become skilled guides who can give thorough explanations about nature, local heritage and seasonal fruit diversity along the jungle trek (Plates 52–55). The community was also taught about the history of Sarawak and its people, the

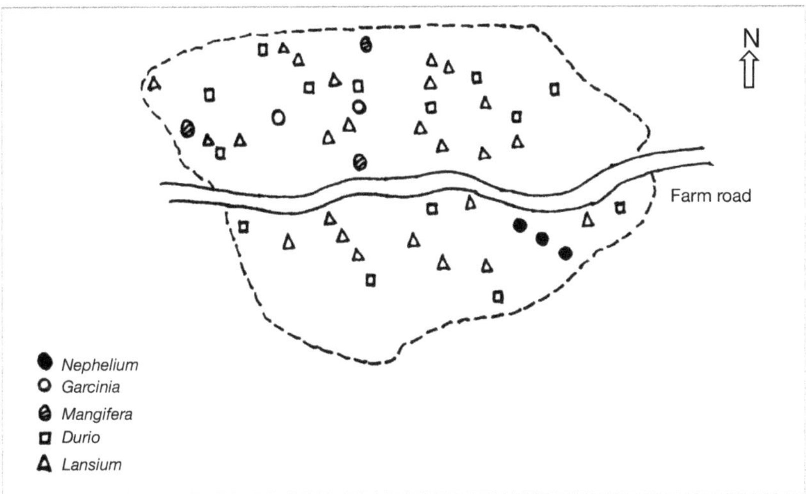

EXISTING TREES (2012)

- *Nephelium* - 3 plants
- *Garcinia* - 2 plants
- *Mangifera* - 3 plants
- *Durio* - 14 plants
- *Lansium*- 25 plants

CURRENT NO. OF TREES (2014)

- *Nephelium* - 23 plants
- *Garcinia* - 7 plants
- *Mangifera* - 56 plants
- *Durio* - 26 plants
- *Lansium* - 25 plants
- *Artocarpus* - 13 plants
- Others - 28 plants

Figure 23.1 Layout of fruit diversity garden in Kampung Kakeng, Serian.
Credit: R. Girid.

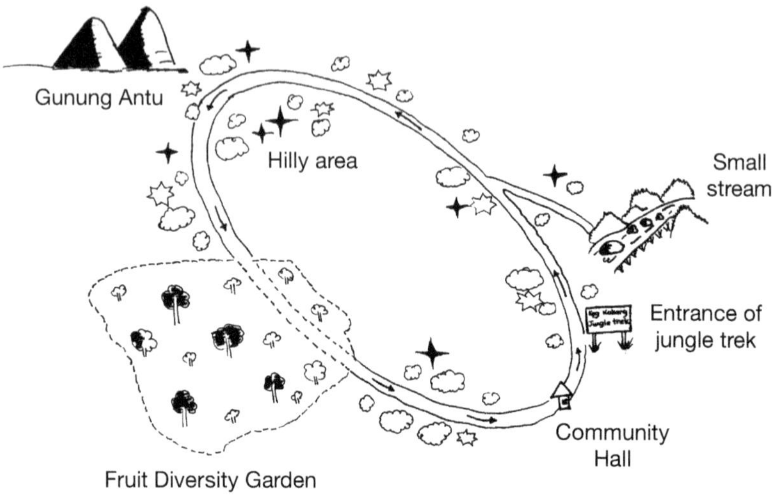

Figure 23.2 Map of trekking route in Kampung Kakeng.
Credit: R. Girid.

thematic interpretation of agro- and ecotourism, communication skills and occupational safety and hygiene for the hospitality industry.

Intervention 3: Developing additional agriculture, culture or heritage based products

A number of agrotourism products, such as the harvesting of pepper berries, traditional paddy planting ('menugal') and harvesting of tropical fruit in the fruit garden, have been developed with the community. Visitors can participate in different operations of fruit, pepper and rice farming activities ranging from planting to harvesting and processing, depending on the farmers' ongoing activities on their farms. Women entrepreneurs are encouraged to venture into handicrafts, antiques, souvenirs and traditional clothing, which can be turned into viable tourism products. For each of the activities identified, one person was nominated to be in charge, and the duration and minimum number of visitors per activity were agreed (Table 23.3). This information, together with the fee requested, was shared with private travel and tour agencies during a familiarization trip to Kampung Kakeng conducted in June 2014.

Impact on livelihood and diversity

Typically the tourism industry in Sarawak employs personnel from outside the region for all but the lowest paid positions, and any entrance fees charged go to the government, not the community. To avoid such a situation, the DoA with the project team had several meetings with women and men of the local

Table 23.3 Agrotourism product information in Kampung Kakeng

Products/services	*Duration (Hour)*	*Availability*	*Person in charge*	*Minimum number of visitors*
Traditional paddy farming	2 to 3	August to February	Mr Gaos	2
Traditional pepper farming	2 to 3	Whole year	Mr Junny	2
Handicraft demonstration	1/2 to 1	Whole year	Mdm Ninges	2
Harvesting and enjoying fruit at fruit diversity garden	1 to 2	Whole year	Mr Jenit	2
Jungle trek	2 to 3	Whole year	Mr Misai Mr Nelson	5

community to develop agrotourism packages (Table 23.1) for a new partnership based on a commitment to hire local residents as managers and use local food and products. If agrotourism is to contribute seriously to conservation and development, the following guidelines have to be agreed upon:

- Provide significant benefits for local residents
- Contribute to the sustainable management of biodiversity and natural resources
- Incorporate environmental education for tourists and residents
- Be developed and managed to minimize negative impacts on the environment and local culture.

Because agrotourism is a new endeavour for many villagers in Kampung Kakeng, its impact on livelihoods is still minimal. Some groups, like the women's handicraft group, have already benefited from tourism activities, although its contribution to their household income is still small. Income generation from tourism-related activities is important as it will motivate community members to raise their living standards and enhance their motivation to conserve the agricultural biodiversity in their area. For instance, the collective decision by the community to establish a fruit diversity garden as one of the tourism products provides a means for them to conserve fruit diversity and use it sustainably in the future. This is proven by the fact that the number of fruit trees conserved in this area increased from 47 to 178 plants in just two years.

Challenges in developing and promoting agrotourism in Kampung Kakeng

Kampung Kakeng has everything it needs in terms of cultural and natural resources to be developed as an agrotourism attraction. However, it became apparent that numerous obstacles impede the development of agrotourism in Kampung Kakeng, Serian.

After several attempts to promote Kampung Kakeng to local tourism stakeholders, there were no efforts by the private or government sectors to facilitate and stimulate the development and promotion of agrotourism in the area. This is probably due to competition from other tourist destinations and the fact that agrotourism is still new in Kampung Kakeng, and Sarawak in general. Promotion of the new products has progressed very slowly because of a lack of skilled and experienced entrepreneurial managers for the tourism activities. Participatory tools have been used to improve the general skills and knowledge of key villagers related to marketing using a theatre play to explain the concept of value chains, participatory value chain mapping to document knowledge on the tourism sector and participatory rapid market appraisal and exposure visits to increase and collect specific market information from tourism agents or other tourist destinations. However, these have not been sufficient to improve the knowledge and skills to a level to enable key villagers to take self-directed actions regarding the marketing and promotion of their tourist activities. A stronger focus on and investment in capacity building of key individuals and staff in the marketing and promotion of tourism activities is thus seen as a key requirement to build up a professional tourism business. Thereafter, a more effective promotion and marketing strategy for both foreign and local markets must be implemented to encourage private sector investment and participation in the agrotourism industry in Kampung Kakeng.

Although various training courses were provided to the community, the lack of commercial attitude and motivation toward agrotourism and the failure to attract young people in the village, as well as a top-down approach in implementing the project, have impeded the success of the tourism activities. This is also due to the fact that there is a lack of awareness among the villagers of the value-added opportunities of tourism and most households feeling comfortable with the current levels of income. Even if they become aware of these opportunities, they do not have the motivation to pursue them as the economic benefits of tourism have not been felt by the community yet. Exposure visits to similar successful agrotourism sites outside Sarawak might be a prerequisite to build the human and social capital of key local stakeholders so that they can learn from others and be self-confident.

The lack of basic infrastructure in Kampung Kakeng such as accommodation for tourists, public toilets, parking spaces, a waste management system and telephone network also decreases the viability of promoting agrotourism in Kampung Kakeng. This is in accordance with Halfacree (1993), in which, in addition to agricultural resources, agrotourism also requires accommodation and other facilities of a similar level to other types of tourism business. In addition, the community lacks financial support for building the renovations necessary for certain agrotourism products. To transform the village into a viable tourist attraction, the village needs an infrastructural facelift and financial assistance from government or other stakeholders. However, while making all these improvements, one should be careful not to disturb the local cultural and agricultural traditions too much and to keep the attractions of natural and agricultural systems intact. During discussions it emerged that the elderly in particular did not want to develop tourist accommodation facilities inside the village as they felt it would disrupt the social fabric.

The sustainability of agrotourism in Kampung Kakeng depends largely on whether it can give clear economic benefits to the community. If tourism is successful in the future, the community will benefit economically from the fees that the visiting tourists pay, including environmental, guide and entrance fees. The sustainability of agrotourism in Kampung Kakeng also depends on the local community taking ownership of the industry. At the moment, the community in Kampung Kakeng is dependent on external parties for funding and other assistance. Levels of participation in decision-making are also low as agrotourism is still an unknown industry to most of the villagers. Akama (1996) explained that local residents not only act as a source for tourism, but they are important assets to make tourism activities more robust and more effective. Therefore, a more participatory approach to managing agrotourism activities needs to be strengthened, nurtured and empowered to make self-directed decision making in the development of agrotourism in the village.

Conclusion

In this chapter, ten potential tourism assets in Kampung Kakeng were successfully identified and several of them have been developed into agrotourism enterprises that can bring economic benefit to the villagers. Despite substantial coordinated efforts, the process was slow due to the lack of knowledge and skills in developing and promoting these agrotourism products, lack of awareness of the opportunities of agrotourism, lack of basic tourism infrastructure and inadequate or insufficient involvement of the local community in decision-making and activities. Not being able to identify an 'agent of change', a skilled entrepreneurial manager locally who can lead all the agrotourism activities, is seen as a key constraint to the progress and adoption of activities. As well as this, it was difficult to involve younger villagers in activities. This is likely because they were not involved and consulted from the start and also because they often have jobs outside the village and could not attend the meetings, which were mostly held during the day. However, despite the various challenges faced in developing agrotourism in the area, the project indicated that agrotourism can live up to its promise of supporting conservation of agrobiodiversity while providing benefits to the community. For instance, a wide range of unique species of fruit trees are now conserved in the fruit diversity garden in Kampung Kakeng, which has helped farmers to see the value of the diversity of species they maintain and its potential as a viable agrotourism attraction in the future. In summary, it is hoped that a proper framework to overcome the aforementioned challenges can be developed to transform the village into a successful agrotourism destination.

Recommendations

The tourism industry is highly competitive, thus it is important to have an effective promotion and marketing strategy to attract visitors to Kampung Kakeng. A range of promotional materials – from leaflets to a website – can be created through partnership with the Ministry of Tourism and other partners such as tour operators and local associations. As the Internet becomes more accessible to the general public, creating a website on Kampung Kakeng could be very useful to promote the range of products and services that the community can offer to potential clients.

Improving facilities and infrastructure to respond to the industry's needs and expectations is vital to transform Kampung Kakeng into a viable tourism attraction. Support and assistance from both the public and private sectors are essential, as the community alone is not yet well placed to initiate and maintain the agrotourism activities.

In order to promote the success of tourism activities as a sustainable livelihood strategy for the people of Kampung Kakeng, it is important to build the capacity of the people first so that they can face the business challenges ahead and make well-informed decisions. This requires the right type of support using participatory methods and a stronger focus on capacity building and training or recruitment of key personnel to strengthen the management of the tourism business and different types of tourism activities. The community should also establish and expand their network with tourism management at other successful sites to foster the exchange of knowledge, best practices and experiences in agrotourism development, which can be facilitated through exposure visits to more successful agrotourism destinations.

Attempts to foster community participation and the active involvement of the younger generation in tourism development in Kampung Kakeng should also be intensified so that they are empowered and are more confident in managing agrotourism. If the younger generation of women and men can take an active role in agrotourism, this might create job opportunities and re-establish the younger generation's ties to their village and traditional culture. This would also reduce rural–urban migration, since the young would no longer have to leave to look for employment in the cities.

Because agrotourism is a relatively new industry for Kampung Kakeng, it may take years before all the efforts to turn the village into a tourism area take root. Therefore, as it is targeting a new market with a new product with subsequently higher risks, this initiative will need strategic investment and support from relevant tourism stakeholders, be they from the private or the government sector, to champion the cause to make Kampung Kakeng a successful agrotourism destination.

References

Akama, J.S. (1996) 'Western environmental values and nature-based tourism in Kenya', *Tourism Management*, vol 17, no 8, pp. 567–574

CBD and UNEP (2007) *Managing Tourism and Biodiversity: User's Manual on the CBD Guidelines on Biodiversity and Tourism Development*, CBD and UNEP, Quebec, Canada; https://www.cbd.int/doc/programmes/tourism/tourism-manual-en.pdf

Ceballos-Lascurain, H. (1996) *Tourism, Ecotourism and Protected Areas*, IUCN – The World Conservation Union, Gland, Switzerland

Halfacree, K. (1993) 'Locality and social representation: Space, discourse and alternative definitions of the rural', *Journal of Rural Studies*, vol 9, no 1, pp. 23–37

Nilsson, P.A. (2002) 'Staying on farms: An ideological background', *Annals of Tourism Research*, vol 29, no 1, pp. 7–24

Veeck, G., Che, D., and Veeck, A. (2006) 'America's changing farmscape: A study of agricultural tourism in Michigan', *Professional Geographer*, vol 58, no 3, pp. 235–248

Weaver, D.B. and Fennell, D.A (1997) 'The vacation farm sector in Saskatchewan: A profile of operators', *Tourism Management*, vol 18, no 6, pp. 357–365

24 Conserving tropical fruit tree diversity by using their products and promoting agrotourism

Lessons from an empowered community in southern Thailand

Montree Issarakraisila, Margaret C. Yoovatana and Songpol Somsri

GPD 'passport'

GPD code:	29
Focus area:	Commercialization and home use Collective action and social networking
Character:	Institutional arrangements and processes incorporating several techniques or methods
Species and varieties involved:	*Garcinia* spp. and *Nephelium* spp.
Name of location:	Kiriwong village, Nakhon Si Thammarat
GIS reference of location(s):	N 08°26′09″; E 99°46′76″ The Kiriwong village is located from 150 to 600 masl altitude at the foot of the 1,835-metre-high Luang Mountain.
Name of farmer (data resource):	Mr Wirat, Mrs Ari, Mr Sontaya, Mrs Gantita Other villagers and members of activity groups

Introduction

Kiriwong is a small village surrounded by the Kao Luang mountain forest landscape of Nakhon Si Thammarat province in the southern region of Thailand, which is known as one of the most important hubs of rich biodiversity in Thailand (Yaimuang *et al.*, 2010). The capital city of Nakhon Si Thammarat is about 780 km south of Bangkok city. The village is known for its very rich natural tropical ecosystem with mixed multispecies tropical fruit tree orchards (Plates 56 and 57). The village economy is based on tropical fruits, mainly mangosteen (*Garcinia mangostana L.*), and the sale of fresh fruits or products made from the plant and tree crops found in their tropical fruit gardens. The village is also well known for tourism, attracting both local and outside visitors to enjoy some of the freshest, cleanest air in Thailand. The tourist activities are organized as 'community-based tourism', a form of ecotourism where the local community has substantial control over, and involvement in, its development and management, and a major proportion of the benefits remain within the community (WWF, 2001).[1]

Kiriwong is located 30 km west of Nakhon Si Thammarat and about 35 km from the ocean. The climate is tropical monsoon with two seasons: the hot season falls from February to April with an average daily temperature in the hottest month of about 29°C, and the rainy season begins from May and ends in January with an average daily temperature in the coolest month of about 26°C. The average yearly rainfall is 2,380 mm with heavy rains occurring from October to December. Soil is mostly sandy loam.

Local context

In 1988, after several days of heavy rains, Kiriwong was devastated by a flash flood that was triggered by the reduced water-retaining capacity of the degraded forests upstream. The flood destroyed houses, temples, orchards, roads and the landscape of the river valley and its feeding streams. This natural calamity prompted the community's concerted effort to protect the fragile and highly diverse local agro-ecosystem of steep mountains covered in natural forests. In order to encourage the participation of all community members in this initiative, the community focused on activities that derive direct livelihood benefits from the indigenous plant species and local resources. The area was already known for its waterfalls, rainforest and beautiful tropical landscape. The farming communities in the area had been growing tropical fruit trees like mangosteen (*Garcinia mangostana L.*), durian (*Durio zibethinus Murr.*), rambutan (*Nephelium lappaceum*) and langsat (*Lansium domesticum*) for several generations, and the subdistrict is known as a major production belt of mangosteen in Thailand. These fruit crops provide the main income source for these communities along with bitter beans called petai from *Parkia speciosa*, a leguminous tree. Both home gardens in the valleys and commercial orchards on the hillsides are traditionally grown as mixed multispecies fruit gardens, called Suan Som Rom by the villagers, which contribute to and benefit from the moist and fertile soil and atmosphere conditions due to the combination of various types and layers of trees and plants (Yaimuang *et al.*, 2010).

Methodology

Kiriwong village is known in Nakhon Si Thammarat for its rich biodiversity and agrotourism activities and was taken up as a site for the GEF UNEP TFT project[2] for the on-farm conservation of tropical fruit tree diversity. Data were gathered through participatory rural appraisal (PRA) and semi-structured key informant interviews and community group meetings held during the project period from 2009–2014. Activities in Kiriwong are organized and implemented through several designated groups. The representatives of four activity groups were invited to explain in more detail about their activities: Mr Sontaya from the herb home group, Mrs Ari from the clothing and natural dye group, Mr Wirat from the agriculture and environment group and Mrs Gantita from the homestay group. Site visits were made to investigate the activities of each group. A detailed study was conducted by a consultant to understand the processes and institutional structure in Kiriwong village; how villagers had organized themselves and developed a range of activities related to the marketing and conservation of local tropical fruit tree species. The village has been an eye opener for many rural development practitioners to see how to facilitate a process of change towards improved incomes and livelihoods by making use of natural resources in a sustainable and environmentally friendly way (Yaimuang *et al.*, 2010). The village particularly shows how the support of local initiatives and activities based on local knowledge and interests facilitated through several semi-independent but collaborating groups can bring about transformative change in the lives of local people. Members of communities participating in the project from India, Indonesia and Malaysia visited the groups and communities in Kiriwong to see what can be possible and to strengthen self-confidence in pursuing similar activities in the commercialization and conservation of local biodiversity.

Description of GPD

Nakhon Si Thammarat province located in the south of Thailand has abundant tropical fruit trees (Table 24.1), and the village has developed a wide range of biodiversity-based products from *Garcinia* and *Nephelium* fruit species, including natural dyes for fabrics, soaps, cosmetics, sweets, juices and health products. Besides the development of various products, the villagers have established several agroecotourism activities such as homestays, cycling or trekking activities and visits to the temple and special natural spots or viewpoints. Tourists often come for half a day to three days, frequently in organized groups, to enjoy the nature, culture, fruits and diversity of Kiriwong village (see Plates 56 and 57). These agroecotourism activities provide additional income and help to bring in customers for the established market outlet that displays the range of processed products that are made by several activity groups. There are now in total nine active activity groups established in the village focusing on specific processing techniques, species or products. They comprise three groups for cloth dyeing, a durian paste group, a herb processing group, a handicraft products group, a group making necklaces from local plants, a money-saving group and a group focusing on environmentally friendly agricultural practices. A range of agroecotourism activities have been developed, from sightseeing to demonstration workshops of the different product groups to see how the products are made. Nowadays, since the establishment of several homestays and small-scale resorts, more tourism activities are provided such as fishing, mountain biking, cultural tours and hiking trails (www.kiriwonggroup.com).

This good practice for diversity management (GPD) can best be described as a process or system where people in the community started to capitalize on their mixed and highly diverse backyards and orchards in this mountainous area for the commercialization of a wide range of products. These activities were organized through several semi-independent and simultaneous operating groups who arrange their own funding and decide their activities based on their own interests. The groups do not work in isolation but collaborate with and strengthen each other in a concerted effort to improve livelihoods while conserving the local highly diverse agro-ecosystem. This practice was initially triggered by environmental concerns after the flash flood to avoid the degradation of the forest and orchards on the hillsides, and was further strengthened with the promotion of agroecotourism by the local government office and several outside donors. It was started more than 20 years ago by a few villagers or so-called 'change agents' and now involves the whole community of the village. Awareness of nature conservation and the active participation of all people in the community were important factors that contributed to the development of this practice and are important drivers for their further development. In the rest of this chapter we outline the activities of four of the activity groups in more detail.

Agriculture and environment group

Mr Wirat strongly believes in environmental conservation and has provided local leadership for social capital building in Kiriwong. In 2001, he organized a group of 30 local women and men farmers and established the agriculture and environment group. The aim of the group was to reduce environmental pollution by using fewer chemicals in agriculture and to follow the idea of a self-sufficient economy. The group received training and shared the knowledge and expertise of members with other villagers about environmentally friendly agricultural practices. The group had financial support from both the Thai government and several non-governmental donors from Thailand and Japan (Nonaka, 1993). One of the first activities, and still the most important, was to set up a nursery in order to produce on a large scale local forest tree seedlings

Table 24.1 Plant species involved and associated activities of the agriculture and environment group

Species or varieties involved	*Canopy structure*	*Activity*
Durian, mangosteen, rambutan, *Garcinia atroviridis*	Tall and large	Seedling propagation, plant and sell saplings to improve mixed cropping orchards, training on production and processing, training on compost usage and seedling propagation.
Lansium domesticum, *Parkia speciosa*	Intermediate	Seedling propagation, plant and sell saplings to improve mixed cropping orchards.
Champak (*Michelia champaca*) and Malacca teak (*Intsia palembanica*), Iron wood (*Hopea odorata*)	Intermediate	Seedling propagation, plant and sell saplings to improve mixed cropping orchards, landscaping and forest restoration; Champak is a valued religious tree in Buddhism.

such as champak (*Michelia champaca*) and Malacca teak (*Intsia palembanica*), and fruit tree seedlings like mangosteen (*Garcinia* spp.) and wild durians (*Durio* spp.) (Table 24.1). The seedlings were distributed to farmers to be planted in the orchards and forests located on the steep hills, which are the origin of rivers and streams, to help improve canopy cover and water-retaining capacity. The traditional practice of growing multiple layers or canopies (tall, intermediate and small perennials) in one area was promoted and demonstrated to maximize light, space and microclimate in tropical environments (Table 24.1; Plates 56 and 57). Other activities developed later included training of other villagers in the production of compost, bioliquid compost and *Trichroderma* (a biocontrol agent),[3] plant propagation training and fruit processing training to promote pollutant-free and environmentally friendly production and processing methods. The group efforts helped to ensure that all surrounding mountains are covered by trees and the area under forest or fruit tree plantation has increased. Some of the group members sell the seedlings every year for additional income, especially those of a popular fast-growing tree, champak (*Michelia champaca*),[4] while other group members focus on producing organic compost for their own use only. Mr Wirat encourages environmentally friendly agriculture through his speeches to farmers and students in nearby schools, which has led to the further promotion and appreciation of traditional mixed cropping orchards.

Tie-dye fabric group[5]

Mrs Ari was one of the founders of a cloth tie-dyeing group in Kiriwong that was established in 1996. The main purpose of the group is to use a traditional dyeing technique to add value to by-products of fruit production and generate

Table 24.2 Plant species which were used as natural colour dyes in natural tie-dyeing group

	Species involved (and part)	*Dye colour*
1	Rambutan (peel)	Grey
2	Mangosteen (leaf)	Orange and pink
3	Jackfruit (bark)	Yellow
4	Tropical almond	Yellowish green
5	*Parkia speciosa*	Grey

extra income for women's groups besides the income derived from their fruit orchards. Table 24.2 shows the number of plant species used for extracting natural dyes. The group received financial support from a private foundation, Komol Keemtong (www.komol.com), to invite an expert from northeastern Thailand to train them in techniques of cloth tying and natural colour dyeing. The expert stayed in the village until the local group had mastered the techniques. Subsequently, the group tested different parts of local fruit trees for suitability for producing natural dyes. By doing this, they developed methods and techniques for dyeing cloth in various shades of different colours. At first, simple products, such as pyjamas and tablecloths, were designed and tailored from these naturally dyed cloths. More products with different designs such as bags, hats, scarfs and shirts were made to expand the products on offer and respond to markets (see www.kiriwonggroup.com/dye) (Plates 58–60).

The number and volume of products have grown steadily since the start 20 years ago, with sales made mostly to visiting tourists. However, for the last four years, growth has stabilized because of the emergence of competing groups and oversaturation of the market. Mrs Ari's tie-dye group is now concentrating on improving the quality of the products to ensure their leading position over other groups. For poorer families, the income obtained can be a substantial contribution to their monthly income. They have been selling products mostly from the village market outlets but are now also exploring sales to distant shops and buyers in Bangkok. As well as the income benefits, the sale of apparel using the traditional dyeing technique has also promoted fruit diversity conservation and the appreciation of traditional Thai culture among tourists and villagers.

'Kiriwong Herbal Home' home processing group

The 'Kiriwong Herbal Home' home processing group was established in 1999 by Mr Sontaya as he was interested in the traditional and medicinal uses of many types of traditional fruits and herbs, particularly mangosteen and its wild relatives (see Table 24.3). Starting with around ten members, this group has grown to about 20 members in 15 years. Each group member produces at least one product, such as herbal soap, natural shampoo and skin balm from

Table 24.3 Plant species that are used as a main ingredient in health products by the herbal home group

Species or varieties involved	*Type of product*
Mangosteen (*Garcinia mangostana*)	Soap, shampoo, conditioner, mangosteen paste, cosmetics, sweets, lip balm
Som Kandarn (*Garcinia atroviridis*)	Soap, shampoo, slimming product (powder), juice, dried sliced fruit (seasoning)
Durian (*Durio zibethinus*)	Durian paste
Herbs and fruits such as lemon grass (*Cymbopogon* spp.), tamarind (*Tamarindus indica*), kaffir lime (*Citrus hystrix*) among many others	Massage oil, lip balm, soap, shampoos

mangosteen, lemon grass and many other herbs and fruit species. Nowadays, more than 20 products are sold and on display in the shop that is located below Mr Sontaya's house in the village (Plates 61–63). Instead of selling low-value fresh fruit with the consequent need to enlarge orchards in the mountainous areas and destroy forests to have more land for cultivation, the home processing group explored how to create high-value products from local herbs and fruit species. It took Mr Sontaya about five years to develop the formula and techniques to use mangosteen peel in combination with a selection of herbs to make a unique herbal soap with proclaimed healthy skin effects because of its anti-bacterial and anti-oxidizing compounds. After that, he learned progressively to improve the packaging and the marketing of the product, mostly by adapting and doing. He now sells the mangosteen soap to local shops as well as star-ranked hotels in Bangkok and even to exporters for overseas markets.

Other health products, such as shampoo and lip balm, are produced from other species such as Som Kandarn (*Garcinia atroviridis*) and kaffir lime (*Citrus hystrix*). All products are promoted on the village website (www.kiriwonggroup.com). Mr Sontaya also demonstrates how to make some of his products to students and visitors as part of his social responsibility. The most highly sold products made by his group are the mangosteen soap and the dried rinds or juice of Som Kandarn fruit, which is popular for its health properties and weight-loss effects (see Chapter 16 for more about *G. atroviridis* production).

Mr Sontaya was the first person in Thailand to develop a soap from mangosteen; hence his nickname Mr Mungkud, which means Mr Mangosteen. In the beginning it was difficult to market the products to retailers and he sold directly to visiting tourists only. In this way he gained confidence in the product, developed his own brand and could finally convince shopkeepers or traders to buy the product although it required substantial initial investment. Now, his shop and demonstration activities are an important part of the attractions for tourists to Kiriwong community (Plates 61–63). Mangosteen soap and other health products sold under the brand 'Mr Mungkud' are among the best rated One Tumbol (Village) One Product (OTOP) products that are promoted by the local government and sold in duty-free shops in Bangkok International Airport. Thanks to these activities and the group purpose, fruit and herb diversity have been conserved and villagers and tourists have regained interest in many of the indigenous herb and fruit species.

Homestay group – 'The Kiriwong Tourism Club'

The Kiriwong Tourism Club was established in 1994 to promote awareness of the natural environment and to manage tourism in the village. Since the flash flood in 1988, visitors had started coming to see the traditional way of life and mixed fruit orchards as well as the landscape in Kiriwong. Visits to tropical fruit orchards during the harvest season, swimming in the river, picnics along the river banks and gardens and sightseeing are the initial tourist activities that mostly attracted local visitors from Nakhon Si Thammarat province. In that period, a few families in the village started to allow visitors to stay overnight in their own houses. This was the beginning of the homestay activities in Kiriwong. With the help of a Bangkok-based NGO, Responsible Ecological Social Tours Project (REST), and government funding, the club developed several community-based tourist projects and activities from 1995 onward, which helped them to market the ecotourism activities among tourist agents and become well-known among domestic tourists outside the province. Kiriwong received the Thailand Tourism Award in 1998 for environment conservation and community self-sufficiency and is considered a model to replicate by the Department of Tourism.

Three types of tourists can be distinguished according to Mrs Gantiti (see Table 24.4). The major and most secure type of visitors are larger groups of government officials, company staff or domestic package tours, who mostly come and visit Kiriwong by bus for half a day to one or two days. They also receive many individual tourists or families from Thailand who often stay for a weekend or a few days. The number of foreign tourists has been increasing also, although this number is still very small. To be able to support all types of tourists, different types of accommodation were required for the specific needs of the different types of guests. At present, there are 48 household members in the tourist club that offer homestays for individuals and small groups of visitors (Plates 64–65). There are an additional ten mini-resorts to accommodate larger groups of visitors. Mrs Gantiti is the coordinator of the tourist club and visitors contact her to book accommodation or to organize activities for them. She provides tours of one day or a few days for groups or individual tourists, mostly to show the history, culture and traditional way of living in the village, or to learn about the various activity and processing groups in the village. Most of these activities support the environment and the conservation of plant diversity directly or indirectly. Information about the homestays and tourism activities can be found on the Internet (www.kiriwonggroup.com), but most information is still in Thai only.

Table 24.4 Types of activity and level of interest in biodiversity and the environment of various types of tourists in Kiriwong village

Type of tourists	*Type of activities*
Individuals, families or groups of local tourists from Nakhon Si Thammarat	See waterfall and go swimming, picnic lunch, exercise, see mixed fruit orchards, visit the activity groups and buy products, participate in religious and cultural events.
	Walking tour at Kiriwong Village – to enjoy walk and eat tropical fruits in the village (e.g. durian, mangosteen, rambutan).
Groups of students, local government officials and private organizations from different parts of Thailand	As above and also often as educational trip to study nature, culture and community management model and experiences.
Foreign tourists (mainly individuals)	To study and enjoy nature, food culture and way of life in Kiriwong. Hire bicycle and bike around villages and orchards and enjoy tropical fruits.

GPD organizational structure and sustainability

In total about 2,500 people live in Kiriwong village, but not all are equally involved in activities. Various groups exist, some officially registered but most of them informal and not registered. With the exception of the savings group, which was established in 1980, most groups were formed in the period 1988 to 2003. Some groups, such as the juice processing group, did not succeed, and have stopped, but those with a strong leadership managed to innovate, adapt and become successful. The finance to establish and conduct group activities came from the members themselves, sometimes supported by grants or loans from local government departments, NGOs or the private sector. External support was most critical during the early years of establishment. Each group has its own independent board to manage their funds and activities. The structure and internal rules of the boards vary depending on the characteristics of a group's activities; usually a leader and a secretary are appointed by mutual agreement among all members. Initially there were no formal links between groups, but they were embedded within the traditional network of close personal relationships within the village, providing mutual support to each other such as requesting an outlet for their products or exchanging their experiences. The groups reached out to the larger community of Kiriwong mainly through events or festivals connected to religious and cultural occasions, which are organized by various activity groups together with the temple schools and local government units. The high ownership of activities by community members and strong but fluid organizational structure contribute to the sustainability of activities as it allows them to adapt easily to new challenges and opportunities.

Impact on diversity and the environment

The benefits that are derived from the mixed cropping/farming system, the development of products from the different fruit tree species by various groups and the promotion of community-based agroecotourism help to maintain the existing diversity in the community as they create both environmental awareness and income opportunities for community members. Community members have also explored enlarging their range of products and activities by using other species and varieties. This practice has helped to conserve unique species like *Garcinia atroviridis*. This species used to be barely known in Kiriwong but has been introduced in mixed cropping orchards, which has led to a substantial increase in the number of trees of this species planted (see Chapters 16, 20, 21 and 25 for more about this species and its uses). Furthermore, with the planting of trees and improved canopy cover on the hillsides, the risk of landslides and flash floods has been reduced. The combined efforts and activities have also reduced the threat of losing globally important biodiversity through conversion of the traditional mixed orchard system into large scale agroforestry estates for rubber, palm oil or the monocropping of mangosteen and durian as has happened in other areas throughout Thailand (Plates 66–69).

Impact on livelihoods

Community members get direct income from the sale of fruits as raw material to the processing groups and through the sales of the higher value products by the various groups. Several community members receive a wage by working for the processing groups and as a shareholder of the group. Finally, members can also earn income by sale of products to the 'Kiriwong local product centre', which is the main outlet of the local products, established in 2005 with financial support from the Department of Industry Promotion, Ministry of Industry, Thailand, and also from the Japan Bank for International Cooperation (JBIC). This outlet is owned by the local government but operated by a private individual. In addition to this they can sell to the numerous small outlets or privately run retail shops in the village. The community also earns income from the agroecotourism activities that provide jobs as guides, drivers or by hosting guests at their homestay. A small portion of the price for every homestay night is donated to a community and environmental welfare fund (Keep Khaoluang Green Fund) to support activities that benefit the whole community and its environment. As a key indicator for the economic and social success of all combined activities, the savings group, which was the first group established in the village, has managed to accumulate over 40 million baht (US$1,300,000) over the last 35 years by the current 2,300 members, which has been used for the various projects and activities.

Conclusion and action plan for scaling up and dissemination

The success of Kiriwong community is attributed to two important factors: first, the self-help (collective action) spirit of villagers through several activity groups; and second, the fact that they have good leaders in the village that functioned as 'change agents' in the whole process. Indicators of empowerment are reflected in how villagers managed to build their asset base of social, personal, natural, financial and physical capitals and made self-directed decisions regarding their livelihoods and environment through several activity groups. In addition, the community groups have managed to link up with external stakeholders such as companies, NGOs and government agencies to leverage funds, expertise and market opportunities for their own benefits.

A strong feeling of ownership and responsibility by the community is the key factor that establishes the balance between the conservation of fruit diversity and natural resources and the improvement of livelihoods. Likewise, strong policy support from the local government to maintain the original ecosystem, and strong linkages with other local agencies and development organizations for community empowerment, are also important factors that have contributed to the success of the good practice, which can be a model for communities elsewhere.

References

Benítez, T., Rincón, A.M., Limón, M.C., and Codón, A.C. (2004) 'Biocontrol mechanisms of Trichoderma strains', *International Microbiology*, vol 7, no 4, pp. 249–260

Nonaka, K. (1993) 'The role of non-government organizations in social development in Thailand', *Journal of National Research Council Thailand*, vol 25, no 1, pp. 1–15

WWF (2001) *Guidelines for Community-based Ecotourism Development*, WWF International, Gland, Switzerland

Yaimuang, S., Wongphan, K., Issarathamnoon, W., Saetang, P., Katanchaleekul, S., Ongla, O., and Boonchaiet, V. (2010) *Handbook of Khiriwong*, Asian Public Intellectuals and Institute of Asian Studies, Chulalongkorn University, Bangkok; http://www.api-fellowships.org/apirp/sites/default/files/kiriwong_handbook.pdf

Notes

1 The terms community-based tourism (CBT) and community-based ecotourism are commonly used to describe the type of tourism that, recognizing the significant social, environmental and economic impacts tourism can have, primarily focuses on tourism's benefits to the local communities.

2 UNEP/GEF funded project on the Conservation and Sustainable Use of Cultivated and Wild Tropical Fruit Diversity: Promoting Sustainable Livelihoods, Food Security and Ecosystem Services.

3 *Trichoderma* strains exert biocontrol against fungal phytopathogens either indirectly, by competing for nutrients and space, modifying the environmental conditions,

or promoting plant growth and plant defensive mechanisms and antibiosis; or directly, by mechanisms such as mycoparasitism (Benítez *et al.*, 2004).

4 In Theravada Buddhism, champak is said to be the tree used for achieving enlightenment, or Bodhi, by the seventeenth Lord Buddha called 'Aththadassi'. It is best known for its strongly fragrant yellow or white flowers. It is, however, primarily cultivated for its timber, and is also used in urban landscaping. Its aril-covered seeds are highly attractive to birds.

5 Tie-dyeing fabric is a very traditional technique created with local imagination and practices to create various designs on cotton fabrics.

25 Value addition of a local food using *Garcinia cowa* leaves through collective action and marketing by a women's group

Samroeng Changprasert, Sombat Tongtao, Chatchanok Noppornphan and Songpol Somsri

GPD 'passport'

GPD code:	28
Focus area:	Commercialization and home use Collective action and social networking
Character:	System
Species and varieties involved:	*Garcinia cowa*
Name of location:	Tambon Khlong Narai, Amphoe Muang, Chanthaburi Province
GIS reference of location(s):	N 12°61′56″; E 102°27′61″ Elevation: 10–15 masl
Name of farmer (data resource):	Mrs Yupa Niyomvanich and various focus group meetings with members

Introduction

Garcinia cowa Roxb. (Guttiferaceae), commonly known as *cowa*, is a lesser-known species that originates from southeast Asia and is found throughout Thailand, where it is known as cha muang. It is one of the 22 *Garcinia* species reported in Thailand, of which mangosteen (*G. mangostana*) is the most well-known (Smitinand, 1980). Cowa is a small- to medium-sized tree whose young leaves and berries are both edible (Yapwattanaphun *et al.*, 2002). It usually grows wild along the margins of forests in various parts of Thailand. This tree grows especially well in coastal areas such as in Chanthaburi, Trat and Rayong Provinces along the east coast of Thailand where the soil is mainly alluvial. Traditionally, it has been used in folk medicine for various purposes (Lim, 2012). For instance, its bark and latex have been used as an antipyretic (anti-fever) and antimicrobial agent (Na Pattalung *et al.*, 1994; Panthong *et al.*, 2006; Ritthiwigrom *et al.*, 2013). The tree produces small, edible fruits that contain hydroxycitric acid (Jena *et al.*, 2002) and are believed to help against fever, stomach ache and constipation. *G. cowa* trees were domesticated in home gardens after farmers discovered their medicinal properties and the potential of their use as a spice or food ingredient in local cuisine at least a hundred years ago. Correspondingly, using the *G. cowa* tree leaves in traditional food recipes has been in practice for about the same length of time. *G. cowa* is now commonly grown in home gardens and orchards by the farmers in Chanthaburi Province. The geography of the region is characterized by short mountain ranges alternating with short river basins that drain into the Gulf of Thailand. The soils in this area are moderately to highly fertile and the climate is characterized by high humidity (72–80 per cent) and warm temperatures (21–35°C). The rainy season begins in May and continues intermittently until the end of October with an average annual rainfall of 2,565 mm (2010–2012). These conditions favour fruit tree production as well as other marketable crops.

Farmers' livelihoods in Chanthaburi province are generally dependent on commercial tropical fruit tree production with occasional additional income derived from non-farm labour. The most popular crops grown commercially are durian (*Durio zibethinus* Murr.) and mangosteen (*Garcinia mangostana* L.), next to less dominant species like rambutan (*Nephelium lappaceum*), salak (*Salacca zalacca*), longan (*Dimocarpus longan*) and langsat (*Lansium domesticum*). During the last three to four decades, Chanthaburi Province has developed from being a region with a traditional farming system with orchards and home gardens combining a wide range of tree species into the major production regions of durian and mangosteen in Thailand. Though monocropping commercial orchards and the number of durian and mangosteen trees have increased sharply, the number of rambutan and mango trees has dwindled. Populations of *G. cowa* have been less affected as their populations have always been small, reflecting its main use as home consumption only. The average farm income in Chanthaburi is about US$300 per month.

Identification of good practice for diversity

The women's group led by Mrs Yupa Niyomvanich was identified as one of the key stakeholders of the UNEP/GEF funded project, 'Conservation and Sustainable Use of Cultivated and Wild Tropical Fruit Diversity: Promoting Sustainable Livelihoods, Food Security and Ecosystem Services' during the initial proposal development phase and baseline survey in 2009. Focus group discussions making use of participatory research tools such as Venn diagrams, Four Cell Analysis (FCA) and Timeline were used to collect information. FCA was used to understand the richness, abundance and trends of local fruit tree diversity. Venn diagrams were used to identify key stakeholders involved in fruit tree production and related value chains. The Timeline was used to document the overview of historical developments that have affected the level of diversity in the area. Mrs Yupa Niyomvanich's process group is participating in the OTOP[1] (One Tambon One Product) programme of the Thai government and is locally well-known for her innovative product *Moo Chamuang*, a traditional Thai dish famous from this region, which she has managed to commercialize and promote among group members.[2] As well as this, the group produces candies or paste made from mangosteen and flakes or chips from durian among several other products made from fruit trees. Several key informant interviews were conducted with her and the group members to document in more detail the process and major driving forces that facilitated the commercialization of *Moo Chamuang* among several other products.

Description of good practice for diversity

The women's processing group was established in 1983 after a major storm damaged the community's durian and mangosteen trees and caused the fruits to fall (Kruijssen *et al.*, 2008). The quality of these fallen or bruised fruits was considered too low to be marketed fresh and therefore some of the women members of the community decided to process the fruits in their homes. In the following years, the women began to counter the low prices of the oversupply in the glut season by using fallen or excess fruits to make paste, candies, flakes and other products to fetch a higher price or to be able to lengthen their shelf life. Assisted by the district-level government, the women established a cooperative, learned to process several kinds of fruits and later acquired a building with processing facilities and a small market outlet (Kruijssen *et al.*, 2008).

In 2004, the Khlong Narai women's group started producing a local food dish named *Moo Chamuang*, a spicy pork curry blended with young leaves of the *G. cowa* tree for sale in local markets. Young leaves of *G. cowa* are traditionally used as a souring spice in Thai cuisine. The Khlong Narai women's group was the first to market this popular dish for which the Chanthaburi region is famous. Mrs Yupa Niyomvanich used her own recipe, which has exquisite taste and quality according to her peers. Plate 70 illustrates the processing steps of this GPD. The spicy pork curry is seasoned with a paste of grilled shallots, galangal rhizomes, dried chillies and crushed *G. cowa* leaves. The leaves add a distinct sourish taste to *Moo Chamuang*, which greatly improves the pork taste, according to local villagers.

Initially the group sold the spicy pork dish directly to consumers at a stand at the local market. Seeing the success, the group decided in 2004 to produce *Moo Chamuang* in a sealed plastic pack to reach out to multiple market outlets, improve shelf life and enable long-distance distribution to a wider group of customers. Seven years later, the women's group began to produce canned *Moo Chamuang* for sale in local and external markets. In the first years they made use of the canning facilities of another processing group, but since 2011 they have used their own canning facility (see Table 25.1). They managed to obtain food quality certification for the product by the Food and Drug Administration of Thailand to guarantee food safety and to attract and strengthen consumer interest.

The value-added products were initially sold to diverse market outlets such as village markets and shop stalls at various exhibitions, fairs and festivals. Lately, the women's group has also been collaborating with the Thailand Post Company to deliver products domestically and globally, thus improving their ability to penetrate more distant markets. Currently, the women's group is able to sell more than 1,200 cans of *Moo Chamuang* per month. In the near future they plan to increase their sales by exploring new market channels, such as exporting products to foreign markets, and expanding production capacity.

Moo Chamuang can be regarded as the most successful product of the Khlong Narai women group. However, before venturing into *Moo Chamuang*, the women's group had explored many other products made from a wide range of fruit trees species as found in their gardens (see Table 25.2). Of this long list of products, at present *Moo Chamuang* generates the most income, followed by mangosteen preserve (paste). Mrs Yupa Niyomvanich explained that they

Table 25.1 Historical timeline of key events of 'Khlong Narai' women's cooperative

Year	*Activity or event*	*Sales turnover of* Moo Chamuang
1983	Group established with the help of the local Extension Department and started producing mangosteen paste, durian sweets and jackfruit flakes among many other products	Establishment phase
1996	Began processing *Moo Chamuang* in sealed plastic bags that were sold in the local market	100–200 packs per month
1997	Entered new market channel by displaying products in city fairs organized by government in Chanthaburi and Bangkok	300–400 packs per month
2003	Obtained equipment to process *Moo Chamuang* in cans with help of professor from Mahidol University in Bangkok and investment grants from government	Started with 600 cans a month
2011	Obtained food safety certification for own canning facility from government	800–1000 cans a month
2014	Started sales through Post Order company. Other women's groups are taking up *Moo Chamuang* as a viable commercial product	1200–1400 cans a month

Table 25.2 Portfolio of products made from fruit tree species by women's groups

Fruit species	*Plant parts*	*Market strategy (Ansoff)*[3]	*Products*
Garcinia cowa (Chamuang)	Young leaves	Diversification	Moo cha muang (pork curry) in cans and in aseptic plastic bag
Durio zibethinus (Durian)	Fruits	Market penetration	Chips, preserves (paste)
Garcinia mangostana (Mangosteen)	Fruits, Peel	Market penetration/ product development	Preserves (paste), candy, juice, cream for skin (lotion), soap
Musa spp. (Banana)	Fruits	Market penetration	Chips, dried whole fruit
Syzygium aquem (Rose apple)	Fruits	Product development	Sweetmeat (preserve)
Nephelium lappaceum (Rambutan)	Fruits	Market penetration	Candy
Citrus madurensis (Som jeed or Som Mapit)	Fruits	Diversification/ product development	Juice

had learned through trial and error. Initially, they focused on mangosteen and durian products, which were easier to make but which receive higher competition, as many other women's groups make these products. By marketing the spicy pork curry, they managed to create a novel product for which little competition existed and which received good demand as it is a well-known dish and recipe that traditionally originates from this region. Now they want to enlarge the market for *Moo Chamuang* by increasing the sales of the spicy pork product to Bangkok. Simultaneously, Mrs Niyomvanich wants to improve some of the other products and explore more novel products such as juice made from a lesser-known Citrus species named som jeed or kumquat (*C. madurensis* Lour) in order to find more products with a good profit margin that can broaden their income base.

Impact on diversity

G. cowa Roxb is a deciduous species and has both male and female trees, which normally results in a genetically diverse population. However, most *G. cowa* trees found in Chanthaburi are female, whose seeds are apomictic, producing plants that are clones of the female parent, in turn limiting the intraspecific genetic diversity of the tree population. However, the creation of a commercial value for *G. cowa* does seem to contribute to interspecific diversity by avoiding and even reversing the loss of this lesser-known semi-wild species. For instance,

FCA analysis between 2010 and 2014 revealed that the number of G. *cowa* increased from 25 to 81 trees in Trok Nong and from 50 to 150 trees in Khlong Narai village. G. *cowa* trees, though initially only found in the wild, are now grown in almost every home garden and orchard in Chanthaburi. The activities by the women's group increased the value of G. *cowa,* resulting in on-farm conservation of this species and increasing the richness in home gardens and orchards by adding another species to the list of their crops. Besides, the start-up of processing activities has enabled the women to explore products for a range of species such as durian, mangosteen, banana, rose apple and rambutan. This helps to trigger interest in maintaining the traditional multispecies orchards and home gardens that did not previously fetch significant income through the sale of fresh fruits and thus were dwindling because of the stronger commercial orientation of farming households on durian and mangosteen only and the growing numbers of monocropping orchards in the province.

Impact on livelihoods

In terms of contribution to livelihood strategies, this practice has created income security for households through the creation of value-added products from traditional food recipes and dishes that were used before only for home consumption. In general, the average income of farmers in this area is around US$300 per month, while the women's group producing *Moo Chamuang* can earn additional income of around US$1,500 per month with the sales of canned (1,200 cans) and packed (600 packs) *Moo Chamuang*. Additionally, they earn about US$260 per month with the sales of mangosteen preserve (300 packs) and US$250 through the sales of mangosteen soap (100 soaps), mangosteen skin cream (50 bottles) and the other products from other fruit species. Around 26 women started to work together to develop the processing activity, which has now been practised for almost 12 years. At present, 30–40 households from Khlong Narai subdistrict are involved and two or three neighbouring women's groups are also in the process of adopting the production of *Moo Chamuang* or similar types of products for the market.

Member households earn additional income through the sale of fresh leaves or fruits to the cooperative and receive salary income for those members who work in the production facility of the women's group. Likewise, local teenagers seeking an employment opportunity collect the fresh leaves or fruits and earn a small wage when free after school. Moreover, local merchants benefit by handling and coordinating the sales of these products. Income generated from canned *Moo Chamuang* or other processed products is more stable and evenly distributed than the sales of fresh fruits. *Moo Chamuang* can be processed all year round, whereas the mangosteen preserve is only processed during June, July and August. This enables community members to diversify and secure their income over the course of the entire year, avoiding sole dependency on fruit sales during the glut season that often brings low and volatile prices.

This practice has also led to the empowerment of the women's group and its members, as the members have been able to make their own decisions as to which products and activities to pursue, have managed to obtain support from government agencies and were able to earn their own income and invest in improved processing facilities such as for canning. In addition, they are proud of their products and activities that have received a lot of attention and won them prizes as best OTOP product. Other women's groups or individuals frequently visit Khlong Narai to learn about the products and operations of the processing group.

Assessment of GPD effect on livelihoods

Effect on livelihood assets

By initiating this local good practice, the women's group members acquired specific skills in the processing and production of canned products such as *Moo Chamuang* and also engaged in other products. They gained insight into the institutional framework of a cooperative, their role as members and shareholders thereof, and the successful management of an enterprising cooperative. It has empowered them to make self-directed decisions regarding their livelihood activities and the use of their own natural, social and financial resources. It has strengthened their linkages and networks with other value chain actors like bankers, traders, retailers, exporters and government departments. Profits made by the cooperative have enabled the women's group to invest in better facilities and improve equipment for canning. The women's group and its members have been able to earn regular income throughout the year with the profits generated from this activity, which has thus provided them with cash outside of the fruit harvest season, when local incomes typically drop. By adding *G. cowa* as a beneficial species to their home gardens and orchards and expanding pre-existing populations, this species will be maintained and secured for future use. This practice provides income to the cooperative through the sales of *Moo Chamuang*, which in the long run contributes to the financial capital of the women in the processing groups, as all group members receive a yearly pay-out based on their share in the cooperative and the profits made.

Driving forces for the success of the GPD

The driving forces for the successful establishment of this activity could be attributed to a push by the Department of Agricultural Extension to establish the women's cooperative to produce mangosteen paste and other products for sale in the first year (Table 25.3). Second, the real driving forces resulted from initiation of the women's cooperative in launching their product of *Moo Chamuang* curry in a simple package that was very well received and accepted in the local market. The increase in demand from the local market drove the women's cooperative to push forward a product development programme.

Table 25.3 Driving forces and conditions favouring or hindering successful functioning of GPD

Good practice and its major components	*Driving forces*	*Conditions favouring success*	*Conditions hindering success*
Organizing women's self-help groups as a cooperative	Financial gains through shares, sale of fresh leaves and employment opportunity to work in processing facility Selection of a good, capable and trustworthy manager	Empowering women's groups through training and support of government organizations Creating financial capital of cooperative by purchase of shares by the members or grants or loans from government or other agencies	Lack of a clear policy support of local government for enhancing community resilience
Value addition of *G. cowa* through canned production of *Moo Chamuang*	Confidence that this recipe would sell to a large market of consumers in Thailand Increase profit margins and income from sales of *Moo Chamuang*	Training of women farmers in food safety and canned processing by government institutions (OTOP programme) Government extension services and subsidies	High investment costs Reliance and dependency on government training and support
Marketing and sales of *Moo Chamuang* and other local products and food culture	Need to find additional ways and channels to market *Moo Chamuang* Find and develop new products based on local diversity and food culture	Guaranteed access to markets for local raw materials and value-added products through diverse outlets, e.g. OTOP and cooperative outlets Favourable conclusions drawn from a preliminary value chain analysis	High competition for similar products Quality maintenance

Finally, Dr Visith Chavisit, an associate professor from Mahidol University, guided the women's cooperative to produce the *Moo Chamuang* dish in a can. Importantly, the women's group is supported by technological help and funding from government offices. First, it established financial capital with the purchase of shares by the members for US$1200. In 2003 the women's cooperative obtained an award of US$73,750 from the central government to buy machinery, build the processing plant and purchase processing equipment for the production of several products. They are highly motivated because of their recognition within the OTOP programme of the Thai government, the establishment of their own cooperative, the creation of jobs for local women and income for shareholders from the profits.

Factors favouring or hindering successful functioning of GPD

Although the local women had the traditional knowledge about how to make this particularly tasty recipe, it was not easy for the founding members to devise a strategy to market *Moo Chamuang* in a form catering to a wide consumer base. The training that was given to them by government programmes such as OTOP helped them to organize themselves. The training on food safety regulations and requirements, simple household-level canning technology and advice regarding the establishment of a cooperative helped the women's group to set up this economically viable enterprise. Later on, financial support from the government together with the successful accumulation of financial capital through profits enabled the cooperative to invest in hardware and an improved production facility. The members stressed that another factor important to their success was the selection of a capable, trustworthy and inspiring manager of the cooperative, Mrs Yupa Niyomvanich, who invented, developed and implemented several product ideas.

Concluding remarks and an action plan for scaling up and dissemination

The principles of this GPD, creating a commercial product from a local popular food dish or product that is made from native fruit species or varieties, can be easily replicated elsewhere. However, mainstreaming this practice may be a challenge, as starting a canning factory, however small, involves substantial initial capital investments that rural communities often lack in part or in whole. Nevertheless, there are other canning techniques that require lower investment costs and may serve as an alternative method for those communities unable to attain initial financial backing.

An action plan for the dissemination of this approach outside the community and beyond Thailand has great potential. Communities from other countries that want to take up this activity can be exposed to this kind of processing activity through exchange visits to the women's cooperative in Khlong Narai. That said, it is essential to first conduct a participatory market chain assessment to evaluate the interest of consumers in other districts and provinces to be able to select which local or traditional fruit species or variety is best suited for manufacturing and what kind of products have the most market potential (see Chapter 22 of this book). The evaluation and selection of the most potentially profitable products and fruit varieties or species should be carried out in conjunction with value chain stakeholders such as traders, exporters and retailers to ensure that market demand is directly taken into account in the value chain analysis process. Stakeholders can also identify which training needs are required from support institutions like the local and national government. As a first step towards establishing this practice, potential and interested communities should develop a business plan to begin raising funds and capital for its implementation.

References

Jena, B.S., Jayaprakasha, G.K., Singh, R.P., and Sakariah, K.K. (2002) 'Organic acids from leaves, fruits, and rinds of *Garcinia cowa*,' *Journal of Agricultural and Food chemistry*, vol 50, no 12, pp. 3431–3434. Available at: http://www.ncbi.nlm.nih.gov/pubmed/12033807, accessed 24 June 2015

Lim, T.K. (2012) *Edible Medicinal and Non-Medicinal Plants*, Springer Netherlands, Dordrecht. Available at: http://www.springerlink.com/index/10.1007/978-94-007-1764-0, accessed 28 April 2015

Na Pattalung, P., Thongtheeraparp, W., Wiriyachitra, P., and Taylor, W.C. (1994) 'Xanthones of *Garcinia cowa*', *Planta Medica*, vol 60, pp. 365–368

Panthong, K., Pongcharoen, W., Phongpaichit, S., and Taylor, W.C. (2006) 'Tetraoxygenated xanthones from the fruits of *Garcinia cowa*', *Phytochemistry*, vol 67, pp. 999–1004

Ritthiwigrom, T., Laphookhieo, S., and Pyne, S.G. (2013). 'Chemical constituents and biological activities of *Garcinia cowa* Roxb', *Maejo International Journal of Science and Technology*, vol 7, no 2, pp. 212–231. Available at: http://www.mijst.mju.ac.th/vol7/212-231.pdf, accessed 24 June 2015

Smitinand, T. (1980) Thai Plant Names (Botanical names-vernacular name), Royal Forest Dept., Bangkok (in Thai)

Yapwattanaphun, C., Subhadrabandhu, S., Sugiura, A., Yonemori, K., and Utsunomiya, N. (2002) 'Utilization of some *Garcinia* species in Thailand', *Acta Horticulturae*, vol 575, pp. 563–570

Notes

1 OTOP stands for 'One Tambon (meaning village) One Product'. It is a rural development and local entrepreneurship stimulus program of the Thai government that aims to create improved rural livelihoods and supports the development and marketing of locally made, unique or traditional products in villages across Thailand.

2 *Kaeng Moo Chamuang* is a Thai curry with a unique sweet and sour taste made from pork belly and a herb called Garcinia leaves or 'Bai Chamuang'. The pork belly is cut into chunks then simmered with curry paste in low heat until it becomes tender. It has an intense but not too spicy flavour, great to eat with warm white rice. http://amazingthaifood.tourismthailand.org/thai-food/thai-regional-foods-eastern.html.

3 The Ansoff matrix is a marketing tool that identifies four alternative types of growth strategy for an enterprise based on new or existing products and for new or existing markets (customers); i.e. market penetration, market development, product development and diversification. See Chapter 22 for a fuller explanation.

26 Value creation for *Garcinia gummi-gutta* and *Garcinia indica* through energy-efficient dryers and product differentiation in the central Western Ghats region of Karnataka, India

Vasudeva R, Narasimha Hegde, B. M. C. Reddy and Bhuwon Sthapit

GPD 'passport'

GPD code:	21
Focus area:	Commercialization and home use Collective action and social networking
Character:	Technique and institutional arrangement
Species and varieties involved:	*Garcinia gummi-gutta* and *G. indica*. Several distinct morphotypes have been identified in each of the species.
Name of location:	The villages of Salkani, Kilara, Menasi and Kallabbe in Sirsi district, Karnataka, India
GIS reference of location(s):	Salkani and Kilara: N 12°61′56″; E 102°27′61″ Menasi and Kallabbe: N 14°42′51″; E 74°40′44″ Elevation of both sites: 546 masl
Name of farmers (data resource):	Mr Girish Bhat, Benagav; Mr Rajesh Bhat, Kallabbe; Mr Manjunath Bhat, Bhairimane; Mr Dattatreya Hegde, Bhairimane; Ms Bhrathi D Hegde, Bhairimane; Mr Eshanna, Amchimani; Ms Lalita V Hegde, Amchimani; and Mr M.B. Nayak, Siddapur

Introduction

Uttara Kannada District, situated in a hotspot of biological diversity, the Western Ghats of India, is one of the largest districts in Karnataka state. It is endowed with rich natural resources. The district has varied geographical features, with thick forests, perennial rivers, abundant flora and fauna and a long coastline of about 140 km. In its 1.025 million ha area, a large amount (0.828 million ha) consists of forest land, and only about 0.12 million ha (roughly amounting to 12 per cent) is under agriculture. The average temperature ranges from 20°C to 33°C. The tropical climate of this region is strongly influenced by the monsoons and moderated by proximity to the sea. During the monsoons, the region receives one of the highest rainfalls in the world. Average annual rainfall in the district is 2,835 mm. However, in the western coastal and crest-line regions it exceeds 4,000 mm. Because of the heavy rainfall, the lateritic soils are rather poor with respect to nutrition. The vegetation in the region is mainly moist deciduous wherein valuable timber wood is found. Evergreen and semi-evergreen formations are fragmented in the crest-line of the Ghats. Deforestation and poaching have been cause for conservation concern in recent years. Arecanut (betelnut) and rice are the main crops in the irrigated region, and cardamom, vanilla and black pepper are also cultivated.

Garcinia is a large genus of the family Clusiaceae (Syn: Guttiferae) that consists of more than 35 genera and more than 800 species, of which about 40 species produce edible fruits, among which G. *mangostana* and G. *indica* are well known. Of the 35 species of *Garcinia* reported in India, seven are endemic to the Western Ghats region and five species are commercially exploited. The rind of G. *gummi-gutta* and G. *indica* fruits is traditionally used by the people of high rainfall tracts of Karnataka and Kerala in culinary preparations as a flavouring agent, and the seeds are also a rich source of an edible fat. The rind of the G. *indica* fruit is used in the preparation of a popular beverage called kokum juice, which is consumed during the summer. Since both the species possess curative properties, especially for stomach and heart ailments, local healers use them in folk medicine. In recent years, G. *gummi-gutta* has become one of the most demanded non-timber forest products (NTFP) species in the region because of its pharmaceutical properties. A chemical (hydroxycitric acid) that has been shown to reduce fat accumulation in humans when consumed (Singh *et al.*, 1995) is extracted from the dried fruit rind. As per the records of Karnataka Forest Department, in Sirsi Forest Division alone more than 2,000 tonnes of dried rinds are produced annually; however, this may be a gross underestimate because of a lack of systematic record keeping (Hegde and Vasudeva, 2010). In recent years, increased commercialization of processed *Garcinia* fruit has pushed this local, subsistence level enterprise into a lucrative business involving international export markets (Vasudeva and Hombe Gowda, 2009).

G. *gummi-gutta* fruits mature during the rainy season. It is essential that the fruit rinds (the economically important part) are processed within three days of collection, otherwise they rot and become useless, causing farmers to lose income. Traditionally, processing is done in makeshift processing units in which the fruit rinds are dried over an open fire in the forest or near or within homes (Plate 71). The rinds are spread over a metal mesh a metre above the flames, which create enough heat to dry the fruit. This processing technique consumes enormous quantities of firewood. Studies have shown that approximately 15–20 kg of firewood is required to obtain 1 kg of dried *Garcinia* rind in the traditional open-fire system (Hegde and Vasudeva, 2010). Firewood for this purpose is usually gathered from the forest where the makeshift processing units are set up. This is a huge hidden cost since it is estimated that more than 46,000 tonnes of fuelwood a year are being used to dry the *Garcinia* rind in Uttara Kannada District alone (Hegde and Vasudeva, 2010).

The study site falls under the high rainfall region surrounded by the forested area. The lifestyle and the culture of the communities are closely associated with the resources of the forest. The communities are involved in the collection of NTFPs such as wild pickle-mango, and the fruits, gums, resins and leaves of various *Garcinia* species. More than 40 per cent of the indigenous communities who traditionally engage in the collection of NTFPs are solely engaged in the harvesting/processing of *G. gummi-gutta* fruit rinds, to be sold to industries, and each of these households earns an average of INR35,000 (US$583) a year solely from such collection. Other species of *Garcinia* such as *G. indica*, *G. morella* and *G. xanthochymous* are equally important to the NTFP harvesters and contribute significantly to their income. These species are not yet completely domesticated and are still mostly found only in forests. Unsustainable harvesting is common and is causing rapid erosion of valuable types.

Methodology

A group of scientists and students from the College of Forestry, Sirsi, interacted with two different groups of progressive farmers and nursery owners to document their good practices in managing mango and *Garcinia* species. These progressive farmers were identified following baseline data gathered from more than 500 households in the focal communities. Semi-structured interviews were conducted, focusing on all four aspects of a good practice for diversity (GPD) management of tropical fruit tree genetic resources (propagation and nursery management, production management, linking farmers with markets and consolidating community roles). Semi-structured questionnaires were combined with unstructured interviews to elucidate the responses. Detailed guidelines for identifying and documenting GPD stories that were circulated by Bioversity International were followed to document the good practices (Sthapit *et al.*, 2008; Chapter 2). The criteria used for identifying good practices included: (1) diversity of target species; (2) sustainability of the practice; (3) contribution to improved livelihoods; (4) impacts on diversity; (5) potential for scaling up; (6) addressing at least one focus area; and (7) applicability to more than one site. Wherever possible, pictures of the practice, method and other details were documented. Local women's self-help groups (SHGs) and youth clubs were involved in the data-gathering process. Women harvesters were interviewed separately to get additional inputs. Each GPD was described using descriptors mentioned in Sthapit *et al.* (2008) and its relevance to the enhancement of livelihood assets was noted for further evaluation.

Description of good practice

The GPD described in this chapter can be categorized as an institutional arrangement with two components: (a) introducing energy-efficient dryers to process *Garcinia gummi-gutta* and *G. indica* fruit rind (see Plate 72); and (b) diversification and value addition of *G. indica* and *G. gummi-gutta* through the commercialization of traditional uses.

Energy-efficient dryers

The introduction of an improved, ecologically designed dryer has reduced the use of firewood. The improved dryers provide the farmers with good-quality processed fruit rinds that fetch a higher price and reduce the risk of household fire hazards and ill effects following smoke inhalation. Most importantly, valuable forest resources in terms of fuelwood are conserved and the emission of greenhouse gases such as CO_2 is reduced. A modern fuelwood-saving dryer designed by the Centre for Sustainable Technologies[1] uses only 4 kg of fuelwood to obtain 1 kg of dried rind, whereas the traditional open-fire system requires about 15 kg of fuelwood to obtain 1 kg of dried rind. Initial trials by LIFE Trust, a local NGO, have validated these results in practice. The scope to reduce the ecological cost of producing dried rinds by scaling up this practice is huge. The GPD essentially involves setting up fuel-efficient dryers on a larger scale to allow the sustainable commercialization of the traditional uses associated with these species. LIFE Trust has helped establish these dryers in at least 12 villages. Karur Village Forest Committee (VFC), which is located in the *Garcinia*-rich forest, has at least three such dryers. In different villages, dryers are owned by individuals, collectively by a group or by the VFC. This good practice represents an integration of modern technology with the basic wisdom of an age-old practice.

Commercialization of traditional uses

The two major value addition processes are the extraction of edible oil from the kernels of *G. gummi-gutta and G. indica* and the powdering of *G. indica* rind.

Butter extracted from the seeds of *G. indica* is good for the skin and is also used as cooking oil. Both varieties of *Garcinia* produce edible fat in their kernels. The fatty acid profile of the oil is similar to oils used in the cosmetic industry. However, so far the seeds of neither species are considered economically important. Butter is extracted from the seeds in limited quantities as a home treatment for sensitive or dry skin and its commercial potential is neglected. Setting up small-scale extraction units at community level could help to commercialize this traditional use and develop additional products like herbal soaps or lip balm from the oil. The traditional method of extracting the rich edible oil from *Garcinia* seeds is difficult. Establishing community oil extraction units encourages the use of *Garcinia* seeds, which otherwise go to waste, and allows farmers to diversify products that can be prepared from this edible oil.

Similarly, powdering the *G. indica* fruit rind makes it easy to use, store, transport and market for diversified uses, such as a spice or an ingredient for kokum juice. Hence powdering *G. indica* fruit rind is considered a GPD.

Impact on diversity

Several farmers and *Garcinia* collectors have opined that collection of fuelwood is quite difficult as it coincides with the rainy season, which has an average of 25 rainy days a month. Because of this constraint, farmers hesitate to undertake the cultivation of these *Garcinia* species. Use of improved dryers has made the processing in terms of fuelwood more efficient, both in terms of labour time and the drying process itself. As a result, many farmers are now interested in cultivating *Garcinia* species.

With time, once the benefits of good-quality products start to become known, there could be increased interest in conserving *Garcinia*, which would help both inter- and intraspecific diversity conservation. With increased use and preparation of different *Garcinia* products, farmers may be encouraged to maintain or start cultivating *Garcinia* trees in their home gardens or orchards for specific purposes. For instance, there may be *G. indica* or *G. gummi-gutta* types that yield higher numbers of seeds or fruits and a greater quantity of edible fat. Such types could be specifically cultivated by farmers. This may help increase the inter- and intraspecific variations on farm (Vasudeva and Sthapit, 2013).

Impact on livelihoods

Use of these wood-efficient dryers has started to slowly improve the livelihoods of rural poor who are dependent on the collection of *Garcinia* fruits through reduced production costs of *Garcinia* rinds while the quality of the dried fruit rinds has simultaneously increased. 'I could directly sell my rinds for a 20 per cent higher price compared with the rinds dried using traditional methods,' says Mr Manjunath Bhat. Because less fuelwood is required to dry the fruit rinds, it takes farmers less time to collect fuelwood and to process the rinds. Wastage in the form of unevenly dried and totally burnt rinds is avoided, thus improving product yield by about 5 per cent and contributing to its higher quality. Further plans for subsidizing more dryers for communities are being made. Dryers are further improved by innovative designs that are more energy-efficient, user-friendly and made of local, easily available construction materials. By increasing the profits for poor households and teaching them about sustainable harvesting practices, rural communities will have more interest in maintaining their valuable forests and fruit trees. Because these improved dryers do not introduce smoke inside the house, illnesses such as asthma caused by smoke inhalation would decrease drastically.

The diversified products from *Garcinia* oil and dried *Garcinia* powder could address demand in wider markets and create additional income. The sales of final products to specific markets and the creation of several income-generating activities will create regular and more secure income with higher margins for the farming households.

Impact on sustainability

Until quite recently, farmers have had no other alternative than the open-fire method, but the recently introduced simple energy-efficient dryer can efficiently process and dry *G. gummi-gutta* and *G. indica* fruit rinds. The importance of the GPD in improving the quality of the product and its versatility in utility must be highlighted. The improved dryer could be effectively used for drying an array of agricultural products such as banana, coconut, cardamom, betelnut and products derived from jackfruit, turmeric and nutmeg that are a part of the agro-forestry systems. Hence the GPD contributes directly to the sustainability of agroforestry systems (Vasudeva *et al.*, 2013).

Assessment of GPD

Effect on livelihood assets

Through processing activities and through diversifying the product range of *Garcinia*, women's groups could earn regular income throughout the year, providing them with regular cash income. This may strengthen their linkages and networks with other value chain stakeholders and service providers such as banks, traders and co-operatives. It also demonstrates to communities different ways of improving traditional methods of processing NTFPs with some external inputs. These activities help to build up a social network that has a direct interest in the conservation and sustainable management of natural resources. The skill levels of women's groups in preparing various products would increase.

Opportunities to produce diversified products increase the product portfolio from which a family can derive cash income and increase the profit margins that families receive from their labour and natural resources. Moreover, they create income on a more regular basis throughout the whole year. Hence this practice reduces economic vulnerability. As several families in this region face difficulties in the repayment of agriculture-related loans due to fluctuating harvests and prices, these additional livelihood activities could create a better economic position for obtaining loans. Because the improved dryer has an overarching influence on environmental stability by reducing fuelwood consumption, it contributes tremendously to the reduction of ecosystem vulnerability when up-scaled in substantial numbers. Furthermore, agricultural wastes such as coconut shells, coconut husks, dried areca leaves and paddy straw, as well as other wastes could be easily used as fuel in the improved dryer without the problem of smoke. Hence this practice reduces the risk of health problems as well as fire hazards, which is obvious when compared with the open-fire method.

Driving forces for the success of the GPD

The major driving force that contributes to the success of this practice is the apparent efficiency of these improved dryers and the reduction in the drudgery of fuelwood collection. Besides, a sense of reducing environmental degradation among communities also contributes to the success. Improved cooking stoves that adopt a similar principle to that of the improved dryers have been popular among communities for the last two decades. This has also contributed to the success.

For the successful functioning of the GPD it is essential that household level dryers are designed and established. Today these dryers are a little bulky and thus need to be established at common community places. They also require a continuous and large volume of fruit rinds each season. Because the initial cost is on the high side (about INR50,000 or about US$833 for each dryer), there is a need to provide more subsidies to individual farmers. Similarly, for the extraction of edible butter from the seeds, smaller extraction units need to be designed that can be used at household level. For a successful functioning of the GPD, good coordination of several different production groups and committed collaboration from a trustworthy NGO partner is required.

Conclusion and action plan for scaling up and dissemination

The importance of energy-efficient dryers is well appreciated in the locations where the rainy season coincides with harvesting time. Popularizing and creating awareness of the usefulness of these dryers to users, policymakers and other donors is necessary so that there is more demand for dryers that can be installed for local community organizations. Scaling up of improved dryer usage can be effectively carried out if officers or policymakers of many departments such as departments of forests and horticulture as well as national banks are informed of the importance and urgency of this practice.

References

Hegde, N.R. and Vasudeva R (2010) 'Ecological cost of drying uppage (*Garcinia gummi-gutta*) fruits in the Western Ghats', in Vasudeva R, B.S. Janagoudar, B.M.C. Reddy,

B. Sthapit, and H.P. Singh (eds), *Garcinia Genetic Resources: Linking Diversity, Livelihood and Management,* Proceedings of National Symposium on Garcinia, College of Forestry, Sirsi, India, pp. 1–188

Singh, R.P., Jayaprakash, G.K., and Sarkaria, K.K. (1995) 'Hydroxy citric acid from *Garcinia cambogia', Biological Memoirs (CFTRI),* vol 21, pp. 27–33

Sthapit, B.R., Subedi, A., Shrestha, P., Shrestha, P., and Upadhyay, M. (2008) 'Practices supporting community management of farmers' varieties', in M.H. Thijssen, Z. Bishaw, A. Beshir, and W.S. de Boef (eds), *Farmers, Seeds and Varieties: Supporting Informal Seed Supply in Ethiopia,* Wageningen International, Wageningen, The Netherlands, pp. 166–171

Vasudeva R and Hombe Gowda, H.C. (2009) 'Domestication of *Garcinia gummi-gutta* (L.) Robson: An Important non-timber forest product yielding species of the Central Western Ghats', in R. Uma Shaanker, A.J. Hiremath, G.C. Joseph, and N.D. Rai (eds), *Non-Timber Forest Products: Conservation, Management and Policy in the Tropics,* Ashoka Trust for Research in Ecology and the Environment, Bangalore, pp. 65–77

Vasudeva R, Hegde, N., and Sthapit, B. (2013) 'Promoting community management of *Garcinia* resources in the Central Western Ghats, India: Initiatives of Tropical Fruits Diversity Project', in W.S. de Boef, A. Subedi, N. Peroni, and M.H. Thijssen (eds), *Farmers, Biodiversity and Resilience: Empowering Communities to Manage Their Livelihoods,* Centre for Development innovation, Wageningen, The Netherlands, pp. 163–169

Vasudeva R and Sthapit, B. (2013) 'Diversity fair of *Garcinia* and aromatic pickle mango: Neglected and under-utilized tree species of the Western Ghats with high potential to improve livelihood', APO Newsletter

Note

1 Centre for Sustainable Technologies, Indian Institute of Science, Bengaluru, India.

Case studies

Working with communities and multi-stakeholder partners

27 Community forests

Utilization and informal regulation for tropical fruit tree conservation

Adhitya Marendra Kiloes, Kuntoro Boga Andri, Achmad Rafieq, M. Winarno, Idha Widi Arsanti and Zahirotul Hikmah Hassan

GPD 'passport'

GPD code:	Combined 31 and 32
Focus area:	Collective action and social networking
Character:	Community forestry; combination of system, method, technique and institutional arrangement
Species and varieties involved:	Six species of mango: *M. casturi*, *M. applanata*, *M. foetida*, *M. odorata*, *M. caesia*, *M. laurina* and 11 species/varieties with names in Bahasa – Hampalam Kalambuai, Hampalam Tapah, Hampalam Pisang, Hampalam Nagara, Asam Pauh, Tandui Masam, Mangga Hambuku, Limus, Rawa-Rawa, Rawa-Rawa Humbut and Hampalam Biasa – have been managed by the community in Telaga Langsat, South Kalimantan. About six, mostly local, varieties, especially of *M. indica* (Podang Urang, Podang Lumut, Arumanis, Gadung, Madu and Golek) have been grown in the state forest land in Kediri, East Java.
Name of location(s):	Telaga Langsat, South Kalimantan and Kediri, East Java
GIS reference of location(s):	Telaga Langsat, South Kalimantan: N 02°45′05″; E 115°22′06″. Elevation: 148 masl Kediri, East Java: N 06°48′47″; E 107°36′52″. Elevation: 500 masl
Name of farmer (data resource):	Mr Nahnuddin, Achmad Ridhani (Telaga Langsat), Mr Jemu, Mustari (Kediri)

Introduction

Indonesia is considered a megadiverse country with most biodiversity found in its low-land forests (Mittermeier *et al.*, 1999; Myers *et al.*, 2000; McCarthy, 2002; Persoon and van Weerd, 2006). The country harbours many types of native fruit species of which the majority are still found wild in the forests. Estimates for native fruit species vary greatly: Uji (2007) reported 329 indigenous species of fruit trees in Indonesia, whereas the Indonesian Ministry of Environment quoted a figure of 450 fruit species. The fruit from tropical trees is nutritionally dense, culturally important and its sale contributes to family food and nutrition of local people (Kiloes, 2014). Indonesia's forests have an important role to play in facilitating economic growth through the provision of high-quality timber and other forest products such as benzoin[1] resin and rattan (Michon, 2005; Garcia-Fernandez and Casado, 2005) and rubber (Tata *et al.*, 2008). In addition, they provide ecosystem services such as facilitating a conducive habitat for pollinators, preventing water run-off and soil erosion, and retaining moisture or shade in the local ecosystem.

The communities that live in tropical forest areas and rely on forest resources for food, fruit, medicines, housing and work are often isolated, with small populations and little formal education or social connections with political powers. Communities have little say in what happens to them and their forests. Forest-dependent communities are extremely vulnerable to changes that happen to them and their systems imposed by external agencies. If their tropical forests are threatened, communities struggle, transform or disappear completely. With them is lost their extensive ecological knowledge.

In Indonesia, human activities such as palm oil production, rubber, timber and mining have already led to large-scale deforestation, soil degradation and massive forest fires. Short-sighted economic planning and inappropriate land use have led to severe ecological degradation and acute environmental and economic problems (Supriatna, 2010). FAO (2010) estimated that forest cover in Indonesia declined by 24.1 million hectares between 1990 and 2010 due to mining and plantation crops, and estimates the remaining permanent forest cover at 114 million hectares. Even in this dreadful scenario, certain indigenous communities continue to undertake a significant number of activities that can best be described as sustainable management of natural biodiversity in communal forests.

Community forestry

In recent years, the inability of the state to control forest degradation has been recognized in many countries. Governments have seen the benefits of handing over forest areas to local communities under a variety of community forest management schemes in many developing countries (Murdiyarso and Skutsch, 2006). White and Martin (2002) estimate that around 14 per cent of all forest in developing countries is under this kind of management, and is most likely to be increased up to 25 per cent (Bluffstone *et al.*, 2012). Under such schemes, villagers get the formal, legal rights to use and profit from the forest products, under jointly agreed management plans that ensure that off-take is kept at sustainable levels. Communities organize themselves by setting by-laws and by self-regulation as regards access to forest products. Their motivations to take part in such a scheme can be various: to maintain the forest to ensure future benefits is a clear often cited reason. For some, it is to ensure a continued supply of firewood and fodder; for others, to enable eco-tourism; yet others participate in the hope that the wild animals that have disappeared from the shrinking habitat will return and provide a means of sustainable subsistence in the future. In a few cases, sustainable timber off-take is the aim (Murdiyarso and Skutsch, 2006). Such initiatives to recognize the rights of local communities are often defined as social forestry, community-based forest management or joint forest management.

In general, Indonesian forest management needs urgent improvement as many of the concessions do not have clearly demarcated boundaries, and forest fires, illegal land clearance and shifting cultivation are widespread. Forest management in Indonesia is governed by two laws, The Forestry Law – 1967 and a new Forestry Law – 1999, which empower the Indonesian Forest Corporation (Perum Perhutani) to manage all forests on public lands in Indonesia and can grant the right of exploitation or extraction to concessionaires. Concessionaires are often large-scale private sector companies, but the 1999 legislation also allows for a wider range of concessionaires including smaller ones. Neither law has specific legislation regarding community-based forest management (CBFM) or the rights of indigenous communities over forests (Blaser *et al.*, 2011), but this does not mean efforts have not been made by the Indonesian government to involve communities in forest management. CBFM emerged in Indonesia in the early 1980s. In 1985, Perum Perhutani began implementing 13 social forestry projects on public lands in Java (Perum Perhutani, 1996), the so-called PMDH programme (Forest Village Community Development Programme), which was extended in 2003.

The sustainable use of forests by communities based on informal community regulations or formal agreements with Perhutani constitutes a good practice for diversity management (GPD), as it helps the local people to use community forest for their livelihoods while also conserving genetic diversity of tropical fruits. This chapter will discuss two cases of forest management by communities, in Kediri in East Java and in Telaga Langsat in Kalimantan, which help to secure tropical fruit tree diversity.

Description of GPD

The practice of community forestry in Indonesia is considered a good practice for diversity management (GPD) as it is a combination of a system, organization or process, which over time and space maintains, enhances and creates tree genetic diversity and ensures its availability to and from farmers and other actors for improved livelihoods on a sustainable basis (Chapter 1). Under the auspices of the UNEP-GEF project 'Conservation and Sustainable Use of Wild and Cultivated Tropical Fruit Diversity: Promoting Sustainable Livelihoods, Food Security, and Ecosystem Services', an assessment was carried out by the project teams of East Java and South Kalimantan.

Methods of identification

Preliminary participatory rural appraisal was conducted, followed by focus group discussions with key male and female farmers, local extension officials and local state forestry service staff, to better understand how local people are involved in decision making of community forestry, local rules and customary rights and benefit sharing within the communities. The team interacted with local people and government agencies to facilitate collaborative learning and constant self-reflection during field visits over two years. GPD guidelines were also used to identify good practices that manage wild and cultivated tropical fruit tree diversity through community rules and traditions.

Forest management agreement between Perhutani and community in Kediri

Tiron is one of the villages in Banyakan subdistrict in Kediri district, East Java, where the community has rich diversity of mango species in their home gardens. A baseline survey revealed that there were 26 types of mango belonging to many species in Tiron (Kiloes *et al.*, 2014). The site is a dry lowland area about 500 m above sea level, spread out on the foothills in a state forest area, in the central part of East Java province. The majority of households own 0.25–0.5 ha of farmland, with an average annual income from tropical fruits of Rp.1,500,000 (US$150) per farmer family. The richness (i.e. the total number of different genotypes present in the area) and evenness (i.e. equity in the frequency of genotypes or alleles in the area) of the diversity are 26 and 0.74, respectively. Most of the mango trees are old, often more than 50 years, with an average yield of 150–300 kg per tree per year. Mango contributes 15–20 per cent of family income.

Perhutani is managing about 1,500 ha of forest land in Tiron village, which is classified as production forest, of which 500 ha is jointly managed with the community. A management contract between Perhutani and the community was developed in 2007, and a forest village community organization, Lembaga Masyarakat Desa Hutan (LMDH) Tiron Lestari was established. The LMDH is a Forest Village Community Development Programme developed by Perhutani, and is also a cooperative between the community members to increase their social capital.

The land managed by Perhutani is mainly planted with timber trees such as teak, mahogany and sengon. The management contract stipulates that the community has the right to plant the forest land with fruit trees or timber species with mango as the major component and agricultural crops, as long as a canopy cover of 70 per cent is maintained. At first, because the forest area is hilly, a terrace system was developed and over the last decade many mango trees have been planted. The lower terraces are planted with mango and secondary crops, the mid terraces with maize and long bean among other crops, whereas the upper terraces are planted with perennial forest crops such as teak. About six, mostly local, varieties of *M. indica* (Podang Urang, Podang Lumut, Arumanis, Gadung, Madu and Golek) have been planted by the community in the forest. Chemical fertilizers or pesticides are rarely used for the trees and crops, only manure or compost is applied. A profit share of 50:50 has been agreed by the community and Perhutani. Half of what the community earns goes to the farmer who manages the crop and the land and the other half is directed into a fund managed by the farmers' group. The fund managed by the group is used for development activities, such as saving and loans for group members, developing nurseries, developing processing methods for products and marketing the mangoes.

Since the start of the contract in 2007, an estimated 10,000 trees have been planted and canopy cover has increased. Besides the fruit trees, annual crops cultivated in the community forest have contributed substantially to the income of many households, which can be up to 30 per cent for land-poor households. The LMDH also gives importance to improving the community's capability for better livelihoods. This arrangement preserves the diversity of local mango in Tiron, as many mango seedling trees have been planted, including six varieties and 7,000 saplings, and also other noncommercial varieties that are mainly planted for a hobby and pride in their mango diversity richness.

To face any future strategic changes, this arrangement is renewed every five years by holding a formal meeting between the community and Perhutani. This meeting provides the community with an opportunity to negotiate an increase in share, voice any problems faced and tell success stories.

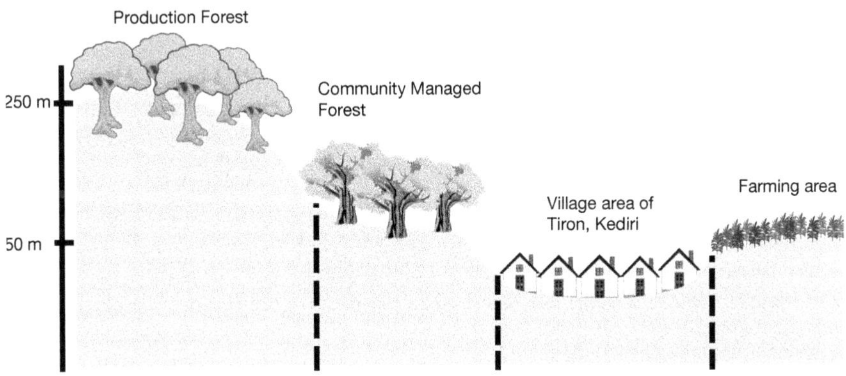

Figure 27.1 Illustration of multi-level landscape in Tiron, Kediri.

Mango diversity maintained by a forest community in Telaga Langsat

Telaga Langsat in Hulu Sungai Selatan District represents a dryland area. The site is located in the central part of South Kalimantan province, at 148 masl. The dry season is from April to September and the rainy season is October to March. It covers an area of 5,808 ha, of which 38.5 per cent is forested, while 26.1 per cent is arable and the rest is housing, village roads and other social facilities. The farming system is irrigated rice-based farming. The total number of households in Telaga Langsat is 2,084 with a total population of 8,780. The average total farm size per household is 0.62 ha with the average home garden 0.06 ha. Every household has at least one mango tree, one to two varieties per household in home gardens and one to three varieties in orchards. The richness and evenness of the mango diversity are 22 and 0.71, respectively. There are 18 types of mango belonging to different species, such as Kasturi, Palipisan, Kuini, Binjai, Hampalam, Hampalam Negara, Limus, Tandui, Rawa-Rawa, Asam Pauh, Hambawang Pulasan, Apel, Golek and Gadung (Kiloes *et al.*, 2014). The major source of income is farming, with main crops being paddy, rubber and vegetables. The average annual income per household is Rp.17,270,500 (US$1,727), of which mango trees contribute approximately 0.6 per cent (Daroini *et al.*, 2013).

The traditional system consists of mainly two elements. This first one is intensive management of tree stands in the buffer zones by allowing preferred trees to prosper in the forest and occasionally cutting or removing undesired trees. The second element is the recognition of 'sacred trees' in the village and surrounding forests that should not be harvested, disturbed or felled.

Managed wild populations

For generations farmers have been planting preferred species or allowing their seeds to grow in the forest buffer zones or community forest surrounding the village, creating a high density of fruit trees, which are harvested based on traditional harvesting rights. Due to this practice in the forest fringes and buffer zones, one will mostly find a higher density of Kuini, Hampalam, Binjai, Asam Pauh and Kasturi compared with the natural forest further away. Harvesting rights are organized as follows. The communities have organized cooperatives to sell the fruit during the fruiting season. The proceeds of the sale are divided among the community members. Any community member passing under a tree can take the fruit and sell it, but they cannot bring a pickup truck or any other form of transportation to carry the harvested fruit to sell it themselves. This practice avoids overharvesting and promotes equality among members of the community.

Sacred trees

The people in Telaga Langsat, South Kalimantan believe that old and large trees are sacred and should not be felled. This belief system has existed in the community for a long time and has been passed down through generations. Mostly this applies to the very large, old trees that play an important role in the forest ecosystem. This study identified at least 20 such trees. Large, old trees of Kasturi, Kuini, Rawa-Rawa, Tandui and some others are some of the sacred trees maintained by the community.

These two elements have been practised for generations and obeyed by the community members. These practices and belief systems help maintain the local ecosystem including the mango tree diversity and have been reinforced by specific informal regulations developed about a decade ago by the elders of the community. The implementation of this informal regulation is controlled by the village security group that was established when the informal regulation was implemented. The fine for breaking these rules is Rp.1,000,000 per tree (about US$100). This fine was agreed by the community members and the money accumulated is used for community nursery development to replant the felled tree. Although the price of felled trees is higher than the fine amount, this practice helps in reforestation through the distribution of new saplings. The regulation is valid both for the mango trees in the village and those in forest buffer zones.

The ownership of the mango trees in the forest buffer zones is determined through an informal agreement between the community leaders and the forest services. Community members have to take care of the mango trees based on the guidance provided by the local agricultural and forestry staff. During the harvest season, they may harvest and sell mango fruits to the local market or to other markets with the assistance of extension staff. As per the informal agreement with the forestry services, the community can have the entire production of the trees, as most of them currently have low commercial value.

The implementation of informal regulations in the community has several positive impacts such as: increased income for low-income community, increased physical security, better food security and health, sustainable natural resource management and participation in the cultural heritage (Kaimowitz, 2003). The practices and benefits derived by the communities in Telaga Langsat attest to this. In fact, in practice this is a traditional spiritual belief (especially regarding the sanctity of trees) that has been converted into a community agreement or informal regulation by the local people.

Driving forces for the success of the GPD

The need to ensure a better life for community members with the available resources, the fear of damage to their environment and livelihood by cutting down trees, profits obtained from using the forest buffer area and additional income and support from the local institutions are the major driving forces for the success of the GPD. A current national programme to grow one billion trees and a local government programme to develop rural small–medium processing units for agricultural crops can support the better future community management of the state buffer forest towards sustainable *in situ* and on-farm conservation of TFT genetic resources.

On the other hand, factors hindering success are the demand for land for rubber and oil palm plantings, an increase in the price of mango main-stem wood for construction and a decrease in spiritual beliefs of younger generations.

Impact on diversity

The practice in Tiron, Kediri, helps considerably to maintain intraspecific diversity of *M. indica*. The six commercial and other non-commercial varieties grown on the lower terraces currently have moderate commercial value both for fresh and processed products in the local market. Increasing demand for fresh and processed products of different varieties in the regional market motivates the community to grow more varieties and trees that in turn can help maintain some of the diversity that may otherwise be lost.

The practice in Telaga Langsat, South Kalimantan, maintains the existing interspecific and intraspecific diversity, even though the mango types they grow currently have low commercial value. The informal village policy not to cut down large, old trees in and around the village and in the buffer zone is one manifestation of consolidating the community role in management of tropical fruit tree genetic resources.

Impact on livelihoods

Farmers with small farms can apply to the village leader to join the community group that manages the state forest in Tiron, Kediri, for growing mango. As a member of the group, farmers can increase their income through mango cultivation. Indirectly, group membership also enhances human capital in Tiron. Farmers gain capacity in managing forest land to cultivate mango for improving their livelihood. As part of the produce is used for home consumption, it can directly improve the nutrition of their household. In Telaga Langsat, the income benefits have been limited. Most of the noncommercial mangoes harvested from species and varieties in Telaga Langsat are marketed locally to obtain additional income. From the profit share the community has started to establish group marketing and processing of local mango fruits and forest produce to add value.

Both arrangements and practices build upon and strengthen the social capital of the community, through traditional beliefs as in Telaga Langsat or through profit-sharing arrangements as a group with the government in Tiron, Kediri. Social capital is a fundamental requirement when pursuing the conservation of a common or public good such as fruit tree diversity, to be able to share the costs of conservation and to avoid overharvesting or destruction by free rider attitudes. Local or national government along with interested nongovernmental agencies can provide policy to assert community resource management.

Future prospects and sustainability of the GPD

One major achievement could be made if specific legislation regarding community-based forest management were developed, which provides forest communities with ownership and harvesting rights and strengthens and recognizes their commitment and involvement in forest management. For the time being, we should encourage the practices as described in this chapter as a practical way forward to create better representation of forest communities in forest management. Incorporating the conservation dimension and ensuring a strong focus on tropical fruit tree diversity in the Forest Village Community Development Programme would be a major incentive for local communities to safeguard their unique mango diversity.

To support the use of state forest and community forests, some technologies that have been developed by the Indonesian Agency for Agricultural Research and Development, such as propagation by grafting and marcotting, are instrumental to have a supply of saplings to be planted in the forest area. Also, development of processing technologies that add value to the mango diversity can be used to support the GPD.

Community participation in meetings of the local government and local forestry services may be promoted to discuss and develop effective ways for improved use of state forests and to further refine how to cooperatively manage forests by the community and the forestry services with an increased share of profits going to the local communities. This can have a positive impact on maintaining TFT diversity (and diversity of other species as well, depending on the context) and also improve livelihoods and environmental services in the sites. Increasing the benefits gained from local TFT resources can further empower local communities and decrease their dependence on support from public and other agencies. To support better use of available resources, conservation efforts of fruit trees can be usefully combined with training in various processing techniques describing the process step by step, including photos and details of the equipment required. Farmer-to-farmer training or Farmer Field Schools (FFS) on community forest management can also be formulated. Exchange visits among farmers in the communities to other locations where such activities have been successfully implemented is a good way to further expand the familiarization and use of this kind of GPD. They can learn about what other villages have done to improve their skills, some of which may be relevant and feasible to try out in their own community.

References

Blaser, J., Sarre, A., Poore, D., and Johnson, S. (2011) *Status of Tropical Forest Management 2011*, ITTO Technical Series No 38, International Tropical Timber Organization, Yokohama, Japan

Bluffstone, R., Robinson, E., and Guthiga, P. (2012) 'REDD+ and community controlled forest in low income countries: Any hope for a linkage?', Environment for Development (EfD), Sweden and Resources for the Future, USA, Discussion Paper Series, October 2012, EfD DP 12-11, University of Gothenburg, Sweden

Daroini, P.B., Dewi, H.A., Andri, K.B., Rafieq, A., Pribadi, Y., Fakhrina, Saderi, D.I., Lesmayati, S., and Lamers, H.A.H. (2013) *Baseline Report Indonesia for the UNEP-GEF Project Entitled: Conservation and Sustainable Use of Cultivated and Wild Tropical Fruit Diversity: Promoting Sustainable Livelihoods, Food Security and Ecosystem Services*, GEF, UNEP, Bioversity and ICHORD, Indonesia

FAO (2010) *Global Forest Resources Assessment 2010 Country Report: Indonesia*, Food and Agriculture Organization of the United Nations; http://www.fao.org/forestry/fra/67090/en/

Garcia-Fernandez, C. and Casado, M.A. (2005) 'Forest recovery in managed agroforestry systems: The case of benzoin and rattan gardens in Indonesia', *Forest Ecology and Management*, vol 214, no 1–3, pp. 158–169

Kaimowitz, D. (2003) 'Forest law enforcement and rural livelihood', *International Forestry Review*, vol 5, no 3, pp. 199–209

Kiloes, A.M. (2014) 'Review about production and trading performance of some of Indonesian tropical fruits in year 2008–2012', Proceedings of a National Seminar, Faculty of Agriculture, University of Sebelas Maret, Indonesia (in Bahasa)

Kiloes, A.M., Winarno, Kuntoro, B.A., Rafieq, A., and Arsanti, I.W. (2014) *Community Fruit Catalogue: Mango and Citrus in East Java and South Kalimantan, Indonesia*, Office of National Project Management Unit of GEF/UNEP Project 'Conservation and Sustainable Use of Wild and Cultivated Tropical Fruit Diversity: Promoting Sustainable Livelihoods, Food Security, and Ecosystem Services', Jakarta

McCarthy, J.F. (2002) 'Turning in circles: District governance, illegal logging, and environmental decline in Sumatra, Indonesia', *Society and Natural Resources*, vol 15, pp. 867–886

Michon, G. (2005) *Domesticating Forests: How Farmers Manage Forest Resources*, IRD, Paris, France, CIFOR, Jakarta, Indonesia and ICRAF, Bogor, Indonesia

Mittermeier, R.A., Myers, N., Gil, P.R., and Mittermeier, C.G. (1999) *Hotspots: Earth's Biologically Richest and Most Endangered Terrestrial Ecoregions*, CEMEX, S.A., Mexico City, Mexico, p. 430

Murdiyarso, D. and Skutsch, M. (2006) *Community Forest Management as a Carbon Mitigation Option*, Center for International Forestry Research, Bogor Barat, Indonesia, pp. 1–7

Myers, N., Mittermeier, R.A., Mittermeier, C.G., da Fonseca, G.A.B., and Kent, J. (2000) 'Biodiversity hotspots for conservation priorities', *Nature*, vol 403, no 6772, pp. 853–858

Persoon, G.A. and van Weerd, M. (2006) 'Biodiversity and natural resource management in insular Southeast Asia', *Island Studies Journal*, vol 1, no 1, pp. 81–108

Perum Perhutani (1996) 'Integration of forest management and forest village development', Perum Perhutani, Jakarta (in Bahasa)

Supriatna, J. (2010) 'International year of biodiversity Indonesia'; http://www.thejakartapost.com/news/2010/01/23/int039l-year-biodiversity-and-indonesia.html

Tata, H.L., van Noordwijk, M., and Marinus, W. (2008) 'Trees and regeneration in rubber agroforests and other forest-derived vegetation in Jambi (Sumatra, Indonesia)', *Indonesian Journal of Forestry Research*, vol 5, no 1

Uji, T. (2007) 'Review: Indonesian native fruits diversity and its potency', *Biodiversitas*, vol 8, no 2, pp. 157–167 (in Bahasa)
White, A. and Martin, A. (2002) *Who Owns the World's Forests?*, Forest Trends and Center for International Environmental Law, Washington, DC

Notes

1 Benzoin was and still is highly valued as an ingredient in incense for burning in rituals and religious ceremonies, not only in all western Indonesian islands, but also in mosques and churches all over the Middle East, North Africa and Europe. The fragrant resin of benzoin has been exploited in the wild and traded from Sumatra for at least 10 centuries, first to China, then to the Middle East and finally to Europe, with an amazing historical continuity. It was and still is used for traditional and modern medicinal purposes, a component valued by the pharmaceutical industry. It was and still is used in perfumery (Michon, 2005).

28 An informal network of grafting experts to help communities conserve and use wild pickle mango (*Mangifera indica*) diversity in the central Western Ghats region of Karnataka, India

Vasudeva R, Narasimha Hegde, B.M.C. Reddy and Bhuwon Sthapit

GPD 'passport'

GPD code:	01
Focus area:	Propagation and planting management Collective action and social networking
Character:	System with institutional arrangement
Species and varieties involved:	Wild pickle mango (*Mangifera indica*) populations restricted to the riverine habitats of the Western Ghats mountains of India. About 40 different types have been recognized and to some extent cultivated by the farmers in and around Sirsi region.
Name of location:	Koligar, Onikere, Vanalli villages in Sirsi district, Karnataka, India
GIS reference of location(s):	N 14°42′51″; E 74°40′44″ Elevation: 546 masl
Name of farmer (data resource):	Mr Dattatreya Hegde, Bhairimane; Ms Bhrathi D Hegde, Bhairimane; Mr Ramesh Hegde, Onikere; Mr Eshanna, Amchimani; Ms Lalita V. Hegde, Amchimani; Mr Krishna Hegde, Onikere; Mr Shantaram, Onikere; Mr M.B. Nayak, Siddapur; Ms Manju Pujari, Kanchikai

Introduction

Wild pickle mango is a delicacy in the Central Western Ghats region of Karnataka, southern India, because of its unique tangy aroma. Wild pickle mango is largely collected from the wild, sold as a commodity and preserved in brine to prepare pickle. It is a major source of income for the rural landless poor (Tesfaye *et al.*, 2015). In this practice, a group of grafting experts, usually three to four, form an informal association and offer themselves as workers in the farmers' fields to graft rare wild pickle mango varieties. Usually the team undertakes work for two to three months between the months of August and October. They visit the farmers' fields to undertake grafting activity and also provide insights into the maintenance of these types. This practice is highly useful in conserving and maintaining the genetic diversity of wild pickle mango as it encourages the increased exchange of grafted plants of unique varieties and, as a result, the on-farm conservation of these types. This practice has undoubtedly increased the diversity of pickle and fruit mango in home gardens. The GPD also contributes to an improvement in total productivity as these experts advise the farmers on other practices of wild pickle mango plantation maintenance, improvising propagation techniques for mass production as well as in popularizing wild pickle mango. In terms of livelihood strategy, the practice encourages communities to rely more on native and home-grown fruit for traditional household uses and provides an alternative income-generating activity.

Context

Uttara Kannada district, situated in one of the world's hotspots of biological diversity – the Western Ghats of India – is one of the largest districts in Karnataka state of India. It is endowed with rich natural resources (see Chapter 11 for full description).

The study site falls under a high rainfall region surrounded by a forested area. The lifestyle and the culture of the communities are closely associated with the resources of the forest. The communities are also involved in the collection of non-timber forest products (NTFPs) such as wild pickle mango and *Garcinia* fruits, gum, resin and leaves. Sizable quantities of wild pickle mango are harvested from the wild and sold in the local market. Because all these resources are being harvested from the wild unsustainably, there has been a rapid erosion of valuable types. However, the adoption of vegetative propagation methods could halt unsustainable harvesting and allow these types to be conserved.

Methodology

The study was conducted in the Uttara Kannada district of Karnataka state, southern India. A group of scientists interacted with progressive farmers and nursery experts of the focal communities to document good practices followed by them to sustainably use mango and *Garcinia* species. The progressive farmers were identified following baseline data gathered from more than 500 households in the focal communities. All four aspects of good practices – propagation and nursery management, production management, linking of farmers with markets and consolidation of the community role in management of tropical fruit tree genetic resources (TFTGR) – were included in an unstructured questionnaire that was adopted to elucidate the responses.

In order to identify and document good practices, the same criteria and method described in Chapter 11 by the same author were employed. This chapter documents one good practice of many from the Sirsi site: the formation of an informal network of grafting experts who help communities conserve and use wild pickle mango diversity.

Description of GPD

Wild pickle mango (*Mangifera indica*), known as Appemidi in the local Kannada language, is a delicacy in the Central Western Ghats region of Karnataka because of its unique aroma, tangy taste and special S-shape (Plates 73–76) (Vasudeva and Rajeshwari, 2014). Locals relish the pickles made of these immature fruits as an integral part of every single meal. The aromatic sap collected from the young fruits is preserved and used to spice up several dishes. This distinctive fruit is bitter and sour with an aroma so unique and variable that it can range from that of cumin seeds to that of camphor. Collected from riverine habitats and sold in local markets, wild pickle mango is a major source of income for the rural landless poor. Fifty years ago, when the wild pickle mango was still undomesticated, people found good-quality trees in sufficient numbers in the forest to meet their needs. However, as the population has grown, demand for the unique pickle mango types has increased. This has often resulted in unscientific harvesting and overharvesting of fruit, occasionally leading to the death of mother trees. Recognizing the danger of losing these unique types and their potential markets, entrepreneurial farmers started to domesticate pickle mangoes as far back as the 1980s. These innovative farmers, through their painstaking efforts, identified wild pickle mango trees with unique aroma and taste in the wild and used them as mother trees (Vasudeva *et al.*, 2011). The scions collected from these recognized mother trees were used to develop different varieties through vegetative propagation.

These farmer-recognized varieties started gaining importance among the local farming communities as more and more farmers started to plant these grafts. This resulted in increased demand for the services of expert grafters.

It took decades of experimentation and trial and error for Mr Eshanna, an innovative farmer, to become an expert in pickle mango grafting, which is more complicated than grafting of many other species. However, what started as a simple desire to grow pickle mangoes on the farm or in the backyard soon expanded into a community-wide effort. 'Whenever two farmers meet, they always have to exchange something,' Mr Eshanna said. 'This exchange could be their best fruit or new information about grafting techniques.' Through this network, farmers were able to collectively identify the hundreds of varieties of pickle mangoes that exist in the forest. They verified the best trees and standardized grafting techniques. Expert grafters would graft pickle mangoes on their friends' and families' farmlands. Over time, this process evolved from being a purely social activity to becoming an informal network of expert grafters who offer their expertise in grafting and maintaining pickle mangoes. Hence this expert group became a locally innovated system that consolidates the role of communities in the conservation and use of rare wild pickle mango diversity.

The GPD is a system wherein a group of experts, usually three to four, form an informal association and offer themselves for work in other farmers' fields to graft the rare wild pickle mango varieties. Usually a team undertakes work for two or three months between the months of August and October. They visit the farmers' fields, undertake grafting activity and also provide insights into the maintenance of this diversity. The cost of their visit is met by the farmers. One group undertakes grafting in about five farms a year. There are several groups in the Sirsi site. For instance, at the Salkani cluster, Mr Dattatreya Hegde, Mr Eshanna and Mr Ramesh Hegde have formed an informal group. For the last 10 years, this type of system has operated only within a close circle of farmers. However, the farmers opined that this practice could be scaled up further with some interventions. Other Western Ghats farmers, from the districts of Uttara Kannada and Shimoga, have expressed their interest in such a practice.

Impact on diversity

This institutional mechanism is highly useful for conserving genetic diversity of wild pickle mango. As a result of the activity of these informal groups, the availability of grafted plants of unique varieties has increased and, as a result, on-farm conservation of these types has taken place. As a testimony of this, today several of the most important wild pickle mango types, such as 'Ananthabhatta Appe' and 'Haladota Appe' have been safely conserved on farm, although the original mother trees have been lost in the wild. Certainly this practice has increased the diversity of pickle mango and the same practice could be used for fruit mango in home gardens. The GPD also contributes to the improvement in total productivity as these experts also advise the farmers on other practices of wild pickle mango plantation maintenance and improvising propagation techniques for mass production, as well as methods for popularizing wild pickle mango.

These informal networks are important as, not only do they conserve the genetic resources for pickle mango, but they also conserve the knowledge about its flavours, multiple uses and tolerance against certain extreme weather conditions such as drought or heavy rains during flowering. Growing pickle mangoes in orchards protects varieties that were in danger of becoming extinct in the wild. Furthermore, these informal groups have also contributed to the increase in on-farm interspecific diversity, as grafting techniques can be extended to other economically important trees as well. For instance, a rare 'white-type' of *Garcinia indica* has also been propagated and popularized by these groups (see Chapter 11 in this book).

Today every wild pickle mango type found on farm in the central Western Ghats is a result of such innovative farmer efforts. In the process, these groups have developed special skills to identify and categorize newer pickle types, thus contributing to the richness of on-farm conservation.

Impact on livelihoods

Wild pickle mango and the food culture of these communities is tightly linked. Various kinds of dishes are prepared regularly in every household using the wild pickle mango preserved in brine solution. Because of their excellent flavour traits, wild pickle mango types are always preferred by the communities for home consumption. People get good-quality pickles year round that are free from any pesticides and chemicals. Furthermore, the grafters obtain a small amount of cash income by undertaking such an exercise. For instance, for Mr Eshanna, the pickle mango business makes up the bulk of his livelihood. He goes to a minimum of 100 orchards and home gardens a year and makes his living by training people and grafting pickle mangos. However, for Mr Dattatreya Hegde from Bhairimane village, the network is primarily a social activity. He refuses to be compensated for his travels. Grafting pickle mangoes on other farmers' lands gives him an excuse to travel and visit friends. However, he earns a substantial income from selling at the local market the pickle mango fruit as well as the grafted plants of these unique wild pickle mango types. There is good demand in the local markets for the grafted plants of these types. Hence in the long run it contributes to the financial capital of the farmers.

The GPD would create a network of farmers with increased capacities to independently set up private nurseries of wild pickle mango, White *Garcinia* and other traditional and economically important species. Further, it would strengthen the linkages of farmers with local markets, retailers and with other nursery experts, which may provide opportunities for marketing. The network of nursery experts has the potential to enhance human capital by capacity building and mobilizing awareness about the market potential of wild pickle mango, thus contributing to livelihood asset creation.

In terms of livelihood strategy, the practice encourages communities to rely more on native and home-grown fruits for traditional household uses. Harvesting a home-grown wild pickle mango for a recipe instead of purchasing from a local market may make a marked difference in the livelihood strategy. In the long run, the practice encourages relying on home-grown fruits for home consumption.

Depending on only one major cash-crop, such as betelnut, for family income is always associated with risk of income loss. The GPD encourages diversification of the fruit crops in the home garden, which in turn helps to diversify the income portfolio of the family through grafting or sale of grafted plants. This practice may, therefore, contribute to the reduction of vulnerability of communities to market risks.

Sustainability and other benefits

This practice can support conservation and use of several traditionally important species in farmers' home gardens, thus directly contributing to the conservation of valuable genetic resources for posterity. Easy availability of traditionally important varieties through these networks would encourage the farming community to raise diversity-rich multispecies gardens. Indeed, such gardens are an important feature of the landscape of the central Western Ghats, and they support several ecosystem services such as being a refuge for native pollinators, and nurturing birds and small animals.

Factors favouring or hindering successful functioning of GPD

The major driving force for setting up this network has been the flair of farmers for recognizing, innovating and exchanging genotypes of several indigenous fruit tree species. Fundamentally this GPD is a system innovated and practised by local farmers, hence the likelihood of it being easily adopted by other farmers is high. However, such innovations have been sporadic and restricted to a few smaller groups. Providing institutional support would encourage greater success of the GPD. The most important issue that needs to be addressed while scaling up is that of sharing the benefits derived from the GPD as a network. For a successful functioning of the GPD, good coordination of several different production groups and committed collaboration from a trustworthy NGO partner is required.

Recommendations for a way forward

Farmers have indicated their interest in scaling up the production of preferred pickle mango types through this network of grafting experts. Because of the local cultural importance of pickle mango, they want to spread the use of Appemidi in popular dishes to surrounding cities. Together, the farmers could make around 10,000 grafts every season and train 50 people to start growing pickle mango in their home gardens and orchards. The group of grafting experts could be key to the spread of this local delicacy, while contributing to livelihood and conservation benefits for the communities.

References

Tesfaye, A., Vasudeva R, Sthapit, B.R., Parthasarathy, V.A., Reddy, B.M.C., and Ramanatha Rao, V. (2015) 'Domestication potential of aromatic pickle-mango

(*Appemidi*) types in the Central Western Ghats, India', *Indian Journal of Plant Genetic Resources*, vol 28, no 1, pp. 106–116. DOI 10.5958/0976-1926.2015.00014.5

Vasudeva R and Rajeshwari, N. (2014) 'Appemidi: Wild aromatic pickle mango of the Central Western Ghats, India', information brochure, UNEP/GEF project on Tropical Fruit Trees

Vasudeva R, Abebe, A.A., Hegde, N.R., and Deepa, G.S. (2011) 'Documenting fruit trait diversity of aromatic pickle mango in the humid tropics of central Western Ghats', in *Proceedings of 'Augmenting Production and Utilization of Mango: Biotic and Abiotic Stresses'* organized by the Society for Development of Subtropical Horticulture, Central Institute for Subtropical Horticulture (ICAR), Lucknow, India

29 Social capital building for tropical fruit tree diversity management

M.P Vasimalai, M. Palanisamy and M. Kiran Kumar

GPD 'passport'

GPD code:	20
Focus area:	Collective action and social networking
Character:	System and institutional arrangement
Species and varieties involved:	Citrus, Garcinia and Mango
Name of location:	Amravati (Warud), Chittoor, Malihabad, Pusa and Sirsi sites, India
GIS reference of location(s):	Amravati (Warud): N 21°46′; E 78°26′ Chittoor: N 13°21′; E 79°10′ Malihabad: N 26°91′; E 80°70′ Pusa: N 21°49′; E 78°35′ Sirsi: N 14°61′; E 74°80′
Name of farmer (data resource):	The leader Mr Rajendra N Karale and members of SHG based in Warud in Maharashtra State The leader Mr Krishna Murthy and members of SHG based in Chittoor in Andhra Pradesh State The leader Mr Sanjith Kumar and members of SHG based in Malihabad in Uttar Pradesh State The leader Mr Ajit Kumar and members of SHG based in Pusa in Bihar State The leader Mr Shantaram Gouda and members of SHG based in Sirsi in Karnataka State

Introduction

Helplessness is the worst form of rural poverty. Collective action allows resource-poor, marginalized men and women farmers to build their social capital and generate new assets for their families' well-being. Social capital can be defined as the features of social organizations that facilitate co-ordination and co-operation for mutual benefit of the members and society as a whole (Putnam, 1995; 2001). Social capital consists of social networks, social connections, relationships and ties, groups, trust, access to wider institutions, the ability to demand and others. These assets and their uses are affected by elements of vulnerability and by processes, institutions and policies. Social capital, like other forms of capital, can also be affected or influenced by the legal system and judicial rules, property rights, political systems, gender perceptions, equity, civil society, trade barriers, cultural norms and values, informal networks and formal institutions (Sajise and Sthapit, 2006). Social networks and connections, reciprocity, trust and mutual benefit encourage collective action to achieve more sustainable development (Bowles and Gintis, 2002; Pretty and Ward, 2001). Government and donor funded projects seek to improve their effectiveness and efficiency by supporting the formation of social capital in the form of groups. Self-help groups (SHGs) or Farmers' Groups (FGs) provide an entry point for efforts to work with community members. Microcredit initiatives (and similar practices such as saving and credit schemes, Community Biodiversity Management fund, etc.) are considered a good practice for mobilizing social capital by helping poor farmers to access financial, human, physical and natural assets.

Empowering farmers and their institutions, creating space for social learning and innovation and a dynamic system of small-scale innovation might be a sustainable way to mainstream good practices for livelihood options and conservation actions using local biodiversity (Chapter 3; Sajise and Sthapit, 2006). Through collective action, it is assumed that natural resources can be conserved and the flow of benefits from them can be shared more equitably among participants. Enhanced social capital can improve environmental outcomes through decreased costs of collective action, increase in knowledge and information flows, increased cooperation, less resource degradation and depletion, more investment in common lands and water systems, and improved monitoring and enforcement (Anderson *et al.*, 2002). Social capital, embedded in participatory groups within rural communities, has been central to equitable and sustainable solutions to local development problems (Pretty and Frank, 2000). Where social capital is well-developed, local groups with locally developed rules and sanctions are able to make more of existing resources than individuals working alone or in competition (Pretty and Ward, 2001).

DHAN Foundation[1] assessed and developed a process to improve social capital in the five Indian sites of the UNEP-GEF project on the 'Conservation and Sustainable Use of Cultivated and Wild Tropical Fruit Diversity: Promoting Sustainable Livelihoods, Food security and Eco-system Services' in partnership with Bioversity International, the Indian Council of Agricultural Research (ICAR), the College of Forestry, Sirsi, the NGO Life Trust and EcoWatch in the Sirsi site. This chapter describes the experiences encountered during the TFT project to build social capital within communities in India through a process of establishing and strengthening SHGs for the benefit of their livelihoods and the conservation of local diversity in fruit tree species and varieties.

Learning and building on successes

Over the last 20 years, DHAN Foundation has promoted the improvement of social capital among marginalized communities through the development or strengthening of localized community organizations (SHGs or other existing organizations) around microfinance. DHAN has worked with more than 1.5 million poor women, small and marginal farmers, landless labourers and fisher folk across India, and has rich experience in organizing more than 55,000 SHGs and networking them into 395 independent self-governed community organizations. The SHG members gain access to and control over financial capital through their own savings and subsequently develop linkages with formal banking and financial institutions. During 2013–14 alone, the SHGs promoted by DHAN Foundation mobilized INR4,386 million (US$70 million) from local commercial banks, at an average annual interest rate of 12 per cent. The same members, if unorganized and borrowing from informal markets, would have to pay interest between 36 per cent and 60 per cent.

These SHGs function through mutuality, trust and self-regulating mechanisms and decision-making processes. They make collective decisions and work in cooperation for a common purpose. Through the structure of a federation of SHGs, members can enhance human capital by participating in training courses that build appropriate production and marketing skills, and create and strengthen physical capital in the form of assets such as farm inputs, storage and post-harvest processing infrastructure. Besides, within an organized structure of a federation, marginalized groups can increase their political, social and economic influence, recognition and bargaining power. A federation of SHGs also serves as peer pressure and a resource for solving the problem of defaulters.

Methodology of social capital building

DHAN Foundation has refined a process for social capital building through SHGs following nine actions grouped into three phases (Table 29.1). The duration of each of these phases varies with the capacity of the SHGs, community leadership and interest, and the experience and community mobilizing skills of the community organizers. Completing all three phases takes at least three years but may exceed that in some contexts. Once the desired number of SHGs is formed and capacity building achieved, phases 1 and 2 are generally assumed to be completed. In general, phase 3 takes a longer

Table 29.1 Phases of social capital building

Preparatory phase (6–12 months)	*Group formation phase (about 12 months)*	*Consolidation phase (about 3–5 years)*
1. Collect secondary and primary data (e.g. using Participatory Rural Appraisal) to understand local situation and identify potential Farmers' Groups 2. Identify and promote self-help groups: select potential members and number of groups for the location based on the family data collected and other criteria. Screening of villages selected for the project	3. Set up institutional modalities of savings and credit programmes 4. Enhance community awareness through regular meetings, orientation and interactions 5. Capacity building of SHGs on governance of fund and develop norms, rules and regulations for group functioning 6. Introduce notebooks and accounting system to the groups, open savings account in a nearby commercial bank branch and arrange loan for the group and insurance for members and their spouses	7. Consolidate roles of SHGs in planning and implementation of community action plans 8. Promote Farmers' Association (3–4 associations per site) 9. Promote Farmers' Federation (one per site)

Notes: SHGs = self-help groups.

time, and mostly includes promotion of the federation, opening an account in a commercial bank, registration, leadership building and mainstreaming linkages with other government service providers.

Phase 1: Preparatory phase

In the context of the TFT project, analyzing the existing situation was the first step for promoting primary groups of farmers. DHAN Foundation carried out a diagnostic study (using Participatory Rural Appraisal) to assess the status of social capital among male and female farmers and orchard labourers involved in tropical fruit tree cultivation among small and marginal fruit growers in five sites within India (Chittoor in Andhra Pradesh, Sirsi in Karnataka, Amravati in Maharashtra, Pusa in Bihar and Malihabad in Uttar Pradesh). Various participatory tools were used to assess groups' interest, relationships and collective actions (Plates 77–80). The purpose of the activity is to understand local dynamics, needs, opportunities, potential and challenges. Both men and women were interviewed to ensure a more gender-equal approach and to facilitate the inclusion of the poor such as labourers and smallholder farmers with orchards in the project.

Table 29.2 Site characteristics and key socio-economic constraints

Site	*State*	*Communities*	*Conservation priority*	*Constraints/barriers*
Amaravati	Maharashtra	Jarud, Nagziri and Bargaon	Citrus Mango	Access to credit for resource-poor farmers Lack of collective marketing despite presence of Ministry of Agriculture Inclusiveness lacking
Chittoor	Andhra Pradesh	Polakala, Talupulapalle Bangarupalyam	Mango	Single variety dominant for commercial production Water stress Small and marginal farmers are yet to be organized Limited women's participation in decision making
Malihabad	Uttar Pradesh	Kasmandikalan, Mohamednagar Talukedari, Sarsanda Gopramau	Mango	No local organizations exist Credit access through money lenders at high interest rates Limited women participation in decision making
Pusa	Bihar	Mahamda, Jagdishpur, Dhobgama Murliyachak	Pomelo Mango	Unorganized farmers and farmers' groups Limited women participation in decision making Inclusiveness lacking
Sirsi	Karnataka	Koligar, Gonsar Kulibeedu	Appe mango *Garcinia*	Out migration is main issue; informal networks of farmers and groups exist

For this diagnostic study, DHAN Foundation teams visited all the sites and the fields of custodian farmers and interacted with them to understand their needs and priorities. Using a structured questionnaire, teams interviewed farmers representing all the villages and conducted focus group discussions with them. They also reviewed secondary data of socio-economic well-being status and land use. The diagnostic study brought out key lessons for each of the project sites with respect to social capital among the farmers and labourers.

Preliminary site characteristics from the diagnostic survey are presented in Table 29.2. In all sites, farmers expressed constraints in accessing credit from banks, and several were under pressure of loans from moneylenders and intermediaries. The informal groups that existed in a few of the villages lacked institutional structure, systems and methods.

These exploratory PRA activities aim to build rapport, quickly assess the agricultural and social needs of the community and identify interested FGs in the community. During this process potential groups were also identified by the community organizer and site management team. A *Padayatra* (a mass procession of local stakeholders and villagers) was organized to spread the message to a larger audience that facilitated formation of more groups in each site.

The group formation process takes place after identifying a cluster of villages with the potential to promote 15 to 20 FGs, which could be managed by a local field associate. DHAN Foundation professional staff and local field associates lived within the community for each cluster of villages. The purpose of this is to build the capacity of local field associates to form the groups, nurturing them and gaining the confidence and trust of the villagers. Professional staff went with the local field associates and demonstrated the process of promoting a group.

A community mobilization team led by DHAN Foundation identified potential members of small and marginal orchard farmers including the landless labourers working in the orchards. Focus was given to include all the small and marginal farmers, including both men and women involved in fruit production. In this project, 35.8 per cent of the 3,276 total members were women. Different knowledge held by women and men, as well as the importance of ensuring equitable benefits and cultural sensitivity, require not only that information be disaggregated by gender but also that meetings be held separately by the same gender. Initially 2 mixed, 14 women and 45 men SHGs were formed and later, by 2015, a total of 11 mixed (115 men and 57 women members), 89 women groups (1,117 members) and 147 men groups (1,987 members) totalling 247 SHGs (3,276 members) were organized. This reflects interest shown by the community women as well as conscious attempts made by the project staff and DHAN to ensure participation of women in the project. The local field associates gathered people in a common place and conducted formal meetings with villagers. The aims and objectives of the TFT project, the importance of biodiversity conservation and the role of communities in conserving the biodiversity were explained. The proposed activities of the project and potential benefits for the community from the project were also discussed. A video about the project, group formation process and activities was screened. Pamphlets about the project with photographs were shared with the members. Following this, potential families to be grouped in the village were identified through a wealth ranking process to identify a marginalized and low-income group of farmers.

Phase 2: Group formation phase

Each group elected office bearers, namely a President, Secretary and Treasurer, among their members. By-laws for the group were adopted with the consensus of all members (see Box 29.1). The by-laws broadly cover the objectives of the group, timings and frequency of meetings, amount of savings by members, interest to be paid on savings, loans and purposes, rate of interest for the loans and repayment period.

Box 29.1 Salient points in the by-laws of Kampalli Mamidi Rythula Sangam (Kampalli Mango Farmers' Group)

1. The group will be formed with farmers of Kampalli village only and they should be mango growers.
2. The meeting will be conducted on a fixed date of every month at the village temple.
3. Each member agrees to save minimum Rs250 per month in the group.
4. Members agree to an interest rate of 12 per cent per annum for any amount of internal loan.
5. Members must repay the loan amount as per the demand fixed by the group.
6. The group agrees to open a savings bank group account in Sapthagiri Grameena Bank.
7. The account will be operated by three leaders of the group, President, Secretary and Treasurer.
8. All members agree to enrol under a life insurance scheme with Life Insurance Company (LIC) to receive benefits from government scheme.
9. Only one member per family should be part of the group.
10. Each member has to pay INR20 as an entrance fee when joining.
11. Members of the group should attend all capacity-building programme and conservation efforts organized by the TFT project on a rotational basis.
12. For external loans the interest rate will be decided based on the bank rate of interest but will not be more than 24 per cent per annum.
13. Leaders will be rotated once every three years.
14. The group should undergo auditing every year.
15. Group should try to avail of benefits from mainstream departments mainly for mango.

DHAN professionals facilitated detailed discussion among the members on the roles of leaders, duration of their leadership positions, accounting system, auditing and appointing an accountant for book keeping. Once agreed upon the group norms, all the members came forward to pay a nominal entrance fee to confirm their membership in their group. The group also decided on how much to save each month with the flexibility that any member who wished to save more could be encouraged to do so.

The group meeting system was then introduced to all the members. The purpose of group meetings is to facilitate discussion on issues related to cultivation, marketing and technology in addition to conducting savings and credit activity in the group. It is conducted in a standard sequence to make it a routine. At the end of the meeting, proceedings are recorded in the minute-book and all the members sign the minutes. The groups have conducted the meetings thereafter following the same following steps:

- Group meetings start with prayer. The members sit in a circle, so that everyone can see the proceedings.
- The president presides over the meeting.
- Discussion on agenda for the meeting follows.
- The minutes of the previous meeting are reviewed.
- Attendance of the members is recorded in the minute-book.
- Financial transactions like savings, lending to members, payment to banks and others are carried out.
- Documentation and disbursal of loans is carried out.
- Awareness and participation in community action plans of tropical fruit tree diversity.
- With resolutions taken, recorded and signed by all the members, the meeting ends.

After the first meeting, a team led by a professional from DHAN Foundation did a quality check of the group. The quality check focused on conceptual clarity with all the members on the by-laws and norms, meeting process and transactions in the group. In the subsequent meeting, baseline data of all the members, their family details, occupation, income and expenditure and asset and credit position were collected.

After the quality check, all the group members were provided with a set of notebooks and passbooks to keep transparent accounts. Field associates were trained in group account writing and the field associates also helped the office bearers of the group to keep the accounts in good shape.

Three months after forming the farmers' group, a savings bank account was opened in the group's name in the nearest branch of a commercial bank, with the President, Secretary and Treasurer as signatories based on the resolution of the group. In the fourth or fifth month, the members were given orientation on bank linkage, quality of operations in the group including attendance, savings and repayment. The purpose is to make people treat the bank's money as their own money and handle it responsibly. In the sixth month, with proper conduct of meetings and transactions, the group can apply for a bank loan. DHAN has always advocated and promoted linkages between the self-help groups and local commercial banks as poor smallholder families have often been kept away from the banking system due to its high transaction costs and thus failed to benefit from it. At the same time, by working together with banks, they gain confidence and courage to demand more services with quality, i.e. feeling of self-respect and being empowered. Capacity building such as this puts the SHGs and banks on an equal footing to negotiate and exert pressure mutually. They are also encouraged to participate in community action plans related to biodiversity-based livelihood programmes.

Phase 3: Consolidation phase

Farmers were taken for an exposure trip to other successful SHGs and custodian orchards and given an orientation about the projects and on people's institutions. Clusters of 15 to 20 FGs were organized into Farmers' Associations representing the office bearers of each group. All the groups in a Gram Panchayat[2] were promoted as Farmers' Associations. Farmers' Associations meet once a month to discuss their progress. They are a forum for cross learning, and decisions taken in the Farmers' Association meetings are shared in the respective group meetings.

The purpose of the Farmers' Association is to bring sustainability and stability to the groups. This kind of nested institution model has been developed and replicated (Figure 29.1) by DHAN Foundation across India during the last two decades to link several local level public sector service providers and leverage technical and financial resources and mainstreaming linkage with diverse actors in the district.

The rationale for federating or creating an overarching structure of several SHGs can be political, economic, social and environmental as it helps to improve marketing and engaging in value chains but also to increase collective action for biodiversity conservation efforts and to be able to change policies:

- Discuss issues that need collective action and common agreed rules and regulations such as biodiversity conservation
- Ensure aggregation of supply that is often required and requested by larger buyers who pay a better price

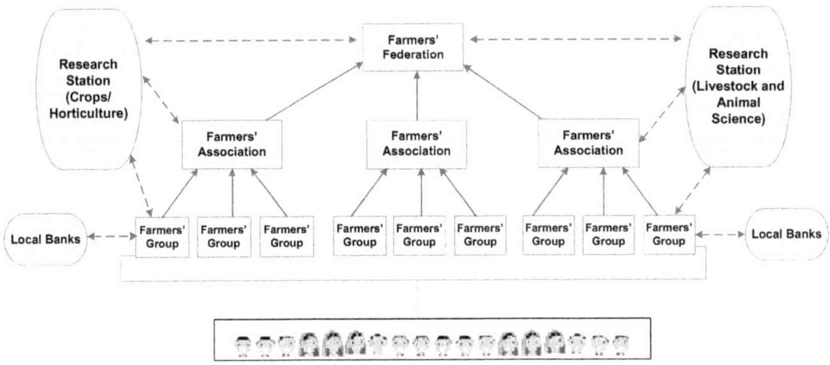

Figure 29.1 Structure of people's institutions for collective action.

- Enable investment in processing equipment required for food safety standards such as HACCP, which is often too big an amount for a single SHG
- Create bargaining power for price negotiations, as dedicated staff (marketing officers) can be hired to continuously search and be in contact with potential buyers
- Enable creation of added value through certification such as fair trade or organic, which is mostly too difficult and expensive for a single SHG
- Ask for funding or help from government organizations or NGOs
- Link up with government agencies and increase their political bargaining power for the needs and demands of all its members

A professional manager with entrepreneurial skills and expertise can be hired from within or outside the community to support the entrepreneurial skills of the members, which is often low (see Chapter 22).

Recent developments in India show that the relatively new legal entity 'Producer Company', which is similar to a federation of SHGs or a cooperative but with a stronger enterprise focus, is quite successful. They are all structures to promote FGs, Farmers' Associations and a Farmers' Federation in the project (collective action) and are seen as an exit strategy to maintain the sustainability of interventions (Srinivasan and Tankha, 2010).

Federations of FGs were promoted in each of the five sites a year after forming the groups. The need to promote Farmers' Federations was discussed in special meetings organized with office bearers of the FGs and in all the group meetings. Then Farmers' Federations were promoted in each site. The Farmers' Federation develops links with related institutions and helps members to avail themselves of government schemes for conserving biodiversity. They are slowly starting to address development issues.

In all of the five project sites of TFT project, a total of 247 FGs (147 men, 89 women and 11 mixed groups) have been promoted and federated into five Farmers' Federations, one for each site (Table 29.3).

To create awareness about biodiversity and organize diverse activities on livelihoods and conservation, systematic efforts were undertaken in all five sites. The importance of biodiversity conservation was discussed during the village meeting, group meeting, and during events such as diversity fairs and farmer exposure visits. As a result, farmers' participation increased in all events from training courses to implementing various biodiversity conservation activities.

Mainstream linkage of FGs in all of the five sites was initiated. Banks, the Life Insurance Corporation of India (LIC) and other ministerial departments responded positively to this initiative. As a result, FGs have been linked with agriculture and horticulture departments at Sirsi. Farmers have received agricultural inputs from the department at a subsidized price and FGs have received horticultural inputs distributed to their members free of cost. The Farmers' Federation of Sirsi collected a life insurance premium from its

Table 29.3 Farmers' Groups and financial details

Name of the site	*Villages*	*Potential number of groups*	*Total number of groups*	*Members*			*Total savings in INR*	*Loan outstanding with members in INR*	*Community Biodiversity Management Fund in INR*
				Men	*Women*	*Total members*			
Amravati	10	115	51	147	454	601	391,080	381,100	200,000
Chittoor	8	100	22	301	0	301	1,007,950	1,503,717	200,000
Malihabad	17	125	61	597	190	787	388,905	220,700	200,000
Pusa	17	110	46	539	205	744	868,925	790,432	200,000
Sirsi	19	140	67	518	325	843	598,040	359,382	2000,000
Total	**65**	**590**	**247**	**2102**	**1174**	**3276**	**3,254,900**	**3,255,331**	**10,00,000**

Note: 1 USD = approx. INR 63

members and insured its members with LIC. The Federation of Pusa explored, with NABARD (National Bank for Agriculture and Rural Development), ATMA (Agriculture Technology Management Agency), KVK (Krishi Vigyan Kendra), RAU (Rajendra Agricultural University) and the district horticulture department, training options and the possibility of availing of various schemes for development. Resource people were engaged from these institutions to provide relevant training to the farmers on aspects of conservation, cultivation, consumption and commercialization of fruits.

Needs-based and income-oriented training programmes such as pickle making, *Garcinia* jam preparation and jackfruit biscuit preparation were organized at the Sirsi site, whereas in the Pusa site, as well as a special type of mango pickle making, training on good agricultural practices for mango, grafting methods and efficient utilization of fertilizer were given. In Chittoor, training was given on mango pest management and roadside mango marketing. Farmers were taken for exposure visits and to diversity fairs to raise awareness and market connections. At the Amravati site, training was given on *Phytophthora* control, orange dish preparation, and pest and disease management in *Citrus*. Access to elite planting materials has been arranged for the farmers.

The Farmers' Federation developed a proposal and submitted to NABARD for the Sirsi, Pusa and Amravati sites to promote a Farmers' Producer Company. A Farmers' Producer Company would arrange orchard and agricultural inputs for farmers. It would also carry out collective marketing and value addition of fruits so that it could obtain a better price for the fruits produced by its farmer members.

In each site, a Community Biodiversity Management (CBM) Fund was established (see Chapter 3) with farmers' monthly savings supplemented with seed money provided by the project. The goal of the fund is to address both development and conservation objectives of households and the community (Shrestha *et al.*, 2013). Local groups should set their own income-generating activities and work together to support community-based conservation efforts. To continue the biodiversity activities through the FGs, a CBM Fund was set up at project sites to which an amount of INR200,000 (about US$3,210) per site was released to each group's bank account. The group gave the money to its members as loans, mainly for activities related to conservation. Any group that is in need can approach the federation for support through the CBM Fund. This rotation of funds generates a small income every year and grows, which in turn acts as capital for the federation. If needed, this CBM Fund can be used as a capital investment for taking up marketing initiatives by the federation. The federation can raise more funds from other sources such as matching grants and/or revolving funds by showing the CBM Fund as collateral and plan for implementing biodiversity conservation activities. For all transactions of the federation, a separate book of accounts is maintained.

Impact on livelihoods and biodiversity

Though the process of promoting social capital was initiated just a year before the closure of the TFT project, DHAN Foundation attempted to promote the needed structures and mechanisms that could strengthen social capital. Some of the features of the social capital built in the project sites were:

- FGs, Framers' Associations and Farmers' Federations were promoted in all five project sites including men and women farmers and landless labourers. Thereby the project created opportunities for small and marginal farmers, and women and landless labourers, who were often not included in conventional agriculture research and extension activities.
- The entire course of social capital formation was highly process-oriented at all stages of promotion starting from diagnostic study, Participatory Rural Appraisal, group formation and development, day-to-day functioning of groups and creation of subsequent structures like Farmers' Associations and Federations.
- A pool of leaders has been identified at the FG level, and at the next levels of Association and Federation. Out of the total members in FGs (3,276), about 950 members (609 men and 341 women) have been chosen by their fellow members as leaders to run their groups. This has created ample opportunities for the leaders to acquire leadership capacities through training.
- Savings, lending and insurance activities introduced in the FGs helped create a favourable environment for mutual cooperation (deciding on whom to lend to), peer pressure (exerting pressure on members who had wilfully not repaid), honesty and transparency (in handling group money), collective decision making, and so on.

All these determinants played a critical role in building trust and mutuality among the members, which led to enhanced social capital. As a result, these farmers were able to mobilize INR3.25 million (US$51,587) in the short period of one year. An equal amount to this they lent to all the members as well, for various farm-related credit needs.

The collective strength of the FGs, their Associations and Federations was helpful in influencing local commercial banks to connect with them, lending to groups and securing insurance services offered to the members. These would not have happened if the farmers had been unorganized and had no institutions to represent their interests. The case studies in Box 29.2 and Box 29.3 give a flavour of the livelihood benefits and biodiversity benefits supported through the project.

Box 29.2 Trust fund to promote collective action on tropical fruit tree (TFT) species – case study

Sri Laxinarasimha Women's Group (LWG) in Salkani village and the Village Forest Committee (VFC), Kilara, in Sirsi site, India, received a modest fund of INR18,000 (about US$285) each as a trust fund under the aegis of the College of Forestry, Sirsi to encourage on-farm and *in situ* conservation of tropical fruit tree diversity. It was deposited and operated through a cooperative bank as a separate account. Guidelines for the effective utilization of the trust fund were developed by the members and they prioritized the activities to be undertaken, which were facilitated under the TFT project. The members of LWG Salkani utilized this fund to get training on pickle making and marketing as well as to visit a famous mango pickle factory. The members of VFC Kilara chose to collect rare 'appe midi' (aromatic pickle mango) types and to establish a diversity plot of TFT species. After three years, LWG was successful in producing about 400 kg of pickles in 2014 from zero production in 2011. The VFC, on the other hand, has established a diversity plot in 2 ha of land, consisting of 70 different varieties of the 'appe midi' types and 10 species of *Garcinia*. LWG focused on making pickles from famous varieties of appe midi such as Malanji and Nandagara and marketed them through a co-operative society in Sirsi as a brand (today the turnover is about US$2,000). Today the diversity plot established at VFC Kilara has become perhaps the largest *ex situ* conservation area of very rare varieties of appe midi and *Garcinia*, from which interested nursery experts around the village can get scion material for their grafting work and further spread these types.

Source: Vasudeva R.

Conclusion and challenges

It took the national TFT project management team in India substantial time to realize that social capital is a pre-requisite for the sustainable on-farm management of tropical fruit tree diversity. Bringing civil society on board in the midst of the project can be considered a positive transformative change amongst a hard core research system. The experience of social capital building for the TFT project implementation was very encouraging, even though not initiated from the outset. Major challenges have been lack of technically sound rural institutions on the ground in most of the sites. Researchers are used to working independently on technical matters without sensitizing the community or creating a common platform for networking civil society, the private sector

Box 29.3 Biodiversity conservation through collective action

Mr Krishna Murthy, a 37-year-old farmer, lives in Kampalli Village of Talupulapalle community in the Chittoor site of the TFT project. His family is very small, with only three members. In the TFT project, he is one of the most active participants. He attended the following training programmes after he became a member of the FG:

- Pest management
- Registration of local, 'naati' varieties
- Post-harvest technologies in mango for processing and export
- Sensitization workshop for mainstreaming mango diversity conservation
- SHG concept and book keeping

After attending these training courses, he slowly started practising many new methods in his orchard such as basin formation before the rainy season, application of tank silt, and safeguarding local varieties.

He came forward to promote Kampalli Mango Farmers' Group after attending a concept seeding meeting organized by DHAN Foundation and was able to succeed in motivating 11 other like-minded farmers. As a leader of the group, every month he conducts meetings, collects savings from all members, ensures entry of books of accounts, etc. From the group CBM Fund he took INR5,000 as a loan for procurement of fertilizers and repaid the entire amount to the group within 6 months. He was able to motivate all the members of the group to enrol under 'Aam Aadmi Bheema Yojana' (a subsidized life insurance scheme) by paying INR100 per member as a premium and 'Universal Health Insurance Scheme' by paying INR500 per member as a premium. He was able to mobilize kitchen garden kits from research stations, fruit fly traps, mango harvesters and mango diversity boxes and supplied them to all the group members to promote the importance of tropical fruit tree diversity in the community.

and government actors. Farmer or self-help groups organized around a Community Biodiversity Management (CBM) Fund have started off well with social capital building practices but require long-term nurturing and support for sustainability and effective use.

References

Anderson, C.L., Locker, L., and Nugent, R. (2002) 'Microcredit, social capital, and common pool resources', *World Development*, vol 30, pp. 95–105

Bowles, S. and Gintis, H. (2002) 'Social capital and community governance', *The Economic Journal*, vol 112, pp. F419–F436

DHAN Foundation (2009) *Resources Mobilization for Rehabilitation of Tanks with People's Participation. Policy Brief 9. Translating Policy into Practices and Practices change*, p. 28; http://dhan.org/cpp/pdf/policybrief9.pdf

Pretty, J. and Ward, H. (2001) 'Social capital and the environment', *World Development*, vol 29, pp. 209–227

Pretty, J. and Frank, B.R. (2000) 'Participation and social capital formation in natural resource management: Achievements and lessons', plenary paper for International Landcare 2000 Conference, Melbourne, Australia

Putnam, R. (1995) 'Bowling alone: America's declining social capital', *Journal of Democracy*, vol 6, no 1, pp. 65–78

Putnam, R. (2001) 'Social capital measurement and consequences', in J.F. Helliwell (ed.), *The Contribution of Human and Social Capital to Sustained Economic Growth and Well-being*, OECD, Canada, pp. 117–135

Sajise, P.E. and Sthapit, B.R. (2006) 'Farmers, farms and agrobiodiversity: On farm conservation experiences of IPGRI', in N.Q. Ng, J.T. Williams, and D.A. Vaughan (eds), *Proceedings of Workshop On In Situ Conservation of Plant Genetic Resources held at FAO, Bangkok, Thailand*; FAO-RAP Bangkok, Thailand, 29–31 August 2005, FAO, Bangkok, Thailand and Regional Office for Asia and the Pacific; National Institute of Agrobiological Sciences, Tsukuba, Japan, pp. 30–46

Shrestha, P., Sthapit, S., Subedi, A., and Sthapit, B.R. (2013) 'Community biodiversity management fund: Promoting conservation through livelihood development in Nepal', in W.S de Boef, A. Subedi, N. Peroni, and M. Thijssen (eds), *Community Biodiversity Management: Promoting Resilience and the Conservation of Plant Genetic Resources*, Earthscan/Routledge, Abingdon, UK and New York

Srinivasan G. and Tankha A. (2010) *SHG Federations Development Costs and Sustainability*, Access/Rabobank, New Delhi, p. 233.

Notes

1 DHAN (Development of Human Action) Foundation, a not-for-profit development institution engaged in poverty reduction at the grassroots for more than two decades in India, has evolved a model of building sustainable people's organizations around development issues (DHAN Foundation, 2009). The experience of DHAN in building more than 300 independent people's organizations around the themes of microfinance, conservation and development of water resources, rainfed farming development, coastal conservation and livelihoods since 1997 has been useful in the creation of a community organization model for the Tropical Fruit Tree project. Learn more about DHAN at (http://www.dhan.org).

2 Gram Panchayat is a local self-government, which is the third tier in the three-tier federal structure of Indian Government. After the Central and State governments, Gram Panchayat is the bottom level structure that is led by elected representatives and leaders. They execute development programmes at the village level.

Part 4

A way forward

30 Reflections on the challenges of good practice identification, documentation, piloting and mainstreaming

Bhuwon Sthapit, Hugo A.H. Lamers and V. Ramanatha Rao

Linking research to practice

The benefits from and threats to agricultural biodiversity and the urgent need for *in situ* and on-farm conservation have been frequently stressed, and concerted efforts on this front have been called for by various scholars and organizations (Bartlett, 2008; Bélair *et al.*, 2010; Jarvis *et al.*, 2011; Sthapit *et al.*, 2012; Bellon *et al.*, 2015). However, many practitioners, researchers, civil society organizations and government institutions are still grappling with questions such as how to ensure the long-term continuation of on-farm/*in situ* conservation practices and processes (de Boef *et al.*, 2012; 2013). Even more importantly, perhaps, how to ensure that these practices and processes are effective and efficient? This book strives to provide specific guidance to answer these questions by zooming in on 'practices', i.e. those activities undertaken on the ground by women and men farmers, to understand and support what their communities contribute to on-farm/*in situ* conservation.

Chapter 2 defines a good practice for diversity management (GPD) as a system, organization or process that over time and space maintains, enhances and creates crop genetic diversity and ensures its availability to and from farmers and other actors for improved livelihoods on a sustainable basis. The identification, analysis, piloting and mainstreaming of good practices is proposed as a way to improve on-farm/*in situ* management of crop genetic resources in general and tropical fruit tree diversity in particular (see Figure 30.1). The approach has been tested through the UNEP/GEF funded project for the conservation and sustainable use of tropical fruit tree diversity in India, Indonesia, Malaysia and Thailand.

In Chapters 5 to 29 the details have been presented in a total of 23 case studies of good practices for diversity management within their specific contexts across eight countries in three continents. Chapter 4 attempts to provide some answers about who is practising (i.e. custodians), whereas Chapter 3 provides a deeper understanding of how practices play a key role in community

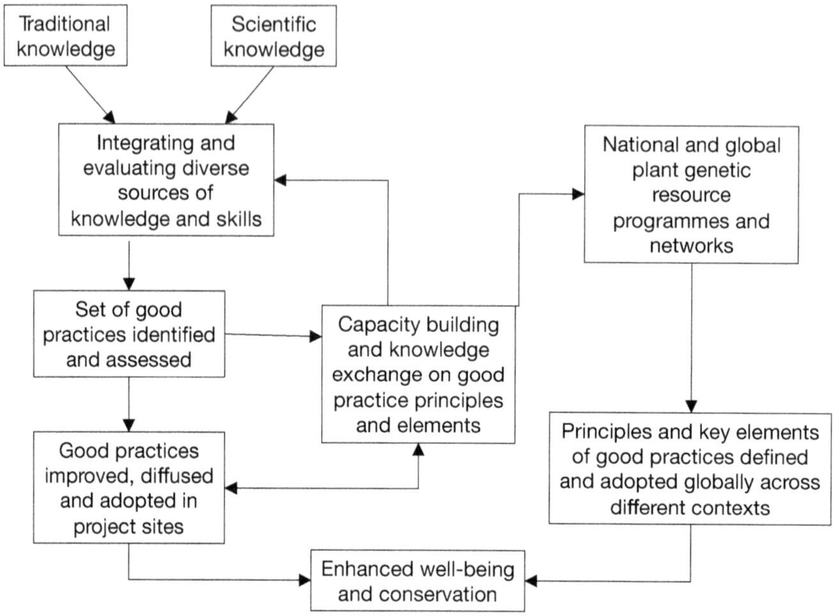

Figure 30.1 Revised impact pathway for improving practices for enhanced community well-being and conservation of tropical fruit tree diversity. The conceptual pathway that led project development (Figure 1.1) was revised following piloting on the ground.

biodiversity management (CBM). We hope these case studies and conceptual chapters give practitioners, researchers and policymakers ideas and tools to help them work with farming communities to achieve on-farm/*in situ* conservation goals while improving these communities' livelihoods.

Our research was guided by two assumptions. First, that many farmers' existing practices are good practices from which scientists could learn. Second, that many practices could be improved by blending the traditional knowledge with scientific knowledge. Such a combination not only leads to synergy between the two sources of knowledge but also to greater adoption and adaption of practices because they fit within farmers' existing mental schemata. Blending traditional and scientific knowledge was sought through platforms where practitioners and researchers could discuss and evaluate knowledge and skills related to GPDs in the light of achieving both conservation of genetic diversity for future use and fulfilment of the farmers' current needs. In this process, a set of GPDs were identified and documented, some traditionally practised by a farming community, others recently evolved through farmer interactions with formal sector interventions. We tested concepts through a process of local, national and regional capacity building, exchange visits and workshops and international conferences. Our theory of change was that identifying and assessing good practices for diversity would lead to their wide adoption and diffusion, thereby contributing to the enhanced well-being of the community and conserved tropical fruit tree diversity for today and tomorrow.

Challenges

In the process of identifying, analyzing and strengthening or mainstreaming GPDs, several challenges and opportunities were identified, which will be of use to others considering similar work.

Developing a learning mindset

It is important to remember that learning involves much more than simply sharing information and knowledge. Learning is an ongoing process that requires access to information and knowledge, gained from personal experience or that of other people relevant to the problem faced. One principle of the CBM approach is that it should provide a platform for collective action and social learning (Chapter 3) to facilitate all activities within a community. However, it is difficult to adhere to principles of ongoing learning and adaptation for those researchers and officials who hail from a classic project management background, in which planning and control are top down and centralized, and who have had little exposure to participatory research. In these cases, one pre-requisite for success is a change in the mindset of implementation partners before practical change can be facilitated on the ground within farming communities. Over the project duration, our experience showed significant changes in the mindset of researchers, extensionists and others involved in the CBM approach applied and described in this book. The people responsible for the identification, adoption and piloting of GPD over time were able not just to perform actions mechanically, but also to understand why it is important, and once that was done, piloting, scaling up and mainstreaming were much faster and more effective.

Terminology

There is no universally accepted definition for 'good practice' or 'best practice' because of the extreme diversity of topics that can be described or defined through the concept. Nevertheless, it is a widely used term in the agriculture and economic development sectors. The wide use and applicability of the term 'good practice' resulted in each individual often having his/her own understanding of the term. Extension agencies in the participating countries are often familiar with 'good agricultural practices' (GAP) that superimpose external inputs and technologies into the farmer's production system (Wannamolee, 2008). Value chain and agricultural development experts often have experience with 'good manufacturing practices' (GMP) as a protocol to be followed with standards and compliance systems that ensure food safety and product quality. However, in the present context, we were looking for those time-tested farmer systems, processes, techniques, methods or combinations of practices arranged in an institutional setting that support the on-farm and *in situ* conservation of tropical fruit tree diversity and its sustainable use. Our focus was practices that combine species and genetic diversity conservation with livelihood benefits, not just focusing on production or productivity (Chapter 2). This requires a conceptual understanding of on-farm and *in situ* conservation of agricultural biodiversity. The first step was to achieve a shared understanding of the definition of GDP among all participants.

Identifying GPD

A key component of creating a shared understanding of what constitutes a good practice across the research community is by specifying how field researchers can identify such practices. After achieving a common understanding of the definition, field researchers were tasked with the difficult job of identifying good practices in their numerous manifestations as well as collecting the data necessary for sharing and evaluating them on a macroscopic level. A wide range of potential good practices were proposed through various data collection methods and then evaluated regarding their fit with the definition and evaluation criteria of a GPD. One of the project research team members, Vasudeva R, compared the search for a GPD with finding a green parrot in a green tree: difficult to identify and demarcate from what surrounds it, but once you know the shape you are looking for the task becomes easier. Often we were sure that a GPD was there hidden in the context, but we still struggled to get a clear view of it.

To complement this method we also adopted a 'positive deviant' approach exploring why some farmers in similar contexts were able to maintain rich TFT diversity with better management and production practices and uses, whereas other neighbouring farmers were not. This led us to identification of custodian households (Chapter 4). With further characterization of such custodian farmers, we found that they have rich traditional knowledge, seek diverse sources of knowledge and skills from various sources and have capacity to innovate (unpacking and repackaging various GPDs) in their own context. They tend to synthesize various sources of information and knowledge to adapt, integrate and coproduce new practices. These farmers are better connected with formal and informal institutions as sources of information and materials. Interviewing these farmers and understanding their practices helped us identify GPDs better.

Full understanding of the GPD required a critical attitude, several cycles of reflection and refinements and sufficient information to facilitate meaningful evaluation. Without this iterative, critical analysis, one can end up with a vaguely described good practice whose contribution to livelihoods and on-farm or *in situ* conservation is not well formulated and determined. If this initial identification process is done only perfunctorily, it complicates all the following steps such as documenting, piloting and mainstreaming.

Documenting GPD

Once potential GPDs were identified, various forms of documentation methods were used, ranging from audio recordings of key informant interviews to videography of selected good practices. Using different documentation methods facilitates triangulation and counters researcher bias or misinterpretation of the information. Digital pictures were found useful to document and phenotype unique elite materials and digital videos or photo series were used to document step-by-step procedures of propagation, cultivation or processing techniques related to unique fruit tree species.

A series of workshops was conducted at different levels, from site level and national level to regional level, to present all potential GPDs and evaluate them. The final regional workshop was used to document selected GPDs, each in a separate descriptive document following a standardized outline. Our approach was to keep the GPD description as short and simple as possible, including only those elements essential to its functioning and that directly contribute to its outcomes, and linking to other GPDs or supportive practices.

Piloting GPD

At a regional workshop dedicated to identifying good practices, held in Thailand on 22–26 February 2011, 33 GPDs were identified. These GPDs were all successful in their community context. However, we realized quickly that good practice in one context might not be replicated in another context in exactly the same way. It was also realized that there was neither time nor funds to pilot all 33 GPDs in different sites across all four project partner countries. Consequently, a few GPDs were selected for piloting in different sites in the same country and in the other three countries, with some adjustments to suit the local context.

The practice of organizing diversity fairs is one of the GPDs that was identified, piloted and adopted widely, repeated many times at village and national levels, and scaled up and mainstreamed in every country. In contrast to the planned 20 diversity fairs, a total of 61 fairs were organized during the project period across four countries, from relatively small but vibrant local village fairs to large exhibitions in major cities targeting high-level policymakers and consumers. A second GPD that was adopted successfully elsewhere was the side-grafting technique, learned from farmers in Thailand, which was adopted by farmers in Sirsi, India. Certain research methods, such as Four Cell Analysis (see Chapter 2), have been taken up by researchers regionally and globally. Similarly, the concept of 'custodian farmers', which emerged gradually through the experiences of project implementation, has been widely embraced as reflecting a useful reality. The concept was explored in a dedicated workshop (New Delhi, 11–12 February 2013; Sthapit *et al.*, 2013) and is informing current research on *in situ* and on-farm conservation.

Though only a few practices could be adopted wholesale across sites and countries, one unexpected outcome of the project was a flourishing cross-fertilization between good practices across countries and sites through the sharing of knowledge, practices and procedures at various meetings, workshops, fairs and exposure visits. A specific knowledge share-fair was organized in Thailand in March 2012 to promote the piloting and scaling of good practices. What happened during and after the event was that farmers, researchers and policymakers unpacked certain GPDs and picked out the most interesting elements and principles, which they then repackaged in their own sites and contexts. In other words, farmers and researchers tend to pick up principles rather than the practice per se. For example, all countries' government partners have taken up work or pledged to work with custodian farmers beyond the TFT project. The Indian government partner (ICAR) has organized several farmer and researcher meetings to identify and decorate custodians across the country for their role in crop diversity conservation. Following this, Indonesian government partners (ICHORD and BPTP) have shown great interest in supporting a network of custodian farmers as an on-farm conservation strategy and as an important link to their *ex situ* conservation mandate and programmes. Similar types of activities, in which custodian farmers were interviewed, invited for key meetings or fairs and received informal or formal recognition for their role in conservation, took place in Malaysia and Thailand, and beyond project countries such as Bolivia and Nepal.

Mainstreaming

Researchers, development workers and donors eventually wish to scale up and scale out GPDs to translate benefits to a wide range of beneficiaries. However, a 'silver bullet' or 'copy and paste' strategy will not suffice for GPDs. Good practices for on-farm conservation of tropical fruit tree diversity are innovations for achieving household food security, livelihood improvements and resilience. This requires a strategy of collective action and networking that enables agencies to digest diverse sources of knowledge, skills and materials, develop the capacity to integrate scientific knowledge into traditional knowledge systems and often co-produce new techniques and practices that solve local problems and collective needs. This dynamic process fits very well in a CBM approach. Greater impacts are achieved by communities who have the capacity to critically analyze their situation, to build complementary collaborations to plug gaps and weaknesses and to develop new approaches from the comparative advantages of contrasting systems. Taking a CBM approach, although it requires a long-term commitment, can create an enabling environment for farmers to innovate GPD over time. The facilitator of this process needs to:

- Build relationships, developing trust and clarity
- Gain understanding and appreciate local reality
- Integrate diverse sources of knowledge
- Develop capacity to unpack and repackage a set of good practices so they are relevant to the local situation

- Facilitate change that enables shifts and breakthroughs
- Review change along pre-set indicators to document impact and outcomes achieved

Lessons learned

Good practices in the management of agricultural biodiversity are highly context specific and sometimes difficult to recognize and disentangle from context. Besides, GPDs should not be treated as good agricultural practices in a general sense, but should be unpacked and repackaged through learning platforms. Understanding the elements, functions and underlying principles is thus more important than replication. The three analogies below illustrate the lessons learned by the researchers during the project, which might help to give readers and practitioners a better understanding of GPDs.

GPD as a house (understanding the concept)

Houses look different all over the world as the result of environmental conditions and socio-cultural traditions; some have flat roofs with very small windows whereas others are made from wood and are built on stilts. Nevertheless, they are built to provide comfortable living conditions for families. In spite of their apparent differences, all houses share several key elements (i.e. doors, windows, roofs and walls) and provide the same key functions to users (i.e. safe place to sleep, eat and live together). Similarly, good practices for on-farm conservation look different across the world because they are embedded within their local socio-cultural and environmental contexts, but they have recognizable key elements and key functions. GPDs are often a constellation of several interlinked key activities, regulations and organizational arrangements. When these key elements are connected, they facilitate the two major functions of a good practice: facilitating conservation of agricultural biodiversity and improving the well-being of its users.

GPD as a green parrot in a green tree (identification)

It is difficult to distinguish a green parrot in a green tree. However, because you can hear the bird, you are sure the parrot is there. Similarly GPDs are sometimes difficult to recognize and describe although you catch some information that makes you sure it is there. It takes a while and several iterations before you can disentangle and distinguish what is part of the good practice and what is part of its context. GPDs are interwoven into their local context and belief system. Once a GPD is identified, one should strive to keep the practice description as concise and simple as possible, including only those elements that directly contribute to its functions.

GPD as LEGO® building blocks (mainstreaming)

Lego® blocks are a widespread toy for children, made of interlocking plastic bricks of different sizes and colours. A Lego® set is a structure such as a car, house or airplane made from several smaller building blocks with precise and exact building instructions included. However, after obtaining several Lego® sets, children often break up all the original constructions, pool all the building blocks together and start making new (imaginary or otherwise) structures. Similarly, GPDs are mainstreamed by unpacking good practices from one context, identifying key elements and principles that can be used to build, innovate or strengthen practices in other geographic regions or sites by custodians, practitioners, researchers or policymakers.

Emerging principles

There are certain basic but essential principles that we should understand before we attempt to pilot and disseminate a GPD to other sites and contexts. Considering the analogy of a house, which has different forms to express the same functions in different environments and socio-cultural contexts, the main characteristics of a GPD are that over time and space it maintains, enhances and creates crop genetic diversity and ensures its availability to and from farmers and other actors for improved livelihoods on a sustainable basis. From this it follows that the key principles of a GPD are that it:

- is embedded in local culture and practices
- is dynamic: it needs to be constantly refined by integration of diverse knowledge sources
- should improve livelihoods (people-centric) and well-being of people
- is contextual, i.e. specific to local situations
- is knowledge intensive and links nature, biodiversity and culture
- needs to be viewed collectively and undertaken together with other practices (although there can be unique, standalone examples)
- aims for empowerment and livelihood improvement and conservation goals
- ensures the maintenance of unique or diverse species or varieties.

A way forward

Our findings suggest that empowering farmers and their institutions, creating space for social learning and innovation and a dynamic system of small-scale innovation might be a sustainable way to mainstream good practices. Experience tells us that approaches, processes, methods and principles are more practical for scaling up than context-specific good practices. Farmers who have better adaptive capacity can pick up principles and adapt practices for their own situation. Such innovation could be more productive, simple, low-cost and sustainable in practice. It is still a debatable issue what kind of approaches create the enabling environment for such small-scale innovation. Research is required to find ways to make such learning platforms effective and efficient in terms of producing a wide range of improved and sustainable GPDs across actors and geographic sites with limited costs. Consolidating roles of custodian farmers in the identification, piloting and mainstreaming of GPDs within a CBM approach might be one way forward. Furthermore, it has been a challenge to formulate general impact indicators and measure the impact of GPDs as they vary so widely in their content and context. Research into learning platforms, custodian farmers, CBM and impact indicators to evaluate the costs and benefits of good practices might pave the way for better results.

References

Bartlett, A. (2008) 'No more adoption rates! Looking for empowerment in agricultural development programmes', *Development in Practice*, vol 18, no 4–5, pp. 524–538. DOI:10.1080/09614520802181269

Bélair, C., Ichikawa, K., Wong, B.Y.L., and Mulongoy, K.J. (eds) (2010) *Sustainable Use of Biological Diversity in Socio-ecological Production Landscapes. Background to the 'Satoyama Initiative for the Benefit of BioDiversity and Human Well-being'*, Technical Series no. 52, Secretariat of the Convention on Biological Diversity, Montreal

Bellon, M.R., Gotor, E., and Caracciolo, F. (2015) 'Assessing the effectiveness of projects supporting on-farm conservation of native crops: Evidence from the High Andes of South America', *World Development*, vol 70, pp. 162–176; http:dx.doi:10.1016/j.worlddev.2015.01.014

de Boef, W.S., Thijssen, M.H., Shrestha, P., Subedi, A., Feyissa, R., Gezu, G., Canci, A., Ferreira, M.A.J.F., Dias, T., Swain, S., and Sthapit, B.R. (2012) 'Moving beyond the dilemma: Practices that contribute to the on-farm management of agro biodiversity', *Journal of Sustainable Agriculture*, vol 36, pp. 788–809. DOI: 10.1080/10440046.2012.695329; http://www.tandfonline.com/eprint/hVuvwUH9HdrZnSQWBiSA/full, accessed 21 August 2012

de Boef, W.S., Subedi, A., Peroni, N., and Thijssen, M. (eds) (2013) *Community Biodiversity Management: Promoting Resilience and the Conservation of Plant Genetic Resources*, Earthscan/Routledge, Abingdon, UK and New York

Jarvis, D.I., Hodgkin, T., Sthapit, B.R., Fadda, C., and Lopez-Noriega, I. (2011) 'A heuristic framework for identifying multiple ways of supporting the conservation and use of traditional crop varieties within the agricultural production system', *Critical Reviews in Plant Sciences*, vol 30, no 1–2, pp. 1–49

Sthapit, B.R., Subedi, A., Lamers, H., Jarvis, D.I., Ramanatha Rao, V., and Reddy, B.M.C. (2012) 'Community based approach to on-farm conservation and sustainable use of agricultural biodiversity in Asia', *Indian Journal of Plant Genetic Resources*, vol 25, no 1, pp. 88–104

Sthapit, B.R., Lamers, H., and Ramanatha Rao, V. (eds) (2013) *Custodian Farmers of Agricultural Biodiversity: Selected Profiles from South and South East Asia*, Proceedings of the Workshop on Custodian Farmers of Agricultural Biodiversity, 11–12 February, New Delhi, India; Bioversity International, New Delhi, India

Wannamolee, W. (2008) 'Development of Good Agricultural Practices (GAP) for fruit and vegetables in Thailand', paper presented at Training of Trainers in Good Agricultural Practices (GAP) and Benchmarking: GLOBALGAP for Fruit and Vegetables, 14–23 July 2008, Kuala Lumpur, Malaysia; FAO, Rome, Italy

Index

Entries in **bold** denote boxes and tables; entries in *italics* denote figures.

For Product Safety Concerns and Information please contact our EU
representative GPSR@taylorandfrancis.com
Taylor & Francis Verlag GmbH, Kaufingerstraße 24, 80331 München, Germany

www.ingramcontent.com/pod-product-compliance
Ingram Content Group UK Ltd.
Pitfield, Milton Keynes, MK11 3LW, UK
UKHW020024200926
475702UK00014B/2045
* 9 7 8 0 3 6 7 1 7 2 7 8 7 *